THE MAN WHO DESIGNED THE FUTURE

THE MAN
WHO DESIGNED THE
FUTURE

NORMAN BEL GEDDES AND THE INVENTION OF TWENTIETH-CENTURY AMERICA

B. ALEXANDRA SZERLIP

MELVILLE HOUSE
BROOKLYN · LONDON

The Man Who Designed the Future

First Melville House Printing: April 2017

Melville House Publishing   8 Blackstock Mews
46 John Street   and   Islington
Brooklyn, NY 11201   London N4 2BT

mhpbooks.com   facebook.com/mhpbooks   @melvillehouse

Frontispiece image courtesy Harry Ransom Center, The University of Texas at Austin, photographer unknown.
Bel Geddes & Co. logo (pg. 121) courtesy: Harry Ransom Center, The University of Texas at Austin.

Library of Congress Cataloging-in-Publication Data

Names: Szerlip, Barbara, author.
Title: The man who designed the future : Norman Bel Geddes and the invention of twentieth-century America / B. Alexandra Szerlip.
Description: Brooklyn : Melville House, 2017. | Includes bibliographical references.
Identifiers: LCCN 2016050576 (print) | LCCN 2016051081 (ebook) | ISBN 9781612195629 (hardback) | ISBN 9781612195551 (ebook)
Subjects: LCSH: Geddes, Norman Bel, 1893–1958. | Designers—United States—Biography. | Design—United States—History—20th century. | BISAC: BIOGRAPHY & AUTOBIOGRAPHY / Artists, Architects, Photographers. | HISTORY / United States / 20th Century. | TECHNOLOGY & ENGINEERING / Industrial Design / General.
Classification: LCC NK1412.G43 S94 2017 (print) | LCC NK1412.G43 (ebook) | DDC 745.092 [B]—dc23
LC record available at https://lccn.loc.gov/2016050576

Printed in the United States of America
10 9 8 7 6 5 4 3 2 1

for Kenneth Rexroth
Autodidact, Man of Letters
1905–1982

The United States did not *happen*, it was designed.

—ARTHUR J. PULOS

We are all swimming underwater and playing hunches.

—DAVE HICKEY

The man who owns whole blocks of real estate, and great ships on the sea, does not own a single minute of tomorrow. Tomorrow! . . . It lies under the seal of midnight—behind the veil of glittering constellations.

—EDWIN HUBBELL CHAPIN, 1861

Contents

Introduction

n 1927, moving pictures became "talkies," the Cyclone Roller Coaster opened at Coney Island, and Sinclair Lewis's brutally satirical *Elmer Gantry* coexisted with "Ain't She Sweet?," one a bestseller, the other the number one pop single on the charts. The world's first underwater vehicular tunnel opened to carry automobiles under the Hudson River, connecting New York to New Jersey, and telephone service was established connecting New York to London, San Francisco to Manila. One of Manhattan's first skyscrapers, an art deco masterwork, was completed,[1] Duke Ellington's band headlined at the Cotton Club, Florenz Ziegfeld opened his eponymous theater (with a mixed-race cast), and, having produced 16 million Model Ts, Henry Ford reluctantly "switched gears," discontinuing them in favor of the more stylish Model A.

America was "potential" writ large, replete with innovators, record breakers, and newly minted film stars. Shakespeare-reading, movie-star-handsome Gene Tunney defeated favorite Jack Dempsey for a second time. George Herman "Babe" Ruth Jr. hit his sixtieth home run. The survivor of a dismal British childhood, Charles Spencer Chaplin had, in short order, crossed an ocean and a continent and put Hollywood on the world map, earning a record $16 million (some $406.5 million today) in the process.

In 1927, Charles Augustus Lindbergh, just two years younger than the century, flew solo across the Atlantic in thirty-three hours, inspiring everything from airplane-shaped piggy banks, a George M. Cohan song, a dance (the Lindy Hop), and a *New York Times* editorial on his hair. A baby elk in the Brooklyn zoo was christened "Lindy Lou," the Pennsylvania Railroad named a Pullman car after him, and a spectacular Broadway parade in the pilot's honor (1,800 tons of ticker tape and confetti) was followed by a celebration dinner requiring 36,000 plates, 300 gallons of green turtle soup, and 12,000 slices of cake.[2]

The subsequent flights of one-eyed Wiley Post, Ruth Elder, Admiral Byrd, and Amelia Earhart were nearly overshadowed. Maybe, wrote novelist F. Scott Fitzgerald—at thirty-one, already on a downward spiral—the "frontiers of the illimitable air" would provide respite from the party fatigue of the earthbound Jazz Age.

"Not Mosquitoes but Aeroplanes." An early Bel Geddes sketch.
(Courtesy: Harry Ransom Center, The University of Texas at Austin/The Graphic, 1920s.)

Less jaded than Fitzgerald and equally, if not more, renowned in his field, was a ninth-grade dropout by the name of Norman Bel Geddes.[3] At thirty-four, he was an internationally lauded theater designer who had, among other things, created sets for the Chicago and Metropolitan Operas and two Gershwin musicals, re-created the massive, full-scale interior of a medieval cathedral (the most extravagant and complex set in Broadway history), and reimagined Manhattan's Palais Royale, transforming it from a moribund Victorian cavern into the city's premier nightclub.

Prior to that, he'd abandoned a successful, hard-won career as an advertising art director. In 1927, he decided to jettison (if temporarily) this second, equally hard-won résumé (set and costume designer, lighting innovator, architect) to embark on a third path—one that would seriously challenge accepted notions of everything from airplanes, ocean liners, automobiles, locomotives, stoves, and circuses to the configuration of factories, homes, cities, and transcontinental highways over the succeeding decades and beyond—a vocation that, as yet, had no official name.

•

THE PATRIOT, BEL Geddes would later write, "was probably the straw that broke the camel's back," breaking theater's absolute hold on his attention. A drama about the assassination of Catherine the Great's son (with John Gielgud in his first stateside role), the play opened at Broadway's Majestic Theatre on January 19, 1927, the culmination of difficult work in challenging circumstances. Bel Geddes had been called in, "after months of worry," to salvage the production.[4]

The play required characters to exit from one of five different locales, mid-speech, and continue, almost uninterrupted, in another. To meet the challenge of almost instantaneous scene changes, Bel Geddes had designed a series of sliding partitions and interchangeable modules, every piece fitting into every other. Rather than cutting the floor into a revolving stage—the standard solution, which reduced the performance area by half and accommodated only two scenes at a time—he'd suspended his five sets on cables so they could be stored high up in "the flies," the space above the stage. Once two half-set platforms pivoted out of the side wings on silent castors to meet center stage, the required "walls" swung into place, and a hinged "ceiling" folded down.

Critics praised Bel Geddes's "patrician" lighting (when a door opens on a monarch's dark bedchamber, a scarlet uniform streaks across the lights), his sound effects (a suspenseful pause marked only by the creaking of boots), and especially his eight elaborate, regal sets, which changed noiselessly, in the twinkling of an eye. "Vivid, unforgettable . . . beyond all praise," wrote the *New York Telegram.*[5]

So ingenious was his interchangeable, raised-and-lowered-by-cable, shifting-back-and-forth-between-five-settings approach that *Scientific American* published a full-page annotated schemata—"How 'Lightning' Stage Changes Are Made."[6]

The play itself was less than stellar. "One wishes," wrote *New York Times* drama critic Brooks Atkinson, "that *The Patriot* moved as expeditiously as its scenery." It closed after a five-day run.

Two weeks later, various newspapers announced that the designer was "deserting" a decade-long theater career and would consider only one annual production hereafter.

I'm "gradually inclining more and more to a state of hermitage," Bel Geddes wrote to a family friend. "I haven't been all together happy . . . The dominating mercenary aspect of theater . . . wears me down."[7] No one, least of all Bel Geddes, could have predicted that, in a few short years, his work on *The Patriot* would inspire a nationwide revolution in, of all things, kitchen appliances.

He wasn't "deserting" theater, Norman insisted. His enthusiasm for the stage remained un-dampened. But the frustration of creating quality work, only to have it fail, through no fault of his own, "combined with the continual demand for my second best instead of my best, for doing the work *the managers* thought the audience wanted instead of the work *I thought* should be done" had caused him to withdraw.

At the same time, he was beginning to think that artists should take a more active role in quotidian things, "useful" things.

Objects for everyday living were being designed by the untalented, or possibly those with talent who lacked the courage to stand their ground against conservative, hidebound bosses. The public was so accustomed to what they saw around them that it didn't occur to them that "function" could be aligned with "beauty." Or that "function aligned with beauty" even mattered, that it could serve to improve the quality of their lives.

Industry, he wrote, was "the defining spirit . . . the driving force" of the age. "Accept it or not, it's a fact . . . It's as absurd to condemn an artist of today for applying his ability to industry as to condemn Phidias, Giotto or Michelangelo for applying theirs to religion."[8] It was a subject he'd have many opportunities to expound on. Michelangelo's Sistine Chapel was "the great moving picture epic of his day." El Greco "might have been a magazine illustrator or a bricklayer, but he still would have been a great artist . . . Gauguin was a banker until he was forty."[9]

Just as he approached scenery, lighting, and theatrical staging as "organic outgrowths" of a script (the traditional method slapped together costumes and sets from previous productions; lighting subtleties were practically unknown), he planned to approach product design as an organic outgrowth of efficiency and ease of use—a natural evolution, with visual integrity following of its own accord.

•

DOUBTS ABOUT THE ephemeral nature of theater, his first love, had been growing for some time. Theater was "marking time," thanks to the rapid progress of motion pictures. Beyond that, no matter how brilliantly conceived and rendered, his sets, costumes, and lighting remained rooted in "make believe." It was *because* of his genuine love of theater that he was taking a hiatus.

A shift was in the air. The Industrial Revolution, scarcely fifty years old, had ushered in an era of new materials, growing mechanized production methods and large-scale capital investment. Imagine the opportunities! Machines were yielding new problems, ones that only artists were capable of solving. "With a logic that seems inescapable," Bel Geddes explained, we were at the forefront of embracing an aesthetic allied with "cogs, cams and crankshafts."

Redirecting his focus from Broadway audiences to an audience of consumers didn't strike him as that broad of a leap. Both endeavors required research, planning, an understanding of available materials and their potential, complicated mechanics, the ability to coordinate with technicians, an adherence to budgets and schedules, and, last but not least, an educated guess as to what would ultimately appeal. Or as Bel Geddes later put it, "keeping one eye on imagination and one eye on the cash register."[10]

The extended timeline of commerce was a lure. Plays were often mounted, rehearsed, and opened in a matter of months, not always to their benefit. In industry, the time lapse between initial sketches and appearance in the marketplace could take a year or more. Bel Geddes suspected, though, that the "human factors" he'd encountered over the years in theater—idiosyncratic authors, egoistical directors, dominating producers, recalcitrant, sometimes petty actors, and troublesome stage hands, all of whom seemed to believe that a demanding temperament was the prerequisite of first-rate talent—were alive and well, if in somewhat different form, in the place he was headed. Still, he anticipated industry to be more pragmatic, less mercurial, better organized, and so, in one sense, "infinitely easier."

His creative peers saw commerce as a sordid, if lucrative, way station, or worse, a corrosive maw from which they'd never return. Why put your soul into something only to have it replicated—by machines—ad nauseam? And how many concessions would they be forced to accept along the way? "Designers may work in this field without irreparable damage," wrote economist Stuart Chase, "but whenever a promising painter, sculptor or writer begins to use his talents for a brisker turnover in soap, teaspoons and silk stockings, the world stands to lose an artist, even as the bank stands to gain a depositor."[11]

But what if working with industry could be a challenge instead of a compromise? A genuine outlet for imagination, but with many times the opportunities, and with a vastly larger audience than, say, painting, or even theater? Bringing the manufacturer, the shopkeeper, and the consumer closer together might be "fun," Bel Geddes told an interviewer, and his theater experience would no doubt come in handy. The public *liked* showmanship. The hardest part would be that industry was less forgiving than the entertainment world.

There were a few precedents.

Factory owner Sir Josiah Wedgwood, of pottery fame, was one of the first industrialists to recognize that "art pays." His factories employed artists as freelancers, even full-time employees.

Alphonse Mucha, the creator of painstaking art nouveau posters and Moët & Chandon champagne ads, was still putting his hand to everything from Job cigarette papers and Lefèvre-Utile biscuit box tops to carpets, wallpaper, postage stamps, and banknotes. Early in his career, Paul Cezanne had designed candy box covers; two summers back, in Paris, his widow had "very kindly" shown them to Bel Geddes in the painter's own house.[12] In Germany, Peter Behrens was busy taking on everything from furniture, clocks, lamps, dentist drills, and cutlery to a steel, glass, and concrete factory.

And at least one respected American had ventured beyond convention. Edward Steichen was busy photographing images to promote Welch's Grape Juice and Fleischmann's Yeast, Pebeco toothpaste and Jergens hand lotion, when he wasn't shooting surreal, mesmerizing photos of matches and mothballs, carpet tacks, weeds, and sugar cubes, casting hard, crosshatched shadows for Stehli, a company that "translated" his images into daring silk fabric designs.

"Practically all artists who do commercial work do it with their noses turned up," Steichen would tell poet Carl Sandburg. "If I can't express the best that's in me through such advertising photographs as *Hands Kneading Dough*, then I'm no good."[13] As a testament to his belief, Steichen often included his commercial photos in art exhibitions. "When I was putting my soul on to canvas and wrapping it up in a gold frame and selling it to a few snob millionaires who could afford it . . . I did not feel quite clean. But now," he said, speaking to advertising executives at J. Walter

Thompson, "I have an exhibition every month that reaches hundreds of thousands of people."[14]

Bel Geddes's childhood hero, Thomas Edison, still alive at eighty, had also "branched out" from his early successes into iron ore processing. (Edison also invented the charcoal briquet for barbecues.) He'd wanted, he said, to do something "so different and so much bigger than anything I've ever done before [that] people will forget my name was ever associated with anything electrical."

•

WHILE SOME SAW Bel Geddes's willingness to partner with corporate executives as a betrayal of artistic principles, others saw it as an act of hubris. A painter or sculptor need only be familiar with a few media, like paint, clay, stone; a writer must be familiar with words; a jeweler, precious metals and gems; a composer with notes and harmonies; but someone aspiring to cavort with industry might need to be familiar with numberless materials, each calling into play a dozen different processes. Neither Cezanne, Mucha, nor Steiche had set his sights on reimagining locomotives, and there was no word that Germany's Peter Behrens was branching out to add ships and airplanes to his already impressive repertoire.

It wasn't that Bel Geddes didn't have doubts. But there was something to be said for intuition. He told himself that as surely as fourteenth-century artists were remembered for their cathedrals, those in the twentieth would be remembered for their factories and the products created in them. "Now is the time," he would write, "for the world is changing, and the fellows on top when the smoke clears will be those who changed with it."[15]

CHAPTER 1

The Boy from Adrian
1893–1912

Accident is the architect of life.
  —GOTTFRIED REINHARDT[16]

Norman Melancton Geddes was born in a large Victorian house with a wide veranda that stretched across its front and bent around the side. The surrounding lawn, covering several acres, sloped gently down to a street shaded with tall, arching elms. Behind the house were trellised flower beds and vegetable gardens, a greenhouse, a carpenter's shop, a sizable stable, a carriage house, and a paddock. The house, kept in running order by half a dozen servants, was as well-appointed inside as out, with crystal chandeliers, marble-topped mantels, and rosewood furniture resting on Aubusson carpets. It all belonged to Judge Norman Geddes, Norman's paternal grandfather, a benevolent widower whose titles, at various times, had included lawyer, professor, church trustee, and bank director.

The Judge's son, Clifton—spoiled, extravagant, a meticulous dresser—was on vacation from the Culver Military Academy when he met his future wife, twenty-four-year-old Flora Luella Yingling,[17] also known as Lulu, a serious brunette beauty studying music at Adrian College, the country's only Methodist institute of higher learning. Despite the school's religious tenets, he'd managed to pursue a courtship that included placing a ladder under Flora's dormitory window. She was charmed by this clever, affluent young man who took her dancing and riding and enjoyed an occasional game of cards (all strictly forbidden in her parents' household) and whose family was gracious and welcoming. She withstood Clifton's advances until graduation, married him on

March 2, 1892, and moved into the Judge's stately Adrian, Michigan, manse.

The evening she went into labor with her first child, April 27, 1893, Grandfather Geddes told his young coachman and stable boy, Will de Haw, to hurry and fetch the doctor. Will ran to the stable and leapt onto "Old Bob," the family's work steed, without bothering to saddle him; later, recounting the tale, he would compare his errand to Paul Revere's famous midnight ride. Clifton was at the Chicago World's Fair receiving blue ribbons for having bred and raised several of the country's finest horses. It was prescient that the future designer, named after both his grandfathers, was born at the start of the Chicago festivities, which introduced the public to Ferris wheels, Shredded Wheat, neon lights, and Scott Joplin's ragtime. Forty years later, he would blueprint elaborate structures for the 1933 Chicago Fair. Six years after that, on the site of a reclaimed swamp, he would create the most iconic World's Fair exhibit of all time.

•

BY THE AGE of four, Norman was following Will de Haw around like a shadow. A full-blooded Cherokee whom the judge had rescued from the notorious Carlisle Indian School at the age of eleven,[18] Will kept his young charge amused by carving people, animals, and ghosts out of every available broom, pitchfork, shovel handle, croquet mallet, and hitching post. Diaper-changing and hair-cutting were also in his repertoire, and he gave Norman his first painting lesson, whitewashing a fence. During the summer months, the two, accompanied by Norman's St. Bernard, Prince, would sometimes walk to a nearby stream to pick watercress and watch the Judge's horses cool themselves down. Soon enough, the boy—gray-eyed, with a thatch of dark blond hair—also had a pet pony, Jocko, and a pony cart to hitch him to.

Dinners in the Geddes's home often included the Judge's "horse" friends or a range of literary, scientific, and artistic people traveling through. Young Norman stayed awake in the dark of his room, listening to the din of voices below. One evening, a man stumbled in, then struck a match. The boy could make out a pair of bushy eyebrows and a shaggy mustache. More curious than frightened, Norman offered the apparition a seat. The apparition declined, but returning downstairs, he reported the encounter as "a most harrowing experience, having steered off course in a heavy fog. An albino seal come to my rescue." The bushy-faced man was Fridtjof Nansen, a Norwegian explorer, scientist, and diplomat destined for the Nobel Peace Prize.

The birth of Dudley Randolph Geddes in 1897 and the discovery that he was partially lame coincided with Grandfather Geddes's rapid decline. The Judge barely let Will de Haw out of his sight. One morning, the seventy-six-year-old fell back onto his bed as Will was fixing his suspenders, and it was over.

Once the funeral was behind them, Clifton and Flora made plans to start afresh. He would resign his various railway commissions, move his family East, and establish himself as an independent businessman. He'd already begun, investing in a company that promoted sturdy "Mission Oak" furniture with clean, undecorated lines, a radical change from Victorian excess, the brainchild of a man named Gustav Stickley.

But before embarking on a new life, a respite seemed in order. The family traveled to New York by railroad car; also on board were Prince, Jocko, and his pony cart. Will de Haw and a groom followed with two carriages drawn by bobtailed horses in silver-plated harnesses. The entourage spent two months traveling south through Mount Vernon, Harper's Ferry, the Shenandoah Valley, and the Blue Ridge Mountains.

The trip's highlight was a visit to the White House, where a congressman from Adrian, one of the Judge's old friends, introduced Mr. and Mrs. C. T. Geddes to President McKinley. "In the party was their son, Norman," reported the *Detroit Journal*, "who caught the favor of the President with his questions to such an extent that he took him through the other rooms to show him the answers . . . As the party was about to leave, President McKinley removed a white carnation from his lapel, pinned it on Norman's jacket, and with a pat on the head the President asked him to come again sometime, to which the six-year-old replied, 'Yes sir, I will.'"

It would take several decades, but Norman kept his word.

More ambitious than Clifton's Stickley patronage was his investment in the development of the new Squirrel Hill section of Pittsburgh, Pennsylvania, where he chose to resettle the family, buying a hilltop home with a terraced lawn, a carriage house, a stable and grazing field, and views in all directions. There was a miniature stable for Prince and Jocko and a small upstairs bedroom for Will de Haw. A large sandbox was installed that Flora dubbed "Atlantic City." A children's nurse was hired, and a butler with the absurdly perfect name of Humphreys. Clifton had eighteen of his prized saddle and harness horses transported from Adrian.

Norman's menagerie expanded to include a quartet of rabbits, and Clifton became director and president of the Pittsburgh Horse Show group, whose membership included ketchup czar Henry J. Heinz. *It runs in our family to like horses*, Norman, age eight, wrote in a school essay. *My father won the cup for the champion saddles of Michigan.* The couple began entertaining well and often, their home a magnet for musical celebrities like Chauncey Olcott. Opera diva Nellie Melba—a Presbyterian with Scottish blood, like Clifton—displayed her famous three-octave range in the parlor, accompanied by Flora on piano.[19]

Will de Haw drove his young charge to school and back in a high-seated, two-wheeled tandem. One morning, as he was dropping Norman off, the lash of his whip touched the rear horse, who jumped, scaring the lead horse, and the pair bolted. Will pulled behind a streetcar in an attempt to stop, and the cart turned over. Dudley and his nurse were also aboard. No one was hurt, but one horse's head went through a streetcar window, its throat bleeding so badly that a policeman shot it. That evening, Clifton asked Will de Haw to leave.

He'd been a part of the Geddes's household for twenty-two years, "the happiest years of my life."[20] Now in his thirties, he had nowhere to go. A Carlisle School alumnus, he likely knew little, if anything, about his birth family, and there was nothing for him to return to in Adrian. Norman mourned the Cherokee's dismissal for months. No one had known him longer than Will or loved him better.

Will's departure was the start of a downward spiral that soon turned the familiar inside out.

It had been a charmed life while it lasted.

One morning in 1902, escorting an overnight guest—C. Victor Herbert, son of the operetta composer—downstairs, nine-year-old Norman saw his parents' faces gray and taut in the dining room. Humphreys could offer only that "something had happened" between dinner and breakfast. Clifton spent most of that day meeting with lawyers behind closed doors in the music room, which was where, returning from school, Norman found him—slumped in a chair, tears streaming down his face.

He'd invested his inheritance, together with funds from friends and relatives, in stocks he'd been told were secure. Everything had been lost, wiped out in a financial panic. The house, its furnishings, everything

Dudley, Flora and Norman (l. to r.), circa 1909. Notice
that Dudley's leg is hidden.
*(Courtesy: Harry Ransom Center, The University of Texas at Austin,
photographer unknown.)*

would be auctioned off. Clifton had never experienced life without money.

Flora was made of sterner stuff, but watching from the back porch,
she wept when Henry Heinz came for the horses. Her husband passed
from stall to stall, stroking the prize-winning steeds he'd bred and raised,
offering each a piece of apple, speaking to each by name. After saying
goodbye to half a dozen, he couldn't continue. As he walked out into the
yard, the horses began whinnying. Heinz looked at Flora. "They know
something's wrong, Lulu." Jocko and his miniature stable were bought
by the Geddes's next-door neighbor, Harry Kendall Thaw. It was unclear
what the dangerously unstable millionaire's son wanted with a diminutive Shetland—perhaps a gift for former showgirl Evelyn Nesbitt, with
whom he was obsessed. Thaw's name would soon be splashed across
international headlines, his murder of architect Stanford White, one of
Nesbitt's former lovers, deemed the "Crime of the Century."

•

CLIFTON SECURED A job in Chicago, and with a loan from Flora's father, the family moved into two small rooms, windows facing a brick wall, in the Ellis Park Hotel. For the first time in her life, Flora was without a piano; it was her habit to play at least twice a day. The instrument's absence struck her eldest as a symbol of the family's low estate.

Undeterred, Flora took advantage of Chicago's Art Institute, introducing her boys to paintings and sculpture. She also took them to the theater. Norman's first play, *Rip Van Winkle*, made an indelible impression. From then on, every story he heard or read led him to imagine how it would look and sound onstage. A production of *Ben Hur*, with its chariot race and chained men pulling the oars of a Roman galley under the lash of an overseer's whip, was so vivid, so instantly graspable, that it seemed as actual and important to Norman as the people in his everyday life. He constructed a miniature cardboard stage fitted out with scenery and cardboard actors. With tiny candles for footlights, he reenacted all the parts in the plays he saw from memory.

Clifton's bosses didn't elaborate on the reasons for his dismissal, other than to say that he'd only been trained for success and that these were not successful times. He'd always had an irreverent manner and light spirit; even Dudley, six, and Norman, ten, noticed a change, a solemnity more dark than dignified. When Norman heard newsboys outside their hotel hawking a special edition—PRESIDENT MCKINLEY ASSASSINATED!—he felt a sense of panic, remembering the kindly old man offering his carnation.

Two months later, Flora and her sons were living with Clifton's sister, Aunt Hattie, in Saginaw. Flora began giving piano and singing lessons. Clifton stayed behind, working a series of dismal salesman jobs, visiting the family most Sundays, to great fanfare. But when he arrived at Christmas loaded down with packages, Flora's delight vanished at the smell of whiskey on his breath; though not nearly as devoted to Scripture as her parents, she was, at heart, religious. Spirits were for sacraments and the occasional glass of sherry or Madeira. Her husband considered religion a balm for the defeated or mentally weak—in Flora's mind, a fair definition of alcohol.

Before returning to Chicago, Clifton told his wife he'd been offered an assistant manager job in Alabama. A cousin owned a brick factory and was hoping to retire soon; if Clifton played his cards right, he could take over before too long. He and Flora had been struggling for almost four years now. The job wasn't as much of a comedown as it might appear, he explained. Adrian had a venerable history of manufacturing bricks from local river clay. Didn't she remember, the year Norman was born, all those headlines about a nationwide financial crisis? Bricks, fence making, and railway lines had rendered his hometown impervious.

Chicago was far, but Alabama was something else entirely. Flora asked him not to go.

·

AS NORMAN DEMONSTRATED his miniature stage to Leon, a Saginaw classmate, one afternoon, one of its "footlights" fell over, setting the cardboard on fire. Leon mentioned two friends whose dad had a barn with a second floor. Why not build a *real* theater up there? The Wicks brothers were willing but had no idea what to do. Norman took charge, rigging up a curtain to run on pulleys. He ran around town on a borrowed bicycle collecting old clothes to be made into costumes and created tickets on a toy printing press. The first row of spectators would sit on porch-swing cushions on the floor, the second row on low stools, the third behind them on chairs. He glued sheets of old wrapping paper together, tacked them onto wooden frames, then sprinkled them with water so they'd shrink taut, ready to be painted. Voilà! Sturdy, movable scenery flats. His friends watched in astonishment as Norman created an entire world out of nothing but his imagination and some local salvage. [21]

A twenty-five-cent piece (lucre from Norman's paper route) got Norman and Leon into Academy Theatre matinees. Sitting up in the cheapest, top-gallery seats, they made notes between the acts to re-create later as half-hour scripts. Their first production was "The Adventures of the Empty House," with Norman as Sherlock Holmes, the second, a Tom Sawyer pastiche, complete with an orchestra trio blowing on tissue paper over combs.

Seven people came to the last show and we made twenty-four cents, Norman wrote his father in Alabama. *We charge five cents for big people and three cents for children under twelve.*

Unwilling to keep leaning on Aunt Hattie's good graces, Flora traveled to Ann Arbor, returning with a twelve-month lease on a large boarding house and enrollment in the University of Michigan's School of Music to pursue a teaching credential. With a great sweep of energy, she moved her family, began cooking three meals a day for her boarders, and started attending classes. When making ends meet proved difficult, she arranged for her boys to stay with her parents in her Newcomerstown, Ohio, hometown.

Eventually, Flora landed a music supervisor's position at New Philadelphia High School, a prestigious job with a decent salary, and Dudley rejoined her. But she saw no point in uprooting her eldest son yet again. He could remain in Newcomerstown, finish high school, and enjoy his new friends.

•

AT FIFTEEN, NORMAN put little stock in the strict German Methodist convictions of Grandpa Melancton and Grandma Catherine Yingling, so different from kindly Judge Geddes. The only book in the Yingling home was an embossed leather Bible so massive that it required its own table in the parlor, and though Norman never once saw anyone reading it, his guardians were quick to quote chapter and verse at the first sign of trouble.

Much as he loved stories, there were no explanations for his mother's unhappiness, the transformation of his once devil-may-care father, and the disappearance, seemingly overnight, of the general ease of life Norman had been unaware of until it was unceremoniously snatched away. Nor could his grandparents' Scriptures explain why Dudley, who'd never done anything to anyone, had been born with a shortened leg. Despite repeated inquiries, he never received a satisfactory answer as to just who or what God was.

Along with being home to M. Yingling & Son: Groceries, Lamps, Glass & Queensware, Stoneware, Newcomerstown was the site of Lock 21 of the Erie Canal, with some dozen barges and twenty trains passing through each day. But despite plank sidewalks and brick-front shops, men in alpaca coats and women with parasols and ostrich-feather hats, the town remained a provincial outpost. In the spring of 1910, when Halley's Comet passed overhead, much of the citizenry disappeared into stockpiled cellars and windowless sheds, anticipating Judgment Day.

For two years, Norman was happy enough, despite the mandatory church services and grisly parables about tortured saints. There were hayrides in the fall, a canal to skate on in winter. Cy Young, the baseball great, was a local; in the off-season, Norman and his friends sat on the fence and watched him train.

Summer meant camping with a few close friends, manning lookout platforms in high trees, searching for arrowheads, swimming the Tuscarawas River at Anderson's Ford, with plenty of frogs and turtles to catch, or paddling a skiff upstream and then shooting back down. After an "Indian" dinner of corn pone and bacon cooked over a campfire, his friends would talk late into the night. How did fireflies make light without a battery? How long had the stars been there and how had they gotten there in the first place? How did fish manage to travel so quickly through water?

Imagining Indian life wasn't that remote a fantasy. Newcomerstown had once been the village of Gekelmukpechunk, home to the Delaware Turtle tribe. There were still expanses of forest where beaver and deer (the bounty that drew French and English fur traders to Ohio) made their home. The Wounded Knee Massacre, which officially ended the era of the American frontier, occurred only three years before Norman was born.

Norman spent hours lying in the grass, studying the trees arching overhead, listening to the sounds of insects, bullfrogs, woodpeckers, and rabbits, finding creatures that lived under rocks and inside rotting logs. (Perhaps he would be a naturalist instead of an Indian. Better yet, an Indian naturalist.) He'd even helped his grandfather rid the henhouse of rats by catching a pair of four-foot-long black snakes and letting them loose inside. But he'd had to be clandestine about it—Grandpa Yingling believed black snakes were manifestations of the Devil.

The child was very much father to the man. Early passions—an elaborate miniature cardboard theater, a Montgomery Ward printing press, a fascination with Native Americans and the hidden physics of the natural world, even his burgeoning collection of reptiles and bugs—would all inform his adult life. Some would provide grist for professional projects, others for ambitious recreational ones—a distinction he would come to view as arbitrary.

•

WITH THE EXCEPTION of geography and history, school was dull. An aspiring newspaper cartoonist, Norman took to decorating the margins of his textbooks during class. The first time he was caught, he did penance standing on a stool in the corner. The second time, his knuckles were whacked with a metal ruler. The third time, his teacher went into a fury, striking his hands repeatedly until a girl in the class began screaming. On his way home that day, he stopped at the post office to write his mother a postcard: *Not going back to school. I never want to learn anything.* But after a few days, and with no word from his mother or admonition from his grandparents, he returned and was sent to the principal's office.

Ordered to bend over a chair, he had his bottom lashed with a doubled leather strap, followed by promises of more to come. Pulling the boy up by his collar, the principal shook him until he began to cry, then ordered him to return to class and apologize. Norman kicked his tormentor in the shin and ran.

Hometown or not, Flora wasn't about to have her son beaten. She summoned her eldest to New Philadelphia to join her and Dudley and managed to get him enrolled in the ninth grade. He'd been demoted a year.

His new school wasn't much of an improvement. More edifying was keeping abreast of the news. A future cartoonist had to be up on events, after all. Spitballs had just been ruled illegal in baseball. Jack Johnson had become the first black world heavyweight boxing champion. A pair of brothers were experimenting with flying machines at a place called Kitty Hawk, traveling an astonishing 852 feet, at 21 miles per hour, through the air.

Edifying, too, was reading up on Thomas Edison. Some of the things Edison had to say helped make sense of the questions Grandpa Yingling could not. "Nature is what we know . . . And nature is not kind, or merciful, or loving. If God made me—the fabled God of . . . mercy, kindness, love— He also made the fish I catch and eat. And where do His mercy, kindness, and love for that fish come in? No. Nature made us—nature did it all—not the gods of the religions."[22] Spurred on by a magazine article on autograph collecting and the rumor that no one could "get" Edison, Norman wrote to the inventor in care of his New Jersey laboratory. Claiming he'd been elected president of a (non-existent) scientific society, the Young Edisons, he asked for a large autographed photo to hang, and he got it.[23]

Then, barely settled in at his new school, he orchestrated what would prove to be the end of his formal education. Having discovered a box of colored chalk in the brightest colors he'd ever seen, Norman took advantage of a ten-minute class recess to sketch an ambitious mural, ten feet long and three feet high, filling the entire blackboard. In it, the high school superintendent, the principal, and his teacher sat on donkeys whose long ears served as handlebars, while various pupils looked on fondly, their fingers touching their chins in saint-like poses. Doves flew overhead, and flowers carpeted the ground.

When no one claimed responsibility, the principal was called in. After several demands that the perpetrator reveal himself, Norman stood up. Weren't schools supposed to encourage promising talent? "I did it," he said.

His gleeful peers were dismissed for the day and Flora was called in. For the remainder of the afternoon, the cartoonist was questioned, chastised, and finally expelled from New Philadelphia High School, told, at fifteen, that he would never amount to anything.

Then the telegram arrived—brutally short—from Mr. F. M. Van Deusen, owner of the Sylacauga Brick Company in Alabama, saying Clifton Terry Geddes was dead.

•

WHEN JAMES HARRISON Donahey, renowned cartoonist for *The Cleveland Plain Dealer*, got wind of the mural incident, he wrote to say that if young Mr. Geddes ever came to Cleveland, he would personally arrange for his enrollment in the Cleveland Institute of Arts. Ten years before, Donahey had been "given the gate" from the same New Philadelphia school under similar circumstances. Flora believed in education. Somehow, they would manage. Norman packed his bags.

Forget cartooning, at least for the first year, and learn the fundamentals, Donahey advised. Avoid shortcuts and everything else will fall into its proper place. But weren't the fundamentals fairly obvious?

A sculpture class, taught by a man Norman initially deemed a third-rate talent, proved a revelation. He began to understand that surfaces were the result of "inner elements," both actual and theoretical, an idea that was to have applications far beyond those his instructor intended. Other classes introduced him to tonal values, showed him how to tackle complicated visual problems by simplifying them, and offered other basics he would build on for the rest of his life.

Arriving to pose for his Live Drawing class one afternoon was a Blackfeet Indian in traditional dress—breechcloth, long braids down each shoulder, and a black-tipped eagle feather tucked into his scalp lock. Ruggedly handsome with high cheekbones, deep-set eyes, a prominent nose, and a long, sinewy neck, "Chief" Thundercloud was the foremost Native American artists' model of his day, a familiar figure at various art schools.[24] He'd posed for John Singer Sargent, spent years in Frederic Remington's employ,[25] and his profile graced the U.S. Treasury's five-, ten-, and twenty-dollar gold pieces, minted the previous year.

Norman could barely wait for class to end so he could introduce himself. Before the day was out, Thundercloud had accepted an invitation to visit his room.

With his new friend's departure, he quickly lost interest in classroom studies. He could learn just as much, he reasoned, from the paintings and sculptures in the Institute's galleries, not to mention from reading up on Indian lore in the library. His tenure at the Cleveland Institute of Arts ultimately lasted a matter of months. But he kept in touch with Chief Thundercloud, sending letters and sketches, and the following year, he headed East for a two-week stay at the Blackfeet's Pennsylvania farm.

Thundercloud spent his summers on the Delaware River, near Dingmans Ferry, with his wife, Henrietta, a German "lady painter," and their daughter, Wanita. During his fortnight visit, Geddes and his fifty-five-year-old host took long walks in the foothills of the Shawangunk Mountains, sometimes making dinner in the woods over a pinewood fire.

The young provincial had found himself an unlikely father figure. What did he have to offer in return, besides his enthusiasm? Perhaps he mentioned that Pennsylvania was where Great-great-grandfather Geddes had first settled after making the long Atlantic crossing from Scotland, and where his father had once had a grand house on a hilltop and prize-winning horses. But mostly he asked questions, as Dominique DePlante—Thundercloud's Christian name—had extraordinary stories to tell: riding with the Pony Express at the age of fifteen, serving as a U.S. Army scout, touring Europe in Buffalo Bill Cody's Wild West show.

Chief Thundercloud, a.k.a. Dominique dePlante

(Courtesy: Harry Ransom Center, The University of Texas at Austin)

Before young Geddes left Dingmans Ferry, his host gave him several traditional garments and a battle shield painted with the image of an eagle-like creature with flashing eyes, a pair of lightning-bolt snakes beneath its wings.

He explained that the myth of the Thunderbird predated the Spanish explorers. With its twenty-foot wingspan, the raptor had long been revered by tribes throughout the western United States, Alaska, and the Japanese islands. It was believed to never set foot on land.

Thundercloud's stories left his young visitor with an ever-growing interest in Native American lore. And the fact that Frederic Remington, his host's longtime employer, had achieved success, despite creating work that flew in the face of what was considered "serious" art, wasn't lost on him.

•

OBSESSED WITH THE idea of portraying the everyday life of Indians prior to European contact, Norman wrote to Uncle Fred, Flora's brother, who had a cabin in Montana's Bighorn Territory, enclosing several sketches. His uncle passed them along to an Indian trader at Sheridan who, in turn, showed them to Joseph Henry Sharp, a well-known painter of Indian life who may have been the first artist to "discover" Taos, New Mexico. Sharp wrote to the curator of anthropology at Chicago's Field Museum of Natural History where a trustee with a particular interest in the Blackfeet offered to pay Norman's travel and living expenses for a three-month visit to Montana's Lame Deer Reservation—an advance against purchase rights, at one dollar each, for any drawings or paintings.

Which was how in July 1912 nineteen-year-old Norman Melancton Geddes found himself on a sooty, three-day train ride, his valise packed with paints, brushes, charcoal sticks, and a stack of sketching pads.[26] Little Beaver, Thundercloud's tall, poised, copper-skinned nephew, met him at the station.

At Browning's General Store, while Little Beaver arranged for supplies and pack horses, Norman outfitted himself with a broad-brimmed hat, a six-shooter, and a hunting knife. The next day, the unlikely pair rode from sunrise to sunset, following trails along vertiginous cliffs, later descending toward lakes noisy with loons and a family of wading bears.

In the morning, looking down on horseback from the crest of a hill, Norman saw a river, a luxuriant meadow, and a vast circle of white tepees. Fifteen summers back, Little Beaver explained, the camp was three times as large, home to forty thousand.[27] Unlike DePlante, who was Canadian by birth, Little Beaver and his stateside tribe were mandated, with the exception of summers, to live on reservations.

Blackfeet religious beliefs, grounded in the natural world, struck Norman as sincere and meaningful. Their dignity made a lasting impression, as did the exceeding generosity of his hosts—gifts of meticulously handmade moccasins, new buckskin leggings, a doeskin breechcloth, a sorrel pony to ride, a new pair of saddle bags to hold his art supplies, even a pristine tepee for his private use.

For the next three months, he shared meals of boiled meat and berries as he sketched and observed. It was the first time he'd ever been absorbed in something he respected. His childhood friend Will de Haw may have crossed his mind. What would the Cherokee's life have been like had he been born a few generations sooner, and where was he now?

Aware that there was more to capture than time allowed, he applied the lessons of his brief Cleveland Institute tenure, simplifying, reducing tonal values, creating semi-silhouettes that, viewed from a distance, appeared almost photographic. His elegant, understated charcoals (eventually totalling more than 400) captured family gatherings inside lodges, meat being smoked after a hunt, hides being tanned and sewn. He sketched night scenes, too: ceremonial dances, fires glowing inside tepees, figures casting shadows on the walls.

When opportunity and his language skills permitted, he asked about the Thunderbird. It arrived, he was told, at certain times of year to build its nests, bringing storms in its wake, and so became known as the Storm Bringer.[28] It fed on young buffalo calves, which it carried up into the air. As the buffalo herds had dwindled, so had sightings of the enormous bird.

An elder by the name of Curley Bear claimed to have found in his youth the mummified remains of a Thunderbird frozen on the peak of Half Dome, 8,000 feet up—perhaps the loser of a midair combat. A bilingual Blackfeet, the mixed-blood daughter of a local white man, volunteered to find the elder and convince him to share the old stories. More than a year later, a series of envelopes filled with long, rambling oral histories would make their way back to Ohio, where Norman was already toying with the idea of a theater production with the myth as its core.

•

WHEN IT FINALLY came, Clifton Geddes's death certificate read "cardiac congestion," but Flora must have had her doubts. At forty-four, her husband had had no history of heart trouble. She knew two things for certain— that they'd had ten good years and that something had died in him after Pittsburgh. The afternoon she watched him say goodbye to his horses, the light had gone from his eyes.

CHAPTER 2

Chicago/Detroit

1913–1916

Show me a thoroughly satisfied man—and I will show you a failure.
—THOMAS EDISON[29]

The director of the Chicago Art Institute offered to waive tuition fees for three night-school classes if Norman would sweep the steps and walks in front of the building every morning at seven. The head of the Toledo Art Museum (another friend of Uncle Fred's) had written to introduce Norman as a young man of talent, as evidenced by his recent Blackfeet reservation sketches, mentioning his precarious finances while skipping over the details of his short-lived Cleveland studies. Now that Norman was slightly older and perhaps wiser, another shot at professional guidance seemed worth considering.

In exchange for a moth-eaten bedroom, he agreed to wash his landlady's third-floor windows. In exchange for two square meals, he got a job hauling cafeteria dishes for an hour and a half at lunchtime, six days a week. Pedometers were the latest craze, especially among marathon dancers. A month after arriving in Chicago, Norman's had already tallied 206 miles.

That left transportation and laundry. He got a job as a gate guard for the Chicago Elevated Railway six days a week for three hours during the commuter rush. That meant free rides plus three dollars and sixty cents a week. When, during his tenth evening on the job, he was thrown down a flight of stairs by a man twice his size who refused to pay his fare, he decided to look elsewhere.

A fellow lodger suggested taking some drawings down to a local engraving company, Barnes-Crosby. To Norman's surprise, he was hired on the spot to copy lettering onto fashion plates for Sears Roebuck catalogs. He was terrified; a single mistake would ruin an entire plate. With only

Norman had been nine then. Everything had happened in a disjointed, underwater-like way after that—moving in with various relatives, the principal's beating, moving again, getting kicked out of school, meeting Thundercloud, and everything that had happened in its wake. Still, when he willed himself to remember back to the good times in Adrian and Pittsburgh, he also had doubts, which he may or may not have shared with his hardworking mother, who would, he knew, see no virtue in dwelling on the irrevocable.

It's unclear how long it took him to write to his father's cousin. A year? Two years? More? Or how long he'd waited hoping for a response, as the letter, when it finally arrived, was undated.

No, your father did not commit suicide, it began. Typed on Sylacauga Brick Company letterhead, it consisted of three astonishing sentences making up one long, unbroken paragraph, without salutation or signature, as if its author wished to remain as anonymous as possible, though *F. M. Van Deusen, Proprietor* was neatly engraved in the upper left-hand corner.

> *He was taken with a pain and went over to the house and laid down, and a Druggest friend of his Bob Edwards heard that he wasnt well and came to see him, he said that he was going up to his Drug Store and would send some medicine that he thought would relieve the pain, which he did, and your Father took it, I think they called it Mayblossam or something like that, well Cliff did not improve and I went over to the house and found him suffering quite a bit, and I asked him what about getting a Doctor, the Doctor came and gave him some other medicine, I remember telling the Doctor that Bob Edwards had sent some medicine here and that Cliff had taken it, in a very few minutes after he had taken the medicine the Doctor gave him he rolled off the couch and died right then . . .*

The "medicines," it turned out, were a fatal mix.

The rest of the letter waxed casual with a story of Cliff riding a horse into a saloon, right up to the bar—a prank that got him arrested. *Well Sylacauga has been good to me,* F. M. Van Deusen concluded, *and I rather think your Father liked it here, for after all his wanderings, he found a home . . .*

The family's abrupt change of circumstances had led, inexorably, to this.

So much was contingent on money which could, for no good reason it seemed, simply vanish.

an hour break at midday, he had no choice but to quit the busboy job, forfeiting his "free" meals. He also quit two of his three night classes at the Institute. After a full day of detail work, his eyes burned and his concentration was shot.

Things quickly fell apart. At the end of his first week at Barnes-Crosby, he was outraged to discover that his paycheck amounted to three dollars. He'd anticipated at least twelve, enough to get by and send a few dollars home. Everyone started at three dollars, the art manager informed him. Adding insult to injury, he learned that the lettering plates he'd worked so hard on (the results were flawless) were the previous years' discards; apparently he was "in training."

Notebook entry, November 6, 1913: Quit Barnes-Crosby. Pedometer—525 1/2 miles in 79 days.

He managed to get the busboy job back and continued washing his landlady's windows. When not in class, he spent his days drawing and painting—in his room, at the zoo, at the Chicago Historical Society, in the Newberry Library. For fun, he found work as a "super" at the Chicago Opera: a mounted picador in *Carmen*, a peasant in *Cavalleria Rusticana*, a spear-wielding Egyptian in *Aida*.

And he met Norwegian neo-impressionist Henrik Lund, in town for a display of his paintings at the Art Institute. Lund gave him an International Exhibition catalog. "The best of Europe's new generation," he said, asking Norman to comment in the margins. "What a poster designer this fellow Matisse could be!" Norman wrote back. "Never saw such big chunks of color put together so wonderfully . . . This Picasso fellow . . . can paint without it being a picture of something." A Brancusi sculpture was "exquisitely beautiful," a Cezanne landscape was mighty close to an ideal rendering of beauty without being "pretty," and as for the Van Gogh—"Wow."

The weather turned extremely cold. The Institute's steps required snow shovels now. He trimmed the frayed edges of his suit and refused his mother's offer of a coat, which would cost fifteen dollars. I'm not outdoors much, he wrote her.

Notebook entry, November 27, 1913: Pedometer—645 miles in 100 days.

In January, he dropped his last class, setting his sights on the Cook County Morgue. School wasn't the only place to learn what he needed to know.

A policeman spotted him standing around the entranceway.

"What's your business here?"

"I'm an art student interested in anatomy."

"Yeah, well get your anatomy moving!"

The following day, with no police in sight, he opened the door, nearly colliding with someone in a stained white coat. "Dunno," the attendant mumbled. "Talk to the fellow at the desk."

How he secured permission to draw in the "library," where the refrigerated bodies were kept, is unclear, but he would always remember the sickening smell of formaldehyde, the exhaust fan hum that changed from room to room, the heavy groaning of hinged doors, and the godawful screams of sirens.

He spent several weeks pursuing his morgue studies and got to know the doctors well enough to gain admittance to the autopsy room to sketch. At the end of his self-imposed tenure, he left with impressions so vivid that, years later, he could close his eyes and draw a nude in any foreshortened position, as if a model were there in front of him.

•

DISTURBED THAT HIS mother was contributing so heavily to his support, Norman headed for Detroit where, talking up his Barnes-Crosby work, he got hired by the Peninsular Engraving Company—fifteen dollars a week for two weeks.

At the end of his contract, he knocked on the company president's door. Having spent some time comparing the color plate fiction illustrations to the chiseled advertising spreads in *The Saturday Evening Post*, Norman was convinced of the former's greater appeal. Peninsular's work was all pen and ink. Weren't they equipped to make color plates? Yes, he was told, but the process was too expensive for general use.

Desperate, he signed on at the Crystal Palace, the glass-roofed Ford factory in nearby Highland Park where a new assembly line system was paying workers "who didn't drink or smoke cigarettes or read or think"[30] an unprecedented five dollars a day for an eight-hour shift (the standard wage was two dollars and forty cents for nine hours) to bolt together Model Ts at the rate of one every ninety-three minutes.

After a week of repetitive, mind-numbing work, he'd had enough.

Rescue came in the guise of a competition. Five of Peninsular Engraving's employees were entering; the president extended the opportunity to young Norman. Four clients—the Detroit Opera House, the Washington Theatre, the Cass Theatre, and the Garrick Theatre—were each offering a prize for the best program cover. "It's not a one-man job," the president explained. "There are four covers and six of you. Talk it over." The deadline was in four weeks.

Being unemployed, Norman had an edge. But four weeks proved to be remarkably short, given that he'd decided to tackle the lot. After a week, he decided to focus on each theater's facade. It took a second week to figure out which elements would bring each one to life. During the final nine days, the night watchman allowed him to work in the quiet of Peninsular's office until sunrise.

He was waiting in the advertising department when the president walked in, beaming. "You did it, Norman. Your design for the Opera House won."

Cheers all around.

And they also gave you the Garrick Prize!

Two out of four. Too good to be true.

The president held out a telegram. And the Washington and the Cass, too, he said, shaking Norman's hand. All four. I've never heard of such a thing.

The next morning, Norman was reinstated at Peninsular, at an astonishing forty dollars a week, with the promise that he could design for color plates from then on.

Miss Olive MacGurn, a kindergarten teacher and fellow tenant where Norman roomed, volunteered to help with his grammar, which was the worst she'd ever encountered. Reading plays aloud to each other—Ibsen, Strindberg, Maeterlinck, Shaw, Wilde—he became fascinated by how scenes and characters built to a climax, how speech, costumes, scenery, and props each played a role. It was like chess, each move designed in relation to future moves, but with no opponent to misalign the strategy.

He could devote only evenings to this absorbing new passion. He was at the office at nine in the morning and seldom left before seven at night. After several months, he was pulling in weekly invoices to the tune of $800.

He could, he reasoned, earn ten times as much in half the time as a freelancer. Knocking on the president's door, he gave two weeks' notice. He left the room with a doubled salary.

Two months later, he took his portfolio of *The Saturday Evening Post* and *Collier's* ads to Peninsular's rival, Apel-Campbell, who offered 50 percent of all profits from his designs in lieu of salary. He took it. Within two years, Bristles (so-called by his Apel-Campbell colleagues for his thick, brush-cut hair) would be Detroit's most talked about and in-demand illustrator.

Now that he had a bank account, he and Olive attended the theater once a week to see everything from Shakespeare to vaudeville. Why, he kept think-

"Bristles." Tribute from a co-worker.
(Courtesy: Harry Ransom Center, The University of Texas at Austin, painted by Gil Spear.)

ing, can't plays be staged with emphasis, even distortions, that reflect the story's psychology and tensions? Why not create scenery, costumes, even sound, that help sustain continuity and mood? He'd never seen that done. And why not graduate the seat rows to allow clearer views of the stage?

He built a model stage on a scale of one-half inch per foot. He purchased two copies each of plays that interested him—*Peer Gynt, Ghosts, Salome*—cut their pages apart, separating dialogue from action, and pasted them up. He populated his stage with small labeled blocks, one for each cast member; folded bits of heavy paper stood in for stage walls and scenery flats. On large sheets of drawing paper, he mapped out the players' principal movements in each scene.

He hadn't forgotten about the Thunderbird. He broke down the legend into four scenarios: the first told entirely in dialogue; the second in pantomime; the third entirely by mood and symbolism; the fourth with music and any sounds, other than words, that would heighten its effect. Then he incorporated all four into a working script, typing the dialogue on the left side of each page, the accompanying stage "business" on the right—a split-page method he would introduce to Hollywood two years later. He made tempera drawings of the sets and costumes as he imagined them and built a three-dimensional model for each scene.

HIS APEL-CAMPBELL REIGN came to an unexpected end when Barnes-Crosby, Norman's former employer, decided to open a Detroit branch and offered Apel's star designer the managerial slot at $125 a week, plus a 5 percent commission on sales. Barnes-Crosby was the largest advertising illustration house west of New York. At the age of twenty-one, the ninth-grade dropout was now an unqualified success, earning the annual equivalent in today's dollars of $160,000.

He began experimenting with mixing his own printing inks. Once he'd mastered the chemicals, he was able to produce unrivaled color clarity. Within three months, his staff had tripled. Barnes-Crosby's poster business was exceeding $5,000 a week and all indications had it doubling again within a year. He rented a small house and arranged for his mother and brother to move to Detroit.

•

IT WAS INCREASINGLY apparent that *Thunderbird*'s scenes would be enhanced if the lighting could be concentrated into single beams. In the theater world of 1914, all special atmospheric effects were painted onto scenery; projecting light in order to cast a shadow was unheard of. Border lights hung above the stage, bunch lights lit from the sides, and footlights sat in troughs downstage. It was all about visibility, with everything uniform, spread out, flat.

Norman imagined something beyond just "lit." Influenced by the drawings (he couldn't read the French and German texts) of Swiss architect Adolphe Appia, who'd pioneered the control of light intensity and color, and with help from the electrician in his office building, he dissected a switchboard, then constructed a miniature one. After two months of long nights and weekends, his theater model had a customized light system powered by drycell batteries. He improvised spotlights out of inch-wide, three-inch-long cardboard tubes, each with the bulb and reflector from a cheap flashlight; ten slots on either side of the stage were equipped with two.

More late-night pondering suggested that the traditional first row of overhead lamps should be in front of, rather than behind, the proscenium; hung at forty-five-degree angles, they'd enhance facial features and add life to actors' eyes and quality to the costumes without spilling over onto the scenery. There was a way, he was certain, to conceal them— hung from the ceiling, the gallery railing, or the balcony balustrade— from the audience's view and prove them as effective as they were in his model. Lighting innovations led to thoughts about the design of the theater buildings themselves, which soon led to half a dozen basic new form concepts, including the "diagonal axis."[31]

Theater stock companies (troupes who alternate different plays, sometimes weekly, in the same venue) were all the rage. One of the most successful was run by Jessie Bonstelle. Though based in New York, Bonstelle maintained satellite companies in several cities, including Detroit. Norman decided she would be the ideal audience for *Thunderbird*. To the surprise of everyone but him, she accepted the invitation.

His model theater was set up on a table in the middle of his bedroom. A geared counterweight raised the tiny curtain; three acts and one hour later, the curtain came smoothly down. Soon after, Bonstelle sponsored an original Geddes one-act on the Garrick Theatre's full-scale stage. *An Arabian Night* called for one special effect: a single shaft of light that faded in and out with the genie's entrances and exits.

By fitting a carbon arc lamp hood into a large tin box, adding a lens, a thousand-watt incandescent tungsten lamp, and a reflector, and then fitting a three-foot extension of stove pipe to the front of the hood's opening, he managed to throw a thirty-foot shaft of light, which in turn cast a ten-foot illuminated circle onto the stage floor. He even managed to wire it to the house switchboard without setting the place on fire.

An Arabian Night limped along for eight performances. It didn't matter. What did was that he'd invented the first thousand-watt spot lamp. It was the first time that theater illumination, controlled by rheostats, could be varied in intensity, color, and angle, allowing a stage to be draped, as *Forbes* would later put it, "in diaphanous shrouds of light."[32]

•

WHEN ONE OF Barnes-Crosby's top clients solicited drawings for *The Saturday Evening Post*, Norman traveled to Manhattan for a series of meetings. It was an opportunity to see Broadway firsthand—"the Great White Way," brighter by night than by day! He sent letters to the city's top producers requesting appointments, and boxed his model stage and accompanying designs and shipped them ahead.

David Belasco wrote back. Belasco—"the wizard of the switchboard." His *Madame Butterfly* and *The Girl of the Golden West* were practically legendary for their poetic visual effects. Just the man to see!

David Belasco's cluttered office, reached by elevator, was atop his eponymous theater on West Forty-Fourth. Multicolored banners hung from the ceiling, photographs and antlers festooned the walls. There was a set of Japanese armor, a chair once used by King Henry the Eighth, a stuffed fish, a lock of Napoleon's hair. Behind a small table desk sat a man in a black suit and black, high-necked vest topped with a clerical collar.

"Your note intrigued me," said the priestly apparition with thick white hair. "What makes you think you have a new idea for the theater?"

"I have a new way to light the stage."

"I'm the foremost proponent of stage lighting."

"I can control the light, illuminating a specific part of the stage."

"All my lamps have dimmers."

"I don't mean controlling the intensity," Norman explained. "Controlling the area of spread."

"Baby spots? I invented them!"

"Two-hundred watters are weak. I use a thousand."

"A thousand-watt *spot*? You've made one?"

Downstairs, in the theater, Norman uncrated and assembled his model. "What are those lights in front of the proscenium?"

"My first border."

"They'll obstruct the view from the balcony and the gallery."

"Not if they're on the face of the balcony railing," Norman countered, "or suspended from the ceiling, out of sight."

"On the railing . . ." Belasco murmured, pursing his lips.

His spots, Norman explained, could light any area of the stage in any color. Hung at forty-five-degree angles, they could contour faces; hung on upper balcony boxes, they created crosslight.

His theories—that the point of art was to lift audiences from "the plane of common life," and that lighting could heighten crescendos, intensify suspense, interpret mood, in short, be *luminous*—failed to convince. As did his plea that with someone else choreographing the atmosphere of a play, the director would be free to concentrate on his actors.

"Bosh!" The Bishop of Broadway's face turned stony. "You're a kid. An amateur. Go home."

•

THE FOLLOWING SUMMER, en route to New York on business again, Norman dropped Belasco a note. The response was curt: he was busy. Broadway, it turned out, was abuzz with Belasco's latest bold stroke—his production of *Boomerang*, a romantic farce, had done away with footlights, replacing them with dimmer-controlled hooded lamps hung *inside the auditorium*. Norman rushed down to buy a standing-room ticket.

The Bishop had strung old locomotive bull's-eye lanterns, attached with small shutters, along the balcony's railing,[33] exactly where they'd discussed putting them *in this very theater*. Norman felt sick. Small solace that the lanterns were restricted to the center, eschewing the crosslight advantage, and that they lacked lenses.

"All the light comes from above, as in nature," Belasco told the press, "by means of reflectors, invented and manufactured in my own shop . . . The glare of the footlights is a thing of the past."[34] Beamed one influential critic, Belasco had "wrought something of a miracle."[35]

There was no proof and no recourse; the only witness to their meeting had been Belasco's electrician. The idea of patenting his thousand-watt spot had never occurred to him. It was the word of an upstart, out-of-town poster designer against that of a Broadway icon three times his age.

•

DURING A VISIT to Uncle Fred in Toledo, Norman made the acquaintance of Miss Helen Belle Schneider, also known as Bel, a young schoolteacher who'd graduated second in her class from Smith College. Her passions were music and poetry; more enchanting, she was a master of bird calls. The afternoon they met, he kissed her. The following evening, he showed up at her house with a copy of his *Thunderbird* script. Before long, they were exchanging letters.

Thanks to Jessie Bonstelle, Norman now had free rein at Detroit's Garrick Theatre, where he spent evenings and weekends observing performances from every angle: one night from the audience, the next from backstage, the next eighty feet up on the greasy gridiron, the steel framing above the stage. The experience of lying stretched out, suspended seven stories above the action, watching the patterns unfold below, would inform Bel Geddes's productions for the rest of his life.

In addition to working a full-time job, and his play dissections, light experiments, and free time spent at the Garrick, Norman decided to publish an illustrated monthly, *InWhich (Being a Book in Which I Say Exactly What I Think)*. It would be unique, he told Bel, because it would keep "Norman Geddes, whoever he is, in the book at all times. Making him *felt*." Bringing his now-considerable printing skills to bear, there were original woodcuts, custom-mixed inks, and an engraved letterhead that folded into its own envelope (a cost-saving measure). Brother Dudley was commandeered as typesetter and printer, working evenings after school on a small hand press. Norman mapped out a subscription base and a profit margin breakdown, secured a standing order from Brentano's, and made plans to acquire a small job press with foot-power and a binding machine should the demand exceed 500 copies.[36]

He and Bel began collaborating on articles about art and signing them *Norman-Bel Geddes*. Along with being a romantic gesture, it was a relief to drop the ponderous *Melancton*. "Norman Geddes, whoever he is" would never be the rightful namesake of a sixteenth-century Lutheran Reformer.

Bel's father, George H. Schneider, was stingy, suspicious, and capable of physical abuse. To his credit, he pledged fifty dollars to *InWhich* in exchange for sixty copies of each issue, but only because he was unaware that his daughter and "the Geddes boy" were courting.

His only interest, beyond his linseed oil refinery, was golf. In an attempt to curry favor, Norman joined him one weekend for a round at Toledo's prestigious Inverness Country Club. By summer's end, he'd converted the Schneider's entire property into a miniature course that began on one side of the entrance walk, circled the house, garage, trees, and shrubs, and finished up on the walk's other side.

It wasn't that he was interested in the game. But once Bel's father had drawn his attention to it . . . well. More than a decade later, a Tennessee hotel owner named Garnet Carter patented the idea. Carter's Tom Thumb Miniature Golf Course, created with a fairyland theme, would initiate a multimillion-dollar national craze.[37]

•

WHEN THE LOCAL newspaper announced that Charles Wakefield Cadman was giving a concert in Ann Arbor, Norman recognized the perfect man to create *Thunderbird*'s score and wangled an introduction through a mutual friend. Cadman had lived among the Omaha, Osage, and Winnebago tribes, transcribing their songs, and he used Native American themes in his compositions. Fritz Kreisler, Nelson Eddy, and Jeanette MacDonald showcased his songs, and Alma Gluck's recording of "Land of the Sky Blue Water" was inching up the top-seller list. Cadman was interested, if and when a production contract was in place. In the meantime, he offered to pass the play along to an heiress he knew who wanted to produce plays on American themes.

Louise Aline Barnsdall, the daughter of one of America's richest oil tycoons, used her inheritance to help fund projects and people she believed in. Margaret Anderson (whose *Little Review* serialized Joyce's *Ulysses*), anarchist Emma Goldman, and birth control pioneer Margaret Sanger all benefitted from a philanthropy that, given the country's anti-Communist hysteria in the wake of the Russian Revolution, had earned Barnsdall a thick FBI dossier.

Of all her interests, theater was Aline Barnsdall's passion, and she found *Thunderbird* "enchanting." Its settings were inexpensive and the script had a "beautiful simplicity," but perhaps, she wrote, the horses, bear, and dogs could be eliminated?[38]

She offered to option the play for a year. She'd pay Mr. Bel Geddes $500 to direct the production, another $500 for use of the music and his stage designs, and $1,500 to cover costumes and stage properties. The only stipulation was that the production be in California (she loathed New York), where she planned to establish a small, year-round theatrical enterprise. She would cover his travel expenses west, plus the usual sliding scale of 5 to 10 percent on gross receipts.[39]

Despite her interest in American plays, Barnsdall was also hoping to debut foreign works, to which end she'd invited Ryszard (Richard) Ordynski to come aboard as company director. A Pole who'd taught theater at the University of Berlin and worked as a drama critic, he'd gone on to manage the Diaghilev Ballet and run two Warsaw theaters, doing "brilliant things" with the esteemed Austrian director Max Reinhardt.

It was a generous offer, but why California when New York was the center of everything? Norman's reluctance only strengthened Aline's determination. "I take big chances with untried plays and designers," she wrote. "The hunger of the artist for self-expression . . . is more real to me than the hunger for food." They would experiment together. Along with mounting *Thunderbird*, she put him under contract as stage and lighting designer for any productions their California troupe mounted, and she opened a $100,000 account for the acquisition of additional plays and actors. Work would begin onsite, in the fall. In the meantime, *Thunderbird*'s costumes would be made under Little Beaver's direction—and no "hideous" tights beneath the breechcloths. "Thousands go to see the Snake Dance in Arizona," Miss Barnsdall observed. "I see no need for the censor to blush." Should he persist, she'd be happy to take him to court.[40]

•

AT SOME POINT, Bel broached the idea of marriage.

At least that's how Norman chose to remember it years later—that being a year older than him, she was in a hurry. Given his responsibilities— ten employees; a mother and brother to help support, not to mention *In-Which*; evenings "studying" at the Garrick; plans for *Thunderbird* and the rest of his career—the timing seemed crazy. Foolish.

In a letter he wrote two weeks before their wedding, a different story emerged:

> *Oh, Bel, my darling . . . It's asking a lot to ask you to marry me when I have not a cent in advance . . . I'd like to get into my own business (with you, of course) if we could only determine which of our endeavors*

we preferred to concentrate on—let's <u>do</u> it . . . No one can hire one of us without the other . . . no one will know which of us did a certain thing that they like—it was done by Norman-Bel Geddes! That name will mean a fortune some day . . .

I've fought and struggled with no foundation all my life. Now I'm going to establish one and grow!

This foundation is you.[41]

By early summer, Norman was deep in the throes of refining his *Thunderbird* script.[42] Arriving late to work after a long night of rewrites, he found the walrus-mustached Mr. Steiz, Barnes-Crosby's business manager, in town from Chicago, waiting. "Don't you realize," Steiz blurted, following a lecture on punctuality, "that you are very much in the wrong, sir?"

"As long as business under my management exceeds your estimates and shows an increase every month, I'm doing my job."

"That is INCORRECT!" The walrus mustache was trembling. "One cannot make a success of a business while nursing a secret desire to make a success of something else!"

He had twenty-four hours to make a choice.

Following a mostly sleepless night, Norman gave two weeks' notice. Apologies, backpedalling, and promises ensued. No one in Detroit could rival young Geddes's technique, Barnes-Crosby's president cajoled. On his last day, Norman was offered more money than he'd ever dreamed of.

Everyone but Flora and Bel (now four months pregnant) told him he was making a terrible, probably irrevocable, mistake. Having hardly settled into middle-class respectability, he was *voluntarily* stepping off a precipice. But leaving his office for the last time, he felt a great weight lifting. *Thunderbird* was about to take wing.

CHAPTER 3

Hollywood

1916–1917

Los Angeles
Los Angeles
the home of the movie star
what kind of angels
are
out there where you are . . .
    —DON MARQUIS, FROM ARCHY & MEHITABEL

B efore Norman left for Los Angeles, Aline Barnsdall wanted him to come to Chicago to meet the man she'd commissioned to build a theater for their new enterprise.

Aline was unconcerned that Frank Lloyd Wright had never designed a theater, that at forty-nine his career was floundering and no one respectable would hire him. It only confirmed that he was a man ahead of his time. He would design a house, too, named for Barnsdall's favorite flower, the hollyhock. There would be guest houses for leading troupe members and promenades among the olive groves. They'd establish a school with fresh teaching methods, establish a profit-sharing system. "I know it will take time, agony of spirit and all that liberty which my gypsy soul adores," she told *The Los Angeles Examiner,* "but it is something I simply have to do."[43]

"You must realize the importance of Mr. Wright to our plan," Aline explained. Working within an architectural masterpiece would be inspirational.

Two years older than Clifton Geddes would have been, Wright struck Norman as fascinating, despite his grand way of speaking and bohemian attire. The architect produced a set of blueprints, one of Hollyhock House and four of the theater, the latter showing an abstract circle within a square. Aline's expression fell. "Aren't these the plans from five months ago?"

The drawings paid little attention to the theater's interior; it could have been a library, a gymnasium, or a hotel lobby. "Mr. Wright," Norman asked, "in what respects will your theater differ from others?"

"At the right time, young man, at the right time."

Donning a broad-brimmed hat, a cape, and a cane, Wright drove Aline, Norman, and Bel out to see some of his work firsthand. Norman didn't see how Oak Park's Unity Temple was any more functional than its Roman, Byzantine, or Gothic counterparts, and the simplicity of the Willits House in Highland Park (simplicity being something Wright spoke of a great deal) was killed by all the parts sticking out. The spacious, gay Midway Gardens was more to his liking; the Coonley House in Riverside was wonderful, he thought; and the exterior of the Frederick Robie House, with its cantilevered balconies and wide, overhanging eaves, was in a class by itself. By afternoon, they arrived at the new Taliesin, Wright's Shining Brow, still in its early stages at the old site, near Spring Green, Wisconsin.[44] Here, Norman was truly enchanted by "rigid materials being pliantly fitted together . . . around the great handsome trees," life being breathed into rock, wood, and burnt soil.

•

CALIFORNIA IN 1916, especially its southern half, was portrayed in print as a God-given, golden, glamorous Eden on a par with Tahiti. Norman and Bel expected orchids and tropical songbirds at every turn. They arrived to a parched summer landscape.

Mountain cats, deer, and coyotes roamed Los Angeles's surrounding hills; the latter could be heard howling under the bright evening skies. Standing in the center of town, one could see snow-capped mountains. (The city's iconic palm trees, some 35,000, would be planted fourteen years later, in preparation for the 1932 Olympics.) The mail was slow; transcontinental telephone lines didn't exist. There was little in the way of public entertainment. The sewage system left a lot to be desired. The October issue of *InWhich* complained of the endless "monotonous" sunshine.

Still, the fruit valleys were well underway, and "one of the great sensations" was to pass through a quarter mile of orange trees in blossom. In season, Santa Monica's high Palisades were covered with a solid blanket of magenta flowers, beneath which lay a wide, clean beach and the prospect of ocean swimming.

Until Wright's theater was built, the Egan Dramatic School at Ninth Street and Figueroa would be home to The Players Producing Company. (Norman suggested calling it The Seagull, after Chekhov. "We must not borrow from foreign countries," Aline reminded him.) The Egan had a well-proportioned stage and seating for 350. Norman quickly hired a carpenter, an electrician, a property man, a painter, a wardrobe woman, and two seamstresses.

He custom-ordered a dozen thousand-watt, lensed spot lamps, made to his specifications, and set to work creating scenery flats, borrowing a page from Seurat's pointillist technique, which he'd seen firsthand at Chicago's International Exhibition. Following his miniature Detroit model, he hung a line of lights, with color gels and dimmers, behind the proscenium, as well as on the chandeliers and balcony rail sides. Connected to a switchboard, all could be operated by a single electrician. *An Arabian Night* had made dramatic use of *one* 1,000-watter. Now, for the first time, barring Belasco's unacknowledged "borrowing," Norman's invention was being applied full-scale—focus lamps as the sole means of lighting a stage, a practice that would quickly become, and remain, a standard.

Thunderbird was scheduled as the Players' opening gambit. Twelve boxes of costumes arrived, courtesy of Little Beaver, as did the rest of the company, twenty-five professionals and amateurs, and a staff to handle the front office. Rehearsals were scheduled to begin in a month. Everything was in place except the casting.

Three weeks later, still without a female lead, the debut was postponed. Aline paid Norman for another option period, promising that *Thunderbird*'s casting would be a priority at the close of the fall season. They'd conduct a nationwide search, if need be.

When Wright arrived in Los Angeles, his blueprints had progressed but the use of space remained, in Norman's view, "basically wrong." The audience sight lines, especially from the balcony, were exceedingly poor; the stage lacked an apron; there were no provisions for handling or storing lighting equipment, scenery, or curtains, and no stage door through which they could even be admitted.

"First the essentials" was Wright's response. "The little things will fall into place later."

If only Wright's beautiful expression—"form follows function"—was in evidence. "If," Norman added, choosing his words with uncharacteristic care, "you'll just provide me with a way of handling lighting equipment and scenery, I'll keep quiet."

The New York Times had been tracking the progress of Richard Ordynski, Max Reinhardt's "brilliant lieutenant," since his arrival on the SS *Rotterdam* in January 1915, blown to American shores "by the ill wind of war" and a well-timed offer to direct Harvard's Dramatic Club.[45]

Norman was excited at the prospect of working with someone so familiar with the kind of theater he'd dreamed of; someone who, like him, had given up a lucrative career, as a university professor, to pursue more fulfilling goals. No wonder Aline wanted a European of his caliber on board. With *Thunderbird* temporarily out of the running, the Players' first season opening was the American premiere of Osip Dymov's *Nju*, with Ordynski directing. Happily, Norman found the Pole easy to work with, cooperative, and sympathetic, his criticisms always sound.

To accommodate *Nju*'s ten rapidly changing scenes, Norman created six ten-foot-tall, hinged screens, in primary colors. Quickly moved from one position to another, folded in different combinations, and lit with colored gels, they provided dramatic intensity, with the added boon of hiding the crew's machinations. No lowered curtain was required, or long waits between scenes.

On October 31, 1916, *Nju*'s opening-night crowd included Cecil B. DeMille (fresh from his *Joan of Arc* epic), D. W. Griffith (*Intolerance*, his masterpiece, had just been released), Samuel Goldfish (later Goldwyn), Erich von Stroheim (new to Hollywood and a new father), Jesse Lasky (he'd just joined forces with Adolph Zukor to create the Famous Players-Lasky Corp.), and Carl Laemmle (his 20,000 *Leagues Under the Sea* just out), along with Theda Bara (fresh from the cover of *Motion Picture Magazine*), "Charlie" Spencer Chaplin (at the top of his form, cranking out a dozen two-reelers in sixteen months), Marie Dressler (a staple in Chaplin shorts), Wallace Beery (basking in the glow of having caught the largest black sea bass in the world off Santa Catalina Island), Wallace Reid (soon to become the country's most popular movie star), Mabel Normand (most recently in *Bright Lights* with "Fatty" Arbuckle), Gloria Swanson (newly arrived in California to work for Mack Sennett), Douglas Fairbanks (who'd just established his own film company), Harold Lloyd (veteran of thirty-four "Luke" comedies that year), and matinee idol Francis X. Bushman.

Los Angeles Times, Examiner, and *News* all ran excellent reviews, praising everything from *Nju*'s casting and the director to its novel lighting and minimal, protean sets.

Norman, Bel, Aline, and Richard Ordynski began dining together once or twice a week. "Ordy" had wonderful tales to tell—about directing Pola Negri in the forgettable *Slave of Passion, Slave of Vice*, about the larger-than-life personalities he'd worked with: Reinhardt, Stanislavski, Diaghilev. He was also, it turned out, a Shakespearean scholar.

Ordynski was the first person on whom Norman could depend for answers about the things that mattered most to him. They were soon spending hours together outside of work, walking the quiet suburban streets (there was a beautiful orange grove on the corner of Hollywood and Vine) or taking a streetcar to the beach, while Norman inundated his new resource with queries.

Still, there were things that took getting used to. Ordynski was a product of prewar European formalities. The striped trousers, short black coat, and vest edged in white that constituted his daily attire; the derby, the cane, and the handkerchief tucked into his cuff; and the ever-present camellia in his lapel offended Norman's midwestern sensibilities, as did his bowing and hand-kissing and his penetrating blue eyes. And that accent. Didn't everyone living in America speak *American?*

Ordy, thirty-eight, and Aline, thirty-four, began seeing each other outside of work (she had a car and a chauffeur; traffic, in the modern sense, didn't exist), taking excursions to Lake Tahoe, out into the desert, up to Riverside's Mission Inn. Everyone was happy for her. For a while.

•

NJU'S DARK, MOODY story was followed by a glittering farce, future Pulitzer Prize–winning playwright Zoe Atkins's *Papa*, a second American premiere and a second critical success. Chaplin sent Norman a telegram: "Congratulations for the first satirical scenery I ever saw and most excellent." The accolades were all the more gratifying because, as with *Nju*, Norman's folding sets had been created for next to nothing.

Papa played through the Thanksgiving holidays, and director Ordynski was feted around town. "I would rather live on crackers and cheese in Los Angeles and do as I please," he was quoted as saying, "than take $5,000 a month in New York and please David Belasco."[46]

On December 2, Bel gave birth to a daughter, Joan. Norman, meanwhile, was building Aline a theater for children between three and twelve in Egan's rehearsal hall, with a special fourteen-by-nine-foot stage, an extended apron, and seven rows of seats.

The season's penultimate offering was *The Widowing of Mrs. Holroyd* by a young D. H. Lawrence. Then, at Ordynski's insistence, and against Aline's better judgment, The Players staged Von Hofmannsthal's *Everyman* as a Christmas offering—a medieval treatise on the evils of wealth and the pointlessness of earthly vanities. Heavy-handed moralizing cloaked in iambic pentameter? Not exactly what Los Angelenos would consider uplifting holiday fare. When Max Reinhardt had staged it on the square in front of Salzburg's cathedral, in front of 5,000 spectators, Ordy explained, he'd timed the performance so that twilight was deepening just as the invisible spirits cried out, warning of Everyman's impending fate, and he'd placed "criers" on every one of the city's church towers, even atop a distant medieval castle, so that the latter's voice arrived, weird and ghostly, five seconds after the others, just as the moon's first cold rays appeared. A sensation! And what better time than the holidays for pageantry? To complicate matters, the elaborate baroque production required the larger Trinity Auditorium in lieu of the group's own home.

Over the Christmas holidays, the Diaghilev Ballet was in town. Norman was thrilled by Bakst's costumes. The orchestrations of Rimsky-Korsakov, Debussy, and Stravinsky struck him as audible blazes of color. And Nijinsky's brilliant, seemingly effortless dancing struck him as an absolute standard of excellence.

As Diaghilev's former stage manager, Richard Ordynski knew the dancer, a fellow Pole, reasonably well. One afternoon, Ordy, Nijinsky, and Norman attended a Chaplin film. Afterward, out on the street, the dancer stuck his toes out at right angles, twirled an imaginary cane, shrugged his shoulders, doffed an imaginary hat, and began an imitation of the Tramp's signature walk, oblivious to a growing collection of onlookers.

Though beautifully staged and well acted, *Everyman* had been a disastrous choice to close out an otherwise laudable first season. The consensus was that the troupe's future as a permanent, pioneering theater colony was being jeopardized by Ordy's growing influence. *Thunderbird*'s casting, meanwhile, had been forgotten.

By late January, Aline's Irish emotions were getting the better of her. She admired Ordynski greatly. More than that, she was in love with him. But he was hardly a reassuring influence. When they disagreed, she assumed she was wrong. After all, he'd worked in theater all his life.

She knew she was plain and thickset. There was always the concern about a suitor's true motives. In February, her father died, enlarging her already vast inheritance. And then there were Wright's protracted theater plans. Did she really want to build it his way and then have to alter it later? And unbeknownst to anyone but her, she was pregnant.

Though she had no interest in becoming "a parasite" (to quote her friend Emma Goldman) paralyzed by marriage, her situation was awkward, at best. Not to mention that, at thirty-four, she was "old" for motherhood. On top of everything else, Ordynski was already married to a young German actress he'd performed with in New York.

Finally, after a series of particularly distressing quarrels, the Pole defected. The company went on to mount several one-acts without him, but Aline's spirit was no longer in it. Despite carefully orchestrated plans for the upcoming spring season, and with another six months on her lease, she paid off the twenty or so troupe members under contract and shut the theater down.

"Is it his?" Norman asked when Aline's condition began to show.

"Did he deny it?" she countered.

Still, she remained loyal to her dream. Her new theater would operate under Norman's direction, subject only to her. He was pleased but dubious. In the meantime, Aline headed for Seattle, home to one of the few female obstetricians in the country, in the company of an old friend who agreed to pose as her husband.

An FBI agent followed.

Louise Aline Elizabeth Barnsdall was born on August 19, 1917, her red hair immediately earning her the soubriquet Sugartop.

•

AFTER EIGHT MONTHS of apartment living in isolated Beverly, Norman had purchased a house, on the installment plan, in a new subdivision half a block from Hollywood Boulevard. By August, when the rainy season announced itself with a relentless, weeks-long downpour, the ceilings began leaking so badly that the rooms filled with mist.

With Barnsdall's theater shut down, a young family to care for, and house payments due, there was the very pressing issue of income.

Hollywood was a new Wild West where enterprising immigrants concocted dreams on makeshift platforms, captured them on celluloid, and, if extraordinarily lucky, transmuted them into gold. With generous boosting from Chaplin, Norman's work at The Players had attracted the attention of Carl Laemmle, founder of Universal Studios, who made an offer: come up with a two-reeler story, on a patriotic theme; if he liked it, Norman would have a shot at directing it, plus a two-year contract.

Norman's scenario on Nathan Hale was rewarded a $933 budget, plus $250 a week, but the experience proved a let-down—slapped-together sets, cameramen focused on mechanics rather than innovation—not *nearly* as interesting as stage work. After three months on Universal's payroll, Norman asked to be released from his contract.

He had a few fallbacks lined up—rebuilding the Denishawn Dance School's stage and installing a lighting system, designing backgrounds for several Tom Ince cowboy pictures at Culver Studios. But by October, he had to admit that after nearly two years in California things weren't exactly working out.

Manhattan

1917–1919

New York had all the iridescence of the beginning of the world.
—F. SCOTT FITZGERALD

Otto Hermann Kahn was often quoted as saying that he had to "atone" for his wealth.

Born into privilege in Mannheim, Germany, he put aside artistic boyhood ambitions to follow in his father's footsteps as a banker. After working for several years in Berlin and London, he sailed for New York, where, as an employee of Wall Street's Kuhn, Loeb & Co. (then one of the largest private U.S. banks[47]), he soon became a preeminent authority on the cutthroat world of railroad financing. Within three years, not yet thirty years old, he married Adelaide (Addie) Wolff, the aristocratic daughter of one of the firm's founders. His rise to senior partner was rapid, his professional status eventually rivaling that of J. P. Morgan.

The most obvious manifestations of Kahn's wealth were his homes. His five-story, seventy-four-room Manhattan townhouse, at the corner of Fifth Avenue and Ninety-First Street, was often compared to a Medici palace. Built in 1918 of imported French limestone, it featured an elaborate vaulted ceiling, a lapis lazuli bathtub, an elevator painted with airy motifs (Addie was claustrophobic), spectacular views of the reservoir and Central Park, and an art collection that included paintings by Rembrandt and Botticelli.

It paled in comparison to his weekend place in Cold Spring Harbor, a chateau worthy of Louis XIV, set on 443 acres; the opening sequence of *Citizen Kane* includes shots of it. Because Kahn wanted it to tower over the rest of Long Island (and the homes of neighbors like the Whitneys, Mackays, and Vanderbilts), workers spent two years piling up dirt with horse-drawn wagons before construction even began. There was also a Palm Beach residence. The grandeur of Kahn's lifestyle would inspire at

Otto Kahn, c. 1909. Arts patron
extraordinaire.
(Courtesy: Library of Congress. Photographer unknown.)

least one misinformed society matron to greet him gushing, "I know your father, the Aga."[48]

What set Otto Kahn apart from his peers was his deep-felt, unapologetic love of the arts, which, he was convinced, had the power to profoundly enrich people's lives. To that end, he insisted that composers, writers, painters, and playwrights were worthy of patronage. It was the civic responsibility—the privilege and "great mission"—of men of means to subsidize their own future cultural heritage.[49] That was what he meant by "atoning" for one's wealth: using money to help create a more substantial form of currency, one that paid immeasurably higher dividends.

In turn-of-the-century New York, no serious financier would dream of associating himself with something as frivolous and effete as what one senior colleague called "this art nonsense." What Kahn deemed "enlightened selfishness" was, to the vast majority of his peers, a dangerously misplaced philanthropy that fell somewhere between folly and professional suicide. One wonders what they thought about Kahn's preference for taking the subway to work, despite owning several automobiles and employing a full-time chauffeur.

Kahn put his money where his mouth was, giving more than $2,000 to Hart Crane (despite the twenty-six-year-old's alcoholism and penchant for waterfront sailors) so the poet could complete his masterwork, *The Bridge;* subscribing to the U.S. legal defense of Joyce's *Ulysses*; and subsidizing the restoration of the Parthenon. When the First World War prevented Isadora Duncan from getting her money out of Europe, Kahn stepped in to help. When Sergei Eisenstein ran out of funds while shooting *Que Viva Mexico!*, Kahn put up $10,000, well aware that the Russian's politics were at odds with his own. And in an era when racism (like anti-Semitism) was politically correct, Kahn helped jump-start both the acting and singing career of a recent Phi Beta Kappa law school graduate named Paul Robeson. He rarely asked for collateral and generally expected repayment if and when the recipient achieved commercial success. "I have," he would write, "a medieval appreciation of dreamers."[50]

Though many sought him out, he assisted others unasked, sometimes finding them by attending obscure venues—always impeccably dressed, rarely without a large pearl stickpin in his tie and a silver-handled ebony cane—that men of his standing would never dream of venturing into. "No one moves with easier grace through weirdly contrasted social spheres," wrote *Newsweek*, "than this suave Maecenas of Manhattan."[51]

One afternoon in the fall of 1917, Kahn's office received a telegram from a fellow who claimed to have designed a Children's Theater for oil heiress Aline Barnsdall, written and directed for the fledgling Universal Studios, created the first completely abstract staging in America, and rubbed elbows with architect Frank Lloyd Wright. Convinced his destiny lay in New York, he wanted to relocate from Los Angeles with his young family.

Norman used almost five of his last seven dollars to send the wire. He hadn't been smart about shoring up his resources. It was bad enough that he'd jettisoned a highly lucrative, hard-won career in advertising. Now he'd backed out of the two-year contract that Carl Laemmle had gone to the trouble of securing for him, despite a mortgage on a house he couldn't afford, "in a subdivided ranch named Hollywood . . . [though] there was no holly in sight . . . a house so new the paint was hardly dry."[52] A sieve, it turned out, masquerading as a domicile. A house that now struck him as a metaphor for every bad decision he'd ever made.

He'd been sitting on a park bench in front of the Biltmore Hotel, confused and discouraged, his face in his hands, staring at the sidewalk. He was far from home, with a wife, a baby, and no savings. He was impulsive, impatient, and opinionated. He was . . .

The fellow sitting next to him on the bench got up, leaving the magazine he'd been reading behind. A breeze ruffled through some of its pages, and a headline caught Norman's eye: "Millionaires Should Help Artists." It was an interview in *The Literary Digest*. Norman had never heard of Otto H. Kahn or Kuhn, Loeb & Co.

Six months before, Woodrow Wilson had finally—officially—declared war on Germany, and even amid the palm trees and scented orange groves, an impossible world away, anti-German sentiment was running high. It was difficult to make sense of the fact that England had taken up arms against its traditional ally—the British royal family was German, after all. It must have crossed Norman's mind that he was grateful to be half Scottish.

But the German in the interview was chairman of the board of the Metropolitan Opera Company, a philanthropist who'd underwritten the Diaghilev Ballet's first U.S. tour. Because of him, Toscanini, Pavlova— and Nijinsky!—had been persuaded to cross the Atlantic.

Norman read the article twice, three times, and then again. How could he convince this Mr. Kahn that he was an artist worth subsidizing? And wasn't what he was about to do typical of the very impulsiveness that had gotten him in this fix in the first place?

He wrote and rewrote the telegram, counting out the words to make sure he could afford to send it across the five thousand long miles that separated his leaky house from Wall Street. After paying for bus fare home, he had two dollars to his name. Fortunately there were already groceries in the house. Nine days later, Western Union phoned. They were holding $200 in his name.

The story of how Bel Geddes's New York career got its start would eventually make its way into dozens of newspapers, magazines, and books, including numerous editions of Dale Carnegie's *How to Win Friends and Influence People*, as an example of persistence against the odds.[53] Within a few short months, Norman would embark on a friendship with a man whose very different story—born in a cholera-ridden city, one of three out of seven siblings to survive infancy, a factory worker at ten, one of the world's first international celebrities by the age of thirty—remains a Dale Carnegie evergreen.

Norman traveled with Bel and baby Joan as far as Toledo, where they would stay while he tested the waters in Manhattan. But first he headed to Detroit to see his mother, brother, and a few friends. One of them gave him a copy of *Music and Bad Manners* by Carl van Vechten, a friend and *New York Times* critic, along with a letter of introduction.

•

THE NEW YORK City Geddes returned to teemed with civil unrest. Thousands of factory workers, dressmakers, barbers, ship mechanics, and railroad conductors were striking for better conditions or being locked out for trying. Eighty thousand New Yorkers were receiving mandatory draft notices (as a family man with dependents, Norman was exempt), and though the press downplayed any antiwar sentiment as Socialist or anarchist conspiracy, many otherwise-patriotic citizens believed that U.S. democracy had no stake in a foreign war. Hadn't civilization progressed to the point that wars were too stupid and expensive to be possible? Meanwhile, long-suffering suffragists had just scored a major Tammany Hall victory, winning the right to vote by a margin of 100,000 ballots.

Other disruptions were afoot. Alfred Stieglitz was busy rethinking his views on photography, moving from painterly composition toward images that reflected the pace and fragmentation of modern life, and French expatriate Marcel Duchamp, intent on de-deifying art, had recently exhibited his shockingly irreverent, readymade *Fountain*, a urinal signed "R. Mutt." Even baseball, it seemed, wasn't immune to controversy. The managers of the New York Giants and the Cincinnati Reds had been arrested for hosting a game at the Polo Grounds on a Sunday, in violation of state blue laws.

With so much convention being challenged, the timing seemed auspicious. Norman was determined to make his mark. He settled into a modest hotel and immediately sent off a note announcing his arrival. Otto Kahn replied in kind, wishing him good luck. When Norman phoned Kahn's secretary, saying that he'd like to thank his benefactor in person, he was told Mr. Kahn was very busy.

Next morning, Carl van Vechten rang his hotel. Over lunch, Norman mentioned Kahn's apparent lack of interest. He must be one of the busiest men in town, Van Vechten explained. "He makes chicken-feed investments in young talent like you, the way a broker does in stocks. If he finds your stock is good, his investment will continue with a zero or two added. He has a reputation for never making mistakes. That's because he never attaches his name to a project until it's proved itself."

"He plays the cello," Van Vechten added, "and not badly, I hear."[54]

On an icy December afternoon, appointment or no, Norman loaded up two wooden boxes filled with his framed watercolors of costume and set designs—*King Lear, Peer Gynt, Pelleas et Melisande, Romeo and Juliet*—and boarded a streetcar for Wall Street.

A uniformed guard at the twenty-two-story Kuhn, Loeb & Co. informed him that deliveries had to be made at the rear. When Norman protested that his boxes contained works of art for Mr. Otto Kahn, the doorman countered that last week's Rembrandt had entered through the back door.

After considerable pleading, Kahn's secretary led him in. The elusive benefactor was on the phone. When he finished, he turned briskly—ramrod straight, his gray mustache carefully waxed, immaculate in a cutaway coat and striped trousers, a rosebud in his lapel—and extended his hand. (Kahn's soldierly bearing dated from his stint, at twenty, in a regiment of hussars.) But he had little time to spare. Weekend guests were awaiting him on his yacht.

Kahn finally acquiesced ("You're my only friend," the young man told him) to a Sunday morning meeting at his townhouse. This time, Norman splurged on a taxi; there were no streetcar routes to upper Fifth Avenue. A butler opened the massive front door.

Seated in the breakfast room over half a grapefruit, the earnest twenty-four-year-old explained that the hyphenated "Bel-Geddes" on his business card was a "firm name," as he often collaborated with his wife, thus bringing to their projects "both masculine and feminine viewpoints." Kahn glanced at the watercolors, then launched into a series of questions. What kind of work did he prefer? Had he ever met Ziegfeld? Seen Diaghilev's Ballet? Had he been to the Met? Did he like opera?

Though his host "scared the gizzards" out of him,[55] Norman didn't want to pigeonhole himself. No, he'd never met Ziegfeld. He'd *like* to, but he didn't have anything *right now* that was Ziegfeld's kind of thing. He'd seen eight Diaghilev ballets, Nijinsky five times. In fact, he'd met the dancer and even gone to a Chaplin film with him! Opera? He'd "supered" in Chicago two seasons back—a picador on horseback, a peasant, an Egyptian soldier.

To Norman's great relief, he and his new (he hoped) mentor knew a couple of people in common. Charlie Cadman's new opera, *Shanewis*, was scheduled to debut at the Metropolitan Opera in March. And rumor had it that Richard Ordynski was the Met's new stage manager. Unbeknownst to Norman, there was also a Hollywood connection. Foreseeing an opportunity in a fledgling industry, Kahn was in the process of convincing his firm to handle Paramount Pictures' first financing.

"Mr. Cadman and Mr. Ordynski both told me," Norman added, "that they'd like me to design the new Cadman opera." It wasn't *entirely* outside the realm of possibility.

His host offered to call up Giulio Gatti-Casazza and arrange an interview. "All I can do is suggest," Kahn said, rising from the table. "I never interfere in managerial matters."

Monday morning, Herr Professor Ordynski was on the phone. Kahn had kept his promise, and word had already gotten round.

Since falling out with Aline, Ryszard Ordynski had continued on in Los Angeles, making his on-screen debut in Fox Studio's *The Rose of Blood*, a lurid spy drama based on an Ordynski story. Kohl-eyed Theda Bara played a governess in the employ of a widowed Russian prince; Ordy appeared as Vassya, a former lover who convinces her to join the revolution. Afterward, Ordy returned to New York to collaborate with his friend Joseph Urban on a reprise of *Nju* at the Bandbox on Fifty-Seventh Street. Critics praised the ingenuity of Urban's shifting flat sets (a concept copied, without credit, from Norman), but the lighting had lacked finesse.

"Listen, Norman, old fellow," Ordynski offered. "Charlie Cadman is in town and would like you to make a sketch for the *Shanewis* powwow scene, but he needs to approve it before leaving town tomorrow. It might get you all of Act Two. Somewhat unorthodox. He knows you don't have a contract, but Mr. Kahn spoke to him . . ."

Already familiar with the libretto, the present-day tale of a well-educated "native" girl, Norman set to work.

Tuesday morning, Norman found his way to Ordynski's office at the Metropolitan Opera House. His blue-eyed friend was at his desk, the signature camellia, ever-present in California, absent from his lapel. Just then, Gatti-Casazza walked by. A large man with deep-set eyes, pale skin, a rectangular salt-and-pepper beard descending from his lower lip, and a drooping mustache, the Met's manager looked very much the impresario. Norman held up his set painting. Gatti glanced at it, then left the room, muttering.

"Come," said Ordynski. "We see Siedle."

The technical director's office was located across the stage and up a flight of stairs. Midway there, Norman stopped. This was a different opera house entirely from the shabby, weather-stained edifice he'd entered half an hour before. Looking out, the house seats formed a lush sea of maroon velvet, crowned by an enormous starburst chandelier. Four levels of balcony lined the walls on either side of the stage. Stretching out below them were the privately owned "Diamond Horseshoe" boxes, where members of the city's elite, Mrs. Astor's "400," still held sway. A heavy gold curtain hung from an elaborate, scalloped valance, set into an equally ornate, gilded proscenium arch several stories high.

"Young man," Edward Siedle began, before Ordy and Norman even got through his door, "I've been in this business for fifty years! My hair is white with worry over it. Now *you* come. What have you got to show me?" Everyone seemed to know who Norman was and why he was there, and no one seemed happy about it.

He held up his painting.

"Do you mean to tell me that there are places in the world with as much color as this?" Siedle asked.

"Plains Indian powwows vary, depending on the tribe. So does the landscape," Norman explained. "The heroine is from a *southern* Oklahoma tribe. Instead of a set that is obviously *scenery*, this will create a *mood*; it will pull the audience in, emotionally and psychologically."

"Why bring in a smart-aleck kid like you for one show, anyway?" Siedle's ears were flushing. "Fox has been in this business all his life, and he's an *artist*." Exasperated, he turned to his stage manager. "Richard, you know the audience won't care if it's Oklahoma or Argentina. I'm on a budget, and I have Fox on salary. You put me in a very awkward position. This Cadman fellow should have told us sooner. I only saw Mr. Kahn's note this morning. Fox has already done his model."

"All right, Edward." Ordynski lowered his eyelids and shrugged his shoulders. "I'll tell Mr. Kahn, Gatti, and Cadman that you don't want Mr. Geddes."

Siedle hesitated. "No need to rush to judgment," he said, then pointed to the offending colors. "Has Gatti seen this?"

"Of course. He approved it," Ordynski lied, "and asked me to arrange things with you."

Siedle took a deep breath, then called in Fred Fox. "It's like this, Freddie," the tech director began, his voice softening. "Mr. Kahn, Mr. Gatti, and Mr. Cadman all want this young Geddes. Would it hurt your pride if we let him do this one job?"

Fred Fox began shifting his weight from one foot to the other. "After all," he objected, "I'm the in-house painter . . ."

"Mr. Fox." It was Ordynski. "We all appreciate the responsibilities of your job, how much you have to do. Thank you for understanding. I shall inform Mr. Gatti of your fine cooperation."

Norman watched in stunned silence as Edward Siedle dictated a half-page contract: Scenery, props, and twenty costumes for Act Two of *Shanewis*, for a fee of $275.[56] He'd been in New York for less than a month and was officially under contract to one of the world's great opera houses.

Fred Fox had assured Norman that the scenery would replicate his design. But the house painters translated his simple lines into literal, naturalized leaves on twigs and branches. Norman attempted to paint a few yards of fabric himself—with thin, sweeping washes—but the technique wouldn't work on vertical surfaces. The Met's drops averaged eighty by sixty feet; the colors ran down.

The morning his set was to be installed, Norman arrived two hours early. Stationed in one of the velvet seats, he watched as Gatti ("an iceberg with a beard") walked up and down the aisles. The master electrician signaled to his assistant, who released the counterweighted pulleys, and all at once the canvas dropped into position.

Richard Ordynski placed a hand on Norman's shoulder. "It's not nearly as bad as you think," he said.

"You saw what I did in L.A." Norman's voice was one long misery. "And this stage is ten times better equipped than Barnsdall's."

"Don't worry. Music critics aren't art critics."

"We open in a month," Norman protested. "What can I do? It's simply awful."

Ordynski smiled wanly. "You should have seen the *other* American operas."

"You don't realize the dreadful standards that prevail," Carl van Vechten offered by way of solace. "You'll probably get better notices than you deserve. The critics will have to say something decent about *somebody*."

•

THANKS TO MUSIC editor Billy Chase, who'd commissioned a portrait a week for the Sunday *New York Times* music section, Norman began spending most afternoons sketching at the Met.

One afternoon in February, during a matinee of *Pagliacci*, he was standing in the wings, attempting to keep out from underfoot, when Enrico Caruso walked by. Resplendent in his white clown suit, with its large pleated collar and conical hat, the singer stopped just upstage, where the lights fell on him.

It was a perfect pose. Norman concentrated, then began a rapid sketch.

He looked up a second time, and then it was done. It was a good one. Caruso was watching him. Grinning.

During the finale of Act One, Geddes stood motionless, listening to the tenor's aching rendition of "Vesti la Giubba."

> *Recitar! Mentre preso dal delirio*
> *Non lo so piu quel che dico e quel che faccio—*

> *To perform! When my head's whirling with madness,*
> *Not knowing what I'm saying or doing—*

Suddenly, the orchestra's organ-like chords stopped. Caruso shouted out—*"Forzati!"* (force yourself) followed by a skittering, maniacal laugh. Then the bitter, angry *"Tu sei pagliaccio"* (You're a clown). The melody rose up the scale, the muscular voice swelling, almost moving beyond language. A few more words, the orchestra's soft refrain, and the curtain slowly descended and the broken-hearted lover rushed into the circus tent.

Caruso exited to thunderous applause, walking through the backstage crowd and wiping his cheeks, which Norman saw were wet with tears. He stopped, looked back, and motioned for Norman to follow him.

In the star's dressing room, a valet, a dresser, and fellow Neapolitan Antonio Scotti squeezed in around a table cluttered with face paint, atomizers, and good luck charms. "Let's see it!" Caruso beamed.

Norman handed over his sketchbook.

"Ah, good! Very good! I like it. Look, Scotti." He waved off the dresser, who was negotiating the cramped space on tiptoes, laying out Act Two's costume. "It is intermission! I would rather draw than sing. Paper!" he boomed. "Now I make you." He proceeded to execute a series of sketches—Norman, himself, and Scotti. When Scotti reached for his, Caruso grabbed it back. "No, no," he said, "Scotti is spoiled, he has too many," adding it to Norman's small pile. "How long it take to draw Galli-Curci?" He'd seen the soprano's portrait in the *Sunday Times*.

Norman was dumbfounded. "How did you know who I was?"

"I didn't. But I know you draw Galli-Curci by the *way* you draw. Your style. You have a strong style, like me. Everyone knows Rico's drawings. Yes? You work for the *Times?*

"Yes and no. They pay me for each portrait."

"How much for Galli-Curci?"

"Fifty dollars."

"How long it take?"

"Fifteen minutes."

"Good pay."

•

BY THE WINTER of 1918, when Norman met him, Caruso was an international superstar, one of the first, his fame fueled, in large part, by some 250 Victor Company phonograph records—his voice was uniquely suited to the new medium—that sold by the millions. Given the increasing difficulty of importing European talent and the ban on works by German composers and on most German-born performers, he was now essential to the Met's success.

A few days after their initial encounter, Norman got word that Caruso had asked for him. Enrico sat enveloped in an enormous white dressing gown, his mouth stretched open, examining his throat with a long handled, dentist-like mirror. "Ah, my new friend. Where have you been?" he shouted. "And tell me. Why you always carry your hat? No place to put it? From now on, you hang it here. Knock and come in. No difference who is here. Punzo, you understand?" His valet nodded.

"And your coat, where is it?"

What would Caruso think, Norman wondered, if he knew that his coat money had gone to buy a pair of opera scores for his mother? "Winter here is a snap compared to the Midwest," he said. "The weather doesn't bother me."

"When it does, it may be too late. You had no drawing in this Sunday's *Times*."

"I drew Titta Ruffo but they didn't use it."

"Tomorrow, you draw at my hotel, yes? Four o'clock is good."

"With pleasure."

"They better use Caruso," he said, grinning.

The following afternoon, sitting at Caruso's dining room table, he sketched a life-sized red charcoal head. His host reciprocated with several drawings of his own. As Norman got up to leave, the tenor held out a heavy winter coat by the shoulders so Norman could slip it on. No one knew better than Caruso that colds could lead to throat ailments and worse. And the winter of 1917–1918 was proving to be particularly brutal, with below-zero temperatures and a litany of blizzards.

Norman resisted; Enrico insisted. "You must," he said. "Don't be a cabbage! When I see you without a coat, I worry."

Despite a devoted entourage, Caruso was a man with few close friends. His sons were living abroad, and Norman was only five years older than his eldest. His natural warmth had been tempered by zealous fans always wanting to touch him, shake his hand, demanding autographs or, sometimes, money. He was interested in Norman's designs, as he enjoyed building sets of his own. Beyond that, both men had been told early on by teachers that they'd never amount to anything, both had found a mentor in Otto Kahn, and Norman had childhood memories of Nellie Melba singing in the family living room. Melba and Caruso had a long professional history.

•

ON MARCH 23, 1918, *Shanewis*'s opening night, Germany's "Big Bertha" was shelling Paris, the Romanov royal family was enduring a captivity that would end badly, and closer to home, the U.S. Post Office was busy seizing copies of James Joyce's *Ulysses*.

Focused on his own imminent demise, twenty-five-year-old Norman stood near the stage, watching the upscale audience settle into their upholstered seats. In the Diamond Horseshoe, box after box of stiff, sparkling, unimaginably wealthy ladies sat examining one another through gem-encrusted opera glasses. The overture began and *Shanewis* limped into being. Though Norman knew little about music, it was obvious that the gilded ticket holders knew even less. Still, the next day's *Times* review noted the "vastness and mystery" of Norman's Indian village, "a mirage that is a part of the true Western landscape."

•

BY NOW, BEL and baby Joan were settled into a one-room flat near Columbia University. For the next few months, when he wasn't sketching at the Met, Norman kept busy trying to arrange meetings with anyone of consequence. In the evenings, when the shows let out, he tracked down the cognoscenti's favorite bars and restaurants, usually falling into bed after midnight.

After considerable effort, William Brady, owner of The Playhouse, was persuaded to sign him up to design a Hector Turnbull play based on Cecil B. DeMille's *The Cheat* (1915), followed by designing lighting and sets for

The Widow's Might, starring Brady's wife, Grace George. Producer Winthrop Ames hired him to redesign the electrical systems in his Little and Booth Theatres, followed by sets and lighting for *The Truth About Blayds*, starring Leslie Howard in his American debut.

It was a start.

Since coming to New York, Norman had heard rumors about a roof garden slated for the Century Theater on West Sixty-Third as a rival to Flo Ziegfeld's New Amsterdam Roof in the heart of Times Square. With efficient air-conditioning still to come, roof gardens were popular summer venues. Ziegfeld's reputed rival was a forty-three-year-old Lithuanian named Morris Gest, who'd earned his stripes under the quixotic Oscar Hammerstein.

Phone calls to Morris Gest's secretary proved fruitless. Norman left messages under assumed names. He wrote notes. He tried endearing himself to underlings hired to protect Gest from people like him. Finally, he began lurking outside Gest's office for three or four hours at a stretch. When the producer finally appeared—in his signature wide-brimmed velour hat and flowing four-in-hand tie—Norman began talking. And talking. Gest closed the door in his face. The routine continued, but at least, Norman reasoned, his quarry never protested.

Undeterred, he unearthed what he could about the man—that he was married to David Belasco's daughter, was famous for murdering English, Russian, and Yiddish with equal fervor, had a penchant for brilliant colors. Then he remembered the sweet long shot that had gotten him to New York in the first place. He went to Western Union and composed a telegram.

> *Leaving for London next week with plans for a new glass dance floor for Charles Cochrane Roof Garden. Would you like to see it before I go? Address me care of Otto H. Kahn, Kuhn, Loeb & Co.*

The next day, Gest's secretary left a message with Bel. Morris Gest wanted to see Geddes *immediately*. (The fact that London was still the target of bombing raids, making Norman's ploy unlikely, didn't seem to occur to either party.) With studied sangfroid, Norman returned the call, saying he didn't have time to squeeze Gest in unless his punctuality could be vouched for. Then he made a point of being half an hour late for their five o'clock appointment. Gest wasn't there.

He waited thirty minutes, then walked out, only to cross paths with the producer on the sidewalk, just coming in. Though not a tall man, everything about him seemed oversized: his huge, sad, luminous eyes, his generous ears and drooping lips. His office was a pasha's fever dream of pillows, Oriental rugs, and stifling incense smoke. "Weren't you in my office a couple of weeks ago?" Gest asked.

"Certainly," Norman lied.

"You know Mr. Kahn?"

"Of course."

"All right, then, how about this dance floor?"

Norman was prepared. A steel frame with two-foot squares of half-inch-thick frosted glass, lit from below. Never done before. Dress the showgirls in thin fabrics. Begin the numbers with overhead light, then bring up the lights beneath the floor and fade the overhead. The girls would look almost nude.

"That's good, that's good."

"And when the show isn't on, people can dance on the glass. It will seem like they're gliding on moonlight."

Morris Gest took him up to the roof, where Norman elaborated on a half-circle foyer that could open out onto the roof terrace, something the New Amsterdam didn't have. The color would be like a set for *Scheherazade*, not the usual stuffy theater décor. After an hour of hemming and hawing, Norman let Morris Gest talk him out of sailing for London.

In exchange for one year's exclusive rights to the glass dance floor design, Gest offered an admittedly small $100 down, plus expenses. Once Norman had proven himself by installing the floor, incorporating a lighting system for the apron and stage, and creating scenery for the opening production—slated to star Ed Wynn, the *Follies* headliner—royalties and additional fees would be determined. The honorarium was all the cash Norman would see.

•

IN SEPTEMBER, NORMAN learned that *Shanewis* was being retained in the Met's repertoire, paired with a second "Native American" opera, *The Legend*. He signed on to design the latter's imaginary Muscovite setting and medieval costumes. In December, Ordy would be directing the world premiere of *Il Trittico*. And it meant crossing paths with Caruso.

Along with their renewed dressing room chats, Geddes may have joined Caruso and his cronies for a post–*Il Trittico* dinner on Mulberry Street (still home to Neapolitan organ-grinders and the occasional trained monkey) for meals accompanied by cigars, poker, and *scopa*. The two continued to make caricatures of each other. Backstage, mid-performance, the tenor would always ask, "What do you think, Normo? Am I getting them tonight?"

Though the winter of 1919 was less harsh than the previous one, Norman developed a persistent sore throat. When Caruso got wind of it, he sent Geddes to his personal physician, who shipped him off to St. Luke's Hospital for a tonsillectomy, followed by a luxurious recuperation—a large corner room with a private nurse in attendance. Rico paid for everything.

·

THAT SPRING, THE Chicago Opera Company offered $15,000 to design lighting, scenery, and costumes for an opera and a ballet, an enormous jump in both salary and responsibility for a twenty-six-year-old. Upon signing the contract, Norman moved his family into a five-room, fourth-floor walk-up in the West Bronx.

Everything in *La Nave*, an opera based on a D'Annunzio story, pointed to the launching, in the finale, of a vast Byzantine battleship, circa 500 B.C. In the original La Scala production, the curtain lowered on a painted backdrop of a ship about to slide down the ways. Old pseudo-realism! Norman decided the climactic launch should leave nothing to the imagination "but the dawn."[57] His design—a towering, seventy-five-foot vessel, its deck twenty feet above the stage—called for four sections to fold at right angles as it slid on a track from the wings onto the stage. The Chicago Opera's technical director, who remembered Norman from his "super" days five years before, approved the sketches and promised cooperation from his crew; the opera's executive committee concurred.

New York was "ground central" for set construction, especially for complex commissions. An unused Greek church was rented, as no Manhattan carpenter shop was large enough. The vessel's bow and fore would occupy most of the stage, its great prow overhanging the orchestra pit so that when *La Nave*'s star, Rosa Raisa, was crucified on the figurehead, she would be *in* the auditorium.

After the vessel was built and assembled to Norman's detailed specs, there wasn't enough room in the church for the sliding mechanism. Norman and his crew would have to wait until everything was transferred and assembled in Chicago, two weeks prior to the premiere, to test-run the launch. A week before the first scheduled rehearsal, all the scenery for *La Nave* and *Boudoir*, the ballet, was packed into a half dozen baggage cars.

A week later, it was discovered that one of the cars—the one containing *La Nave*'s finale—had gone missing. While the railroad company tried to trace it, there wasn't much choice but to proceed with rehearsals. In the meantime, the opera's general manager, Cleofonte Campanini, had returned from Europe with Italo Montemezzi (*La Nave*'s composer would conduct the premiere) and George Maxwell, manager of the New York office of Riccordi & Company, publishers and copyright owners of every Italian opera of note.

Norman took his designs to Campanini's office for the trio's inspection. First came the production curtain, done in mosaic style, showing a ship at sea, to be used during the overture and two intermissions. Next came the prologue scenery and a barrage of heated Italian.

"Impossible!" said Maxwell. "Why didn't you replicate what was done at La Scala? We sent you photographs!"

The set, Norman explained, was entirely in accord with Mr. D'Annunzio's description. If they'd wanted the same old thing, why had they hired him to design a new production?

"Those photographs were sent for you *to follow*. You had no right to deviate from them."

Norman studied George Maxwell's face. "Your business manager assured me they were merely for my information." He took a deep breath. "The whole thing has to meld together—the story, the music, the setting, the action, the lighting—so that the audience loses itself in the story."

"It *must* be a facsimile of the Milan production!"

"My contract imposes no such limitation."

Maxwell exploded. "What impudence, to trample on La Scala's traditions. To think of great artists like D'Annunzio and Montemezzi having their work trifled with!"

Norman uncovered some of his 200 drawings, including 60 costume designs. "Awful! Terrible!" howled Maxwell.

Herbert Johnson, the opera's business manager, was sent for. In a scene reminiscent of Geddes's first encounter with Siedle, Fred Fox, and Ordynski at the Met, he ended the discussion with dispatch. "Gentlemen," he said, "Mr. Geddes was authorized to make the designs as you see them. The scenery has been built. I suggest we stop arguing. We open in five days."

In faraway Italy, D'Annunzio interrupted his political activities long enough to challenge the designer to a duel; Geddes read about it in the newspaper.

Next morning, the prologue scene was arranged and lit. The costumed chorus, moving in front of the mosaic curtain, created a *moving* mosaic, much to the approval of Beatty, Johnson, and stage director Jules Beck. It was the best thing Norman had ever put on a stage.

"Maxwell doesn't know good from bad," explained soprano Mary Garden. "But his company rules Italian opera. They own the copyrights. Even Gatti is terrified of him."

Maxwell tried, unsuccessfully, to get old scenery substituted for Norman's new designs, and a press release was drafted announcing the premiere's postponement. Then, at two in the morning, the phone rang. The baggage car had been located in Indiana. The premiere was back on schedule. "Old World celebrities, a poet warrior, and a great musician are being asked to bow to an American innovator!" a woeful George Maxwell informed the press.[58]

The following afternoon, opening day, tensions were running high. What happened next is unclear. A sandbag *may* have come loose just over Norman's head. Another version had him being slugged by one of the Italians and thrown out into the gutter.[59]

In the fracas, the gravity slide, which had been calculated at a reduced scale, remained untested. There was no guarantee it would work. The hulk of the ship, with forty actors on her deck, at a combined weight of six tons, would be held in place by a chain. Once released, the ship would, in theory, be propelled by its own weight, rolling its way down a sloping ramp on ball bearings set into tracks.

That evening, November 18, 1919, the prologue and first two acts ran smoothly, to considerable applause.

When time came for the finale, Norman walked to the rear of the auditorium to watch. The sun rose over the horizon, thanks to fifty 1,000-

watt Bel Geddes focus lamps bunched together on one side of the stage. Finally, the crash of cymbals came, the cue for the chain's release, then the clink of sledges. Norman clenched the railing in front of him. The enormous ship wobbled. Then tottered. Finally, it began to glide downstage, slowly picking up speed. The mid and rear sections folded into place, and the great prow arrived at its destination, jutting out some seven stories above the front row seats. The cast's final choruses were drowned out by the audience's cheers. The curtain closed to a standing ovation.

Boudoir, the ballet that followed, was something of an anticlimax, though the reviews were generous, and it quickly led to a commission for *Erminie*, a French musical comedy. Norman's *Erminie* staircase—what appeared to be a freehand pen-and-ink drawing down which the play's lead made his thirty-foot slide entrance—caused gasps, and sometimes screams, from the audience. Every performance sold out, and Norman repaid Otto Kahn the $200 that had brought him East.

CHAPTER 5

From Dante to Gershwin

1919–1922

I could speak of a Whiteman rehearsal
At the old Palais Royale when Paul
Played the Rhapsody that lifted Gershwin
From the "Alley" to Carnegie Hall.
   —IRVING BERLIN[60]

J anuary 16, 1920. Prohibition was going into effect at midnight.
Curious as to what all the fuss was about, Norman Geddes had
his first taste of alcohol that night. A Tom Collins. ("God knows,"
Kirk Brice told his workaholic friend as he mixed it, "you need *some*
way to relax.")

Prohibition would create far more drinkers (and drunks) than it dis-
suaded, making the United States the world's greatest importer of cocktail shakers. *Zozzled, scrooched, spifficated, soused, stink-o, embalmed,*
plastered, pie-eyed, loaded to the muzzle, fried to the hat. Dozens of new
synonyms entered the lexicon. Drinking went with smoking (tobacco
or mary jane), which went nicely with jazz and dancing. Thousands of
speakeasies popped up in Manhattan alone. Peepholes. Secret passwords.
You never knew who you might run into. One night during a raid, Texas
Guinan rushed the Prince of Wales into the kitchen of her club, dressed
him as a chef, and told him to start frying eggs, pronto.

Enter Cafe Society. The floodgates of a hidebound social order, dictated by thoroughbred bloodlines and inherited wealth, were about to
crack open, races and classes mixing, helped along by two headline-
making alliances: the weddings of Israel Isidore Baline, also known as
Irving Berlin (son of poor, Russian, pogrom-escaping Jews), to a Roman
Catholic Comstock Lode heiress, and of boxing champ James Joseph
"Gene" Tunney (descendant of poor, Irish, potato-famine-escaping immigrants) to Andrew Carnegie's grand-niece.[61] A decade later, the egg-frying

Prince would make an even more scandalous leap, eschewing the British Crown for a twice-divorced American commoner.

In their zeal, the nation's temperance fanatics had handed over the key to Pandora's box. And Pandora knew how to throw a party.

•

THE CLEVER, LIGHT touch of *Erminie*'s sets led to illustration assignments for *Harper's Bazaar* and *Vanity Fair*.

Editor Frank Crowninshield was the force behind *Vanity Fair*'s many brazen firsts—a full-color Matisse nude, a paper doll of J. P. Morgan in his underpants, the famous "We Nominate for Oblivion" feature. A New England blue blood with Edwardian manners, he ran a fairly loose ship, with a crew that included Edmund Wilson, Dorothy Parker, Robert Benchley, and, later, Clare Boothe Brokaw (the future Mrs. Henry Luce).

"Crownie" was an amateur magician, and Norman sealed their friendship one afternoon by performing a trick he'd perfected during a brief teenage stint as Zedsky, the Boy Magician. "Volunteer" Bob Benchley was ushered into a large burlap sack. The sack was knotted, the knot sealed with hot wax. Norman recited his patter, and voilà! There was Benchley, standing free, the empty bag still tied and passed around for the wondering inspection of all.

Soon after, at a small stag dinner at Crownie's well-appointed penthouse, Norman was introduced to the difference between turtle and terrapin soup "and how much that difference cost."[62] (The once-prolific diamondbacks were teetering on extinction.) The soup had been ordered two days in advance from Crowninshield's Baltimore club, placed on a one o'clock train, transported by valet to the prestigious Knickerbocker Club for final preparation (cream, sherry), then brought to the penthouse in a hot tureen at the appointed hour.

The evening made a lasting impression. Some fifteen years later, Bel Geddes's household would include several terrapins as pets and a butler who served sherry with the soup.[63]

Norman's social circle enlarged when Benchley treated him to lunch in a musty, second-rate hotel on West Forty-Fourth Street, the Algonquin, where on the far side of the dining room stood an entirely unremarkable round table. Norman found himself shaking hands with Heywood Broun, Harold Ross, Marc Connelly, George S. Kaufman, and "all the famous rest."

At first he found their verbal pyrotechnics and acute sense of timing "breathtaking." He would come to know them—the "laughably austere" Alex Woollcott in particular—well. They were "an awfully smarty pants group" who lived for gags, witticisms ("Get me out of these wet clothes and into a dry martini"), and dreadful puns mixed in with scathing bon mots. Groucho Marx, whose brother Harpo was Woollcott's devoted (if unlikely) friend, remarked that the price of admission was "a serpent's tongue and a half-concealed stiletto."

Along with verbal jousting, the so-called Algonquin Round Table had two favorite pastimes, both played as blood sports. After three nearly twenty-four-hour poker sessions and some losses, Norman "got awfully tired" of the relentless bluffs and counter-bluffs. Croquet reached its apogee during summers on Neshobe, Woollcott's private island, to which Bel Geddes was eventually invited—seven rustic acres in the center of a Vermont lake, where the critic played demonic camp counselor to his anointed brood.

At about the same time, Crownie introduced Norman to the Coffee Table Club, where poker and "gold-plated" conversation, the Algonquin's forte, were forbidden. In the course of an afternoon, he met William and Lucius Beebe, Nelson Doubleday, Alva Johnston, cartoonists Don Marquis and Rube Goldberg, photographer Arnold Genthe, Broadway producer Gilbert Miller, conductor Walter Damrosch, painter Rockwell Kent, and the prime minister of Australia. Aware of the designer's finances, Crowninshield quietly covered Norman's initiation fee and first-year dues.

Suddenly, his world was exploding.

•

BOXING IN THE Twenties was even more popular than baseball.

On a humid, overcast Saturday in July 1921, Norman and three friends—Zoë Atkins (*Nju*) and actresses Jobyna Howland and Ethel Barrymore—boarded a special train bound for Jersey City to witness the "Fight of the Century," Jack Dempsey and French war hero Georges Carpentier vying for the world heavyweight crown. There were 80,000 watching; another 100,000 gathered in Times Square around a loudspeaker; still another 500,000 listened to the bout thanks to wireless "technology."

Within a year, the number of U.S. households that owned a radiola set would jump from an estimated 60,000 to 400,000. By 1928, the snappy lilt of dance bands, the tension of quiz shows, the screams of sirens and murder victims in detective dramas, the drone of political speeches, plus the buzz of static would be commonplace in millions of living rooms. Not even movies infiltrated public life as quickly as "talking furniture," some of it eventually designed by Bel Geddes.

•

FOLLOWING HIS WORK on *Cleopatra's Night* the previous January (*The New York Times* found his sets "most gorgeous"[64]), with Ordynski directing, Norman had been permanently banned from the Met thanks to George Maxwell, who'd written Gatti-Casazza a scathing epistle.

Then, in August, came news of Caruso.

Though Norman had visited the tenor's new fourteen-room suite at the Knickerbocker, between the banishment and the tenor's newfound family life (Enrico had remarried and was a new father), they had inevitably seen less and less of each other.

First Caruso contracted pneumonia, then pleurisy. Ever the professional, he'd continued to tour and sing. It didn't help that, for years, he'd smoked two packs of Egyptian cigarettes every day. The previous Christmas Eve, he'd given his 607th Met performance, his sides strapped in bandages beneath his costume to lessen the pain of his infected lungs. Now, seven months later, he was dead—not even fifty!—and with the dubious honor of being enshrined in a glass casket, like an immense wax doll, for the edification of weeping thousands.

Rico of the winter coat, the good luck charms, the private hospital room, the fulsome laugh, the nights of poker and scopa. Rico in an immense pleated collar, alone under the spotlight. *Normo is my friend and you will tell me you like him*, he'd told an astonished Gatti. *Don't be a cabbage! What do you think, Normo? Am I getting them tonight?*

•

ERMINIE'S HUGE SUCCESS was followed by an excruciatingly slow and difficult year. Meanwhile, after decades of eighteen-hour days, Otto Kahn had suffered a fairly severe stroke. He was only fifty-three.

Had he not done *Erminie*, Norman would have considered designing *The Merry Widow* and *The Chocolate Soldier*, the only forthcoming offers. But he was determined to avoid being typecast. No more musicals until he had at least one play under his belt.

Sitting at his desk in his fourth-floor Bronx walk-up, he searched for a project, something, anything. The hours dragged and the days crept. Discouragement, he told himself, was part of the price of admission for trying to create things that didn't exist. After ten months of concentrated idleness, the objects on his desk and pictures on the walls began to annoy him. He finally removed them, pushing the desk into a bare corner. Better.

He stared at the wall day after day. At some point, a section began to throb. *I knew it was my imagination, but . . . the thing became more localized and definite. The wall was pulsating. I told myself the whole thing was absurd . . .*

Maybe he was going nuts. Da Vinci had written something in his notebooks about arbitrary patterns, stains on a wall, ashes in a fireplace, clouds . . . Perhaps he had a fever? Then, a particular spot began glowing *like a coal being breathed on. The harder I stared, the hotter it burned.*[65]

One night, the pulsing, glowing spot began to spiral, like a fiery corkscrew burning a hole in the wall. It seemed to go on for hours. *My body was hot and sweaty, I could feel my hair plastered to my forehead . . . My head was on fire . . . I staggered into the next room and fell headlong into a bookcase. Dazed and scared, I discovered I was holding a book . . .* which he opened at random and began to read.

As flowerets, bent and closed by the chill of night, after the sun shines on them, straighten themselves all open on the stem, so I became with my weak virtue. And such good daring hastened to my heart, that I began like one enfranchised.

It was the Norton translation of Dante's *Divine Comedy*. The lines struck him as prophetic. He immediately read the three volumes from beginning to end. Next day, with the pulsating spot still in place, he began work on a dramatic visualization that would end up occupying half his working hours for the next two years. More than simply a play, it would be an entire world—a total environment of color, shape, movement, and sound, a full-immersion experience shattering the invisible "fourth wall" that traditionally separated viewers from the viewed. It was the beginning of an ambitious and groundbreaking theatrical exploration, one with consequences beyond even Norman's fecund imaginings.

With a copy of every translation he could find, he pasted the pages up on large sheets, comparing the versions line by line. He became convinced that the story's dominant image, like the spinning, burning spot, was a gyre, a widening circle, "like a fly entering a funnel through the spout."

The action would begin slowly, along parallel lines, but as the tempo increased, the lines would bend and arc, the angles vanish. The set would be a deep bowl of an amphitheater—modern, abstract—composed entirely of steps, tiers, and levels (what Thornton Wilder would later call "a Zeppelin hangar"[66]), cut through with uneven planes and framed by four massive, irregular towers. At the back of the stage, a slope would rise up five stories. A twelve-foot ledge would lead down toward the audience, whose chairs would be arranged in a half circle, like ancient Greek theater.

A complex system of lights would glow and diminish, "creep," "slide," "roam," "settle," and "burst," creating texture, blotting things out, throwing them into fantastic relief, sometimes accentuated by artificial smoke. For "The Inferno," all light would emanate from the pit, spreading until the stage and the auditorium were suffused in a whirling, nebulous glow. For "Purgatory," light would come from the rear, in silhouette, lifting overhead, a starry night sky growing brighter and brighter until, with the light coming in from all sides and *directed into the audience*, Dante's words would ring out:

"O abundant Grace, by the Eternal Light, let my sight be consumed!"

Blackout.

Then ever so gradually, a soft glow would return, enveloping the audience in "an abiding sense of the indefinable vastness of eternity." "Paradise."

The extent to which Bel Geddes influenced the development and practice of modern theatrical lighting "can only be estimated," theater arts professor George Bogusch would write decades later, referring to his elaborate concept for Dante, his "spots," and other innovations to come, "but the available evidence suggests it was very great."[67]

There would be no music. He imagined a sonority of vibration and tone, "exquisite, terrible, never common," that obscured, even transcended, the instruments playing it. With electrical sound amplification yet to be invented, he mapped out three large, vibrating chambers. Placed beneath the stage, they would magnify the orchestra's sound waves (created in supplementary rooms) with baffles, then release them via pivoted panels, the results multiplying into a series of echoes. Speakers—traditionally backstage center—would be placed in the bottom of the pit, on the tower tops, below the audience's seats, at either side and at the center rear of the auditorium, and from the ceiling—"surround sound" two decades before the fact.

Actualizing this otherworldly soundscape proved slow going, but progress was made experimenting with compressed air, wire, steel sheets, iron balls, steam, and water.

Music and sound pioneer John Cage was nine years old at the time.

Costumes, worn or carried, would serve as scenery, some ordinary, others highly abstract. He fashioned eighty-six earth forms, long, opaque pointed garments on wire frames that rested on the wearer's hips, strapped onto their waists and thighs, self-balancing. (The longest reached out thirty feet over the pit but was only two feet high. The shortest, nineteen feet long, rose up to twelve.) For "The Inferno," there were gargantuan bats with hinged, telescopic wings, controlled from within by wires. A serpent composed of eight crawling men was covered with a single, slimy-looking garment. In "Purgatory," actors would carry feather-like objects suggesting giant angels. In "Paradise," they'd be dressed in gauze.

Given the enormity of the stage, masks were the only way to convey intense facial expression—from passivity to mortal terror—and all would be constructed to amplify the wearer's voice. Everyone—523 actors and chorus members—would wear them. They could be hooked, three at a time, on an actor's belt. Soldiers would have body masks created for different parts of their physique. The acting would be stylized and exaggerated, limbs figuring as largely as voices. The overall effect—costumes, movement, light, sound—would be hallucinatory and profoundly moving.

People had less idea about theater's possibilities than geographers had about the world in the fourteenth century, Norman would write.[68] The only person he could really talk to about what he was trying to accomplish was Richard Ordynski.

Years later, describing an exhibit of Norman's *Divine Comedy* costume drawings, a critic would write:

> *He is the indispensable visionary who has transcended the practical and profitable present . . . for an impractical, less profitable glimpse into future . . .*
>
> *Color, color, color! Masses of color that weave and mingle in*

*flamboyant patterns: soft colors, contrasting colors, grandiose colors!
But no dissonance . . . Geddes knows the tricks of radiating lines, of
concentric formations . . . an extraordinary concept.*[69]

•

HE BEGAN OFFERING stage design classes, a way to both get out of the apartment and bring in a bit of cash. Every Saturday for five months, he met
with fifteen to twenty carefully selected pupils in a loft space. Tuition was
$200.[70] (Eventually there were some 200 alumni, among them future
industrial designer Russell Wright, future Tony Award–winning cos-
tumer Aline Bernstein, and a Columbia University grad who, by 1930,
was managing Mei Lan Fang, one of the greatest Peking opera performers in China's history.) "Each part must be an essential element," went
one lecture. "The value of the design lies in its homogeneous integrity, in
giving significance to an idea and fulfilling it."

To make his complex *Divine Comedy* tangible, he created a four-foot-
wide stage model, populated by 500 costumed and masked figures. "This
boy is a genius," noted Carl van Vechten in his private daybook, following
a guided walk-through.[71] Eventually, Norman scraped together the funds
to hire Francis Bruguiere, a friend of Alfred Steiglitz. He wanted hyperrealism, drama, *atmosphere*, not simply a photographic record. Henry
Dreyfuss, one of Norman's Saturday students, was commandeered to assist. A quiet, bespectacled seventeen-year-old, he was seated on the floor
beneath the model, instructed to puff away on three cigarettes at once—
the first time he'd ever smoked.

Bruguiere's haunting photos, widely published in magazines, newspapers, and theater journals, appeared decades before photography was
accepted as art. Norman's radicalism was further underscored by what
was passing for theatrical innovation. In Irving Berlin's newly minted
Music Box Review on West Fifty-Fourth, a dozen young women appeared
in pearl-encrusted gowns, headdresses, and shoes. When the lights were
abruptly cut, the pearls, radium-filled and blasted with arc lights minutes
before, outlined the women with an eerie phosphorescence—smoke and
mirrors compared with Bel Geddes's transcendent vision.

The *Divine Comedy* model and its miniature figures constituted the
bulk of America's submission to the International Exhibition of Theatrical Art in Amsterdam, where, together with a book of Bruguiere's photographs, they created more buzz than anything in Norman's résumé. "A

certain excitement and trembling struck me," gushed *New York Times* theater critic Stark Young. "It was as if [Dante's] soul was present and visible . . . in all its strange purity and force."[72]

Suddenly Bel Geddes was spending all his time fielding phone calls and procuring and revising production estimates, including one for Madison Square Garden. Letters poured in from Fordham University, Notre Dame, and as far afield as France and Italy. Catholic societies made proposals. Leopold Stokowski suggested that Dante, Bel Geddes, and Igor Stravinsky would make an extraordinary mix.[73]

Though the vastly ambitious project was, in the end, never produced, it opened doors that would inform and transform Bel Geddes's career for decades to come.

•

PATRONS OF THE Palais Royale at Forty-Eighth and Broadway were expected to pay exorbitant entry fees, followed by a pricey menu that included Green Turtle Royal Consommé (minus the terrapins) and entrees like Reindeer Steak Grand Veneur. Credit for the nightclub's tremendous success belonged, in large part, to an elegant Oliver Hardy look-alike whose pencil-thin mustache, double chin, and deep widow's peak were becoming recognizable trademarks. Paul Whiteman was Manhattan's very hot ticket. His band's orchestrated jazz didn't encourage the "sin" in syncopation, wasn't impudent and demanding with an edge of despair, like the music one found in Harlem.

By 1922, management finally acknowledged that the decor—gilded cupids and heavy velvet drapes, Versailles-like murals, dusty palms and ornate chandeliers—was at odds with a sophisticated postwar world.

Bel Geddes was offered the commission, his first major work outside of theater. He asked for $10,000 in advance (about $141,000 today) and was "flabbergasted" when the owners agreed. Barely skipping a beat, he switched from Dante's thirteenth-century *Comedy* to Fitzgerald's Jazz Age.

The dining room, designed to seat 400 (as per Mrs. Astor's dictum), featured an enormous oval, diamond-parquet dance floor. Norman arranged intimate circular tables around its edges. The rest of the edifice came down, replaced by pale gray walls with simple curves and angles; accents of gold and vermilion were picked up in the chairs. At the far end of the room was an orchestra platform set off by slender colonnades. High, shallow niches housed stylized friezes of flappers and their partners—cut in layered relief (a technique Norman had developed for *Boudoir*)—their exaggerated poses reminiscent of Matisse's dancers but ever so slightly louche. Lit from behind with vivid color, they created sharp, blocky shadows. Reviewers would call them everything from "heroic" to "grotesque."[74]

The Palais Royale reinvented. From Victorian cave to Manhattan's "hot ticket." *(Courtesy: Harry Ransom Center, The University of Texas at Austin, photographer unknown.)*

The high ceiling was an elliptical umbrella canopy. Hundreds of concealed dimmer-controlled spotlights softened its stark white curves, which in turn diffused light through the room. Acoustics, fire exits, air-conditioning, and waiters' access around crowded tables were all carefully thought out and incorporated into the blueprints.

With this latest commission, Norman moved his family out of the boroughs and into Manhattan proper (a place he'd once compared to a pinball machine), into a four-story brownstone at 133 East Thirty-Seventh Street, complete with backyard. While Bel busied herself furnishing the lower floors, Norman set up shop above. Dudley came to stay, giving four-and-a-half-year-old Joan a live-in uncle, Bel an extra pair of hands, and Norman an unfamiliar sense of extended family.

•

COINCIDING WITH THE Palais Royale job came an offer to design the sets for *Orange Blossoms*, a "naughty" French musical scheduled for mid-September at the Fulton on Forty-Sixth Street.

The plot was uninspired. More interesting were a pair of headline-making scandals. The first involved charges of "moral turpitude" between the play's two leads, filed by a real-life cuckolded bridegroom, followed by a deportation summons that threatened to derail the entire enterprise. The second involved irate corset manufacturers descending on the Ritz-Carlton, a scenario in which Norman played an unlikely role.

"Occasionally boys are born with such energy, obstinacy and talent that they are criticized as long as they live and not always forgiven after they are dead," observed Paris-based correspondent Janet Flanner. "They steal ideas from the future . . . [leaving] an historical mark on . . . their decade." The description fit Bel Geddes, but she was speaking of Paul Poiret, one of the world's most esteemed couturiers, a man "always magnificently out of step."[75]

Like Irving Berlin, a brilliant songwriter and self-taught pianist who couldn't read music, Paul Poiret was a brilliant, self-taught draper who couldn't sew. He was the first clothing designer to bridge the gap between fashion and other arts, hiring Raoul Dufy, Erte, Fortuny, and an unknown Man Ray to design his fabrics. He was also the first to embrace marketing in the modern sense, expanding his empire of lavish coats and dresses to include glassware, wallpaper, and furniture and founding a perfume house—Rosine—more than a decade before Chanel.

Poiret loved theater, and the commission to create *Orange Blossoms'* costumes was an opportunity to promote "export" designs in a country "always pregnant with some new thing." As his ship approached the Statue of Liberty (a gift to America from his kinsmen), the newspapers announced he planned to open a stateside establishment. That he hadn't exactly said as much was lost in the stampede of designers and merchants vociferously protesting an infringement on their constitutional rights and their livelihoods, all because this . . . this *Parisian* . . . was trying to avoid import duties.

In some circles, the inventor of "the sheath" and "the sack" was considered a liberator, having freed women from the tyranny of wasp waists. The nation's corset manufacturers—who, to a man, had never been compressed into constraints that made eating and breathing problematic, and sitting, at times, impossible—didn't see it that way. It didn't help that, during a previous visit, Poiret had been quoted as saying that Americans lacked spontaneity and preferred baseball to museums, and that the black satin dresses he saw everywhere "betrayed the influence of the clergy."[76] Wearing black, "serviceable" as it was, *was* "tiresome everlastingly," concurred etiquette doyenne Emily Post.[77] But then, Miss Post was homegrown, not an upstart *ku-toor-e-yeah* foreigner.

Newspapers were inundated with complaints, and when word got out that the sheath menace was staying at the Ritz-Carlton, he was accosted in the lobby. The success of his *Orange Blossoms* gowns was overlooked in the wash of public hostility. Adding to the melee, a contingent of vociferous young flappers took umbrage with the Frenchman's insistence on long skirts. "Legs are made to walk with," and what did manufacturers, who heeded dictators like Poiret, think they were, anyway—"a bunch of jellyfish with no minds of their own?"[78] Within two weeks of his arrival, Paris's lauded fashion-maker was ready to *swim* home.

Norman found himself in the role of ad hoc spokesman. "I have probably been with Mr. Poiret as much as anyone since he arrived," he informed *The Evening Globe*, given their collaboration on the play, "and I can say that he has been quite misrepresented."[79] Norman knew little about haute couture, but he knew what it meant to be disparaged for being innovative. Recalling his own experiences—Belasco's (feigned) disdain for his thousand-watt spots, Wright's intransigence, Urban's "borrowing" of his folding screens—he understood better than most when this elegant, portly man insisted he wasn't a "capricious despot" but an antenna, someone who anticipates and interprets what has yet to happen.

Despite the headlines, *Orange Blossoms* managed a respectable ninety-five performances. Norman's sets were cited for their "beautiful backgrounds of grey and moon color"[80] and a rhinestone curtain "that was impossible to light."[81] Two months later, the reimagined Palais Royale opened its doors, an event its designer would always associate with the birth of his second daughter, Barbara, who "debuted" on Halloween.

Despite concerns about the dampening effect of a recent police crackdown on hip flasks (which, Norman noted, "could not be manufactured fast enough"), the new Royale was an immediate success. "The loveliest room in all the world is the Geddes triumph," crowed *The Evening Globe*.[82] "A style unto itself," wrote *The New York Times*, particularly intrigued by a small elevated stage hidden behind an apparently solid wall that, at the touch of a button, slid noiselessly aside.[83] The atmosphere was electric with short-skirted cigarette and camera girls, tuxedoed waiters, and luscious floor shows. Norman's sexy, indirect lighting, the kind that made dowagers look like debutants, embraced them all.

Regulars included the Vanderbilts, Lord and Lady Mountbatten, Maurice Ravel, Scott and Zelda Fitzgerald, Duke Ellington, and fifteen-year-old prodigy Roger Wolfe Kahn. Otto's youngest could play eighteen different instruments. At seventeen, he'd buy his own orchestra and begin recording songs ("Hot Hot Hottentot," among others). At nineteen, he'd open his own jazz club, a coup that would land him on the cover of *Time*.

Another regular was a young Tin Pan Alley plugger named George Gershwin, who, like Roger Kahn, sometimes sat in with the band. Having heard his "Blue Monday Blues," Whiteman requested a concert piece for a musical experiment he was staking his reputation on. Gershwin pleaded other commitments. "Hot or cold," Whiteman told him, "the body goes out on Lincoln's birthday," February 12, the only day the Aeolian Hall was available. Over the next three weeks, the composer hunkered down over a battered upright. Rehearsals took place at the Royale after hours, starting at about 2:00 a.m. Whiteman's arranger added brass "smears" and "dirty" reed inflections, then in vogue, to justify the "blue" in "Rhapsody in Blue."

It was scheduled for last in the nearly four-hour program. When the opening clarinet solo finally began, its low trill slid playfully up two and a half octaves, a fluttering, almost drunken whoop that rose into an extended wail. The rest, as they say, is history.

The following day, Walter Damrosch commissioned Gershwin's "Concerto in F," which would, like its predecessor, be refined beneath the Royale's umbrella ceiling. Gershwin, in turn, commissioned the nightclub's designer to create sets for his and brother Ira's first musical, *Lady Be Good*, set to open in December.

CHAPTER 6

Prelude to a Miracle

1923–1924

To anyone who knew Broadway in the Twenties . . . [the shows] didn't succeed or fail—they exploded over the clouds or went down like the *Titanic*.
—BILLY ROSE[84]

One late winter afternoon, the phone rang. "Max Reinhardt is in town and wants to meet you. He saw your *Divine Comedy* drawings in Amsterdam and was profoundly impressed." It was Kenneth Macgowan, *The Globe*'s drama critic.

Reinhardt! Richard Ordynski often spoke of him, and Norman had read everything he could find about the man. A former actor, and a very good one. A master of combining ancient and contemporary theater. A producer of everything from Aristophanes to Shaw who'd hired Edvard Munch to create sets, commissioned Oscar Wilde to write *Salome*, Richard Strauss to compose a score. Half of Europe's best actors had worked with him.

Max Reinhardt had made theater *matter.*

"I'm honored," Bel Geddes said. "But does he speak English?"

"He has an interpreter."

"I don't speak German and interpreters are a nuisance. I appreciate the opportunity, but I think we'd better just call it off."

"He's worked in a dozen countries. It won't be a problem, I promise you."

Macgowan argued for a while, then hung up, clearly annoyed. The next day someone by the name of Dr. Rudolph Kommer phoned to say that he'd seen Norman's L.A. productions, back in 1916, and could he possibly call on him?

"Delighted."

"May I bring a friend?"

"Of course."

"It's Max Reinhardt."

Norman began to explain.

"I know about that," Kommer admitted. "But you see, neither Macgowan nor I could tell the professor you won't see him. We'll stay only a short time. He has a meeting with Morris Gest and David Belasco not far from you."

•

THE *DIVINE COMEDY* model was about the size of a grand piano. For the next couple of hours, Norman (thirty) talked his visitor (fifty) through, with help from his 200-plus figures and drawings, working the miniature dimmer board

If anyone could appreciate Bel Geddes's concept of *a single set* that achieves monumentality, enchantment, and transformation through lighting, it was Max. Both were unapologetic, grand-scale dreamers.

"Vat a thrill you vood have had if you stood in Amsterdam and seen how your work attracts people," Reinhardt began, when the run-through was over. "And rightly so. The little people vill oppose you, accuse you of vat ever is, in English, the vord for the disease of thinking big things.' They vill say you are impractical, they vill resent you. Let zem."

He wanted to see more. Norman began digging out drawings he hadn't looked at in years—designs that had been staged or built, others that never got past the drawing board. When there was nothing left to show, Reinhardt beckoned Kommer to the window. After a time, Kommer turned to his host. "Professor Reinhardt requests that you consider designing *The Miracle* for its New York production."

Otto Kahn had been trying to bring *The Miracle* to New York since 1911, when he'd attended a London production (nineteen-year-old Ernst Lubitsch was one of the 2,000 supers), but the Great War had intervened. Now he was advancing $400,000 of the required $600,000. Norman had no idea what *The Miracle* was, nor did he know that Joseph Urban, who'd designed the sets for its German and Austrian productions, was anticipating the commission. But Reinhardt thought it politic to have a native-born designer.

It was a wordless spectacle, Reinhardt began, based on the legend of Sister Beatrice.

Would audiences weaned on Ziegfeld's *Follies* and snappy Irving Berlin tunes embrace a three-hour play about pilgrimage and redemption? In New York, a pilgrimage meant a subway ride to the Village or a basement speakeasy. Otto Kahn or not, the great Reinhardt or no, Dante notwithstanding, this was a medieval Christian play, Grandfather Melancton's meat, not his. But as it happened, Reinhardt, né Maximilian Goldmann, was Jewish.

"In London, eight thousand people came every night to see it."

"Eight thousand?!"

"Ya. The Olympia is the largest covered arena in the vorld, four times the Albert Hall. Had it not been a pantomime, ve vould not have done it there. No one vood have heard us in so vast a space! Since then, it has been staged all over Europe. Seventeen times."

"Then why do you need a designer?"

"Every venue vas different. In Berlin, a circus building; in Stockholm, the Royal Opera House; in other places, a church or hall, or in open air."

"The staging, the music, and the acting of a pantomime are functions of za space," the Austrian continued, pointing toward Norman's model. "It must come together as an indivisible unit, a living organism with its own laws. The smallest variation requires a corresponding change in za music, za actors' movements, za set, everything."

Still, after so many revivals . . .

"Each time zere is more to realize, to be nourished by, like an inexhaustible spring." Kommer's translations, offered through a continual haze of cigarette smoke, were essential now. "Vhat is most interesting is zat zis play, *which requires the performance of Roman Catholic rites*, has succeeded even in Protestant cities—London, Berlin, Stockholm, Greek Oriental Romania. All the vize men of theater insisted it vas lunacy! Madness! Possibly even sacrilege! Such a furor each time, za idea of presenting religion on stage for entertainment. But it has *never* failed. In Vienna, even the Cardinal Archbishop attended—discreetly, at the dress rehearsal. And vye not? Isn't theater a mirror of human life?"

From the floor below came the sound of Bel ringing the dinner bell. They'd come for half an hour and stayed for five.

The only dark note, from Norman's point of view, was that Morris Gest, at Otto Kahn's suggestion, had been chosen to produce. Aside from a begrudging fascination, Norman didn't like the man. That Gest's father-in-law was David Belasco, the unholy Bishop of Broadway, didn't help. But Reinhardt understood that a production of this magnitude had to be sold to the American public, *pre*sold, or risk bankruptcy. In

1923, a big musical might require $50,000; Reinhardt's budget was twelve times higher. It would prove to be the most expensive onstage production ever mounted. Anywhere. Gest, a master of buzz and spin, knew the landscape.

Since Norman's glass dance floor encounter, five years before, Gest and his partner had become two of Broadway's busiest producers, specializing in Russian imports. The *Chauve Souris* circus had been one of the sensations of the 1922 season, inspiring the creation of Russian tea rooms from Midtown to the Village. Then, af-

Max and Norman, circa 1924.
(Courtesy: Harry Ransom Center, The University of Texas at Austin, photo by Maurice Goldberg)

ter seven months of prolonged and costly parlays, Gest had scored the Moscow Arts Theatre's first-ever American tour, complete with director Constantin Stanislavski and actress Olga Knipper-Tchekhova, with all their own scenery shipped halfway around the world—a $200,000 risk. (Kahn was a major "angel" for the enterprise.) Norman, who'd wanted to name Barnsdall's theater The Seagull, must have felt a pang of admiration for the man who'd brought Knipper, Chekhov's widow, and the company that had premiered many of Chekhov's plays, to Manhattan.

It had been Gest, too, who'd pushed to bring *The Miracle* stateside. Reinhardt had suggested several less complicated and financially burdensome options. *I only want your best,* Gest had insisted. At the same time, he was busy arranging a Farewell Tour for the great Eleanora Duse, who was coming out of retirement. Not bad for "an ignorant immigrant" who'd begun life in America hawking newspapers, shining shoes, and, rumor had it, painting sparrows yellow to sell them as canaries.[85]

•

IN LATE MAY 1923, Max Reinhardt sailed for Europe, leaving Morris Gest with firm instructions to conclude a satisfactory contract with Norman and the Century Theatre, one of the only available venues, as soon as possible. The play was scheduled to open Christmas Eve. Meanwhile, Max had determined that the props and costumes from the London production, already in New York, were unsuitable.

The project was daunting. Along with converting a theater into a realistic Gothic cathedral—arguments for an abstract set had fallen on deaf ears—Bel Geddes was responsible for the sound script (singing, prayer chanting, howling crowds, a symphony orchestra, a chorus and organ, plus some of his *Divine Comedy*'s vibrating instruments), action charts, lighting (coordinated with makeup), and now the design of hundreds of props and costumes.

He wanted the same $10,000 he'd asked for the Palais Royale. Gest, whose negotiation method was part poker player, part P. T. Barnum, and part tantrum, kept him hanging for a week.

"It's a single set!" Gest roared. "Two thousand is my price. If you don't want it, someone else will."

"It may be a single set," Norman countered, "but it's going to cover half a city block, and nine imaginary dream sequences have to play out on it. Not to mention everything else I'm in charge of. Five thousand, and not a penny less."

The words weren't out of his mouth when Gest called for his secretary, his press agent, and the photographers who'd conveniently been waiting in a nearby room. Flashbulbs popped as Gest roared about the greatest designer living, his own discovery, and how he'd just agreed to pay five thousand smackers for a single set.

•

NORMAN FELT IMMENSE sympathy for Aline's frustration over her disbanded company and Wright's nowhere-in-sight theater, so when she wrote asking if he and his family would join her and Sugartop on a European vacation as her guests—he was already planning a month in Salzburg, at Max's seventeenth-century *schloss*, to finalize *The Miracle* plans—he couldn't refuse her.

"You understand, don't you, Norman?" she wrote. "I can't give up the theater and I can't take hold of it." She was confident he was the one "to work something out with."[86]

"I'm taking a month out of the busiest time of my life just to settle what you and I are going to do together," Norman wrote back. They'd have a week on the ship and three weeks in France to "develop something fine."

For the next three months, he worked at fever pitch, preparing a wooden scale model of the cathedral's interior with preliminary details fashioned in clay, scale models of the dream sequences, plus miniature brass costumes. Years later, Brooks Atkinson would note that Norman also created more drawings than even Mont Saint-Michel's twelfth-century builders had likely required: 871 structural drawings for architecture, more than 100 *final* color drawings (based on four times as many preliminary studies) of the 36 stained glass windows, 42 color drawings for scenery, 474 watercolors for costumes, 206 for props—all accompanied by specifications on materials and use—plus lighting and wiring diagrams, and action and mechanical scenery change charts. All told, more than 3,000 illustrations would accompany him to Salzburg. Meanwhile, knowing nothing about Gothic architecture, he spent weeks researching in local libraries and museums. There were also estimates to procure and schedules to develop with potential contractors. Rehearsals and construction were slated to begin in early fall.

At Max's insistence, the Century Theatre (proscenium stage, three balconies, and notoriously bad acoustics) had to be transformed into a resonant, all-encompassing Gothic cathedral that extended, uninterrupted, from the stage all the way to the exit doors as one immense *environment*. The stage floor would be extended out over the orchestra pit into the auditorium. Boxes on either side of the auditorium would be transformed into recessed doorways with high pointed arches or rebuilt as cloisters.

There had to be room for a resounding organ; the aisles had to accommodate hundreds of supernumeraries careening around as jesters, beggars, lepers, and lunatics. Everything had to be fireproof, built in portable sections, and most difficult of all, the Century's contract demanded that *no structural alterations be made*—no nails could be driven into walls, none of the theater's pseudo-Renaissance plaster swags and cupids altered or damaged. The solution was encasing the stage and auditorium with architectural scenery—painted on canvas but built of wood.

The plan was for ticket holders to enter a dim, towering 110-foot church, their footsteps echoing on the stone-slabbed aisles (an asbestos composition). As they looked for their seats (pews for 3,100 people), priests, sacristans, and the occasional worshiper would be moving about lighting candles or counting their beads. The smell of incense would mix with the smell of melting wax. The only illumination, beyond the candles (more than 800) and faux candles (834), would be brilliant shafts of artificial sunlight, punctuating the sacred gloom through three dozen Bel Geddes–designed stained glass windows—ranging from 40 to 80 feet in height, made of thin 10,000-square-foot sheets of muslin stretched and painted to appear semitransparent when lit from behind.

The goal was realism through exaggeration. No curtain, no blackouts, and no stagehands—the actors would make set and prop changes themselves. Electric motors, operated at the push of a button, would replace manpower. A 150-foot glass floor (shades of Bel Geddes's first project for Gest), set into the fore stage, would glow diffusely from below. There would be pipes for smoke effects (from a built-in generator), "vampire" traps (actors could suddenly spring up), and "grave" traps (to sink into).

The dream sequences had to be plausible illusions of everything from a primeval forest, a banquet hall, a wedding chapel, a black mass, a coronation throne room, a public square, the inside of a stable, and a wintery roadway, and each had to dissolve seamlessly into the next. Painted with special pigments, certain dream sequence costumes would appear to vibrate under overhead ultraviolet lights.

Some parts of the story would be augmented by beams of light directed through banks of smoke and fog that, at times, selectively masked parts of the set or action, so that scenery could be changed unnoticed. There would be suspended snow boxes and a gossamer matrix of 600 tiny, low-wattage bulbs to suggest stars or fireflies.

Taken all together, it would be a multimedia, virtual reality experience forty years before either concept would go mainstream in the 1960s.

Morris Gest, meanwhile, had left town, though not before securing the brilliant Michel Fokine to choreograph the crowd and mob scenes. He was reputedly headed for Odessa to see his parents and check on his Russian contracts before meeting up with Max in Salzburg. Norman was flabbergasted to read about this French leave, after the fact, in the newspaper. He managed as best he could, parsing out his blueprints (designed to come in $14,000 *below* the agreed-upon $100,000 budget) to a trio of contractors he thought reliable enough to handle the enormous job.

Casting—the two leads in particular, the Madonna and the Nun— was crucial. The former required dignity, tranquility, and the ability to stand statue-still for long periods while holding an audience's attention. The latter required exceptional stamina—three hours of running up and down the aisles, from one temptation to another—and the ability, at the end, to recite the Lord's Prayer, the pantomime's only speech.

During Max Reinhardt's New York visit, he'd been invited to William Randolph Hearst's yacht and offered a $100,000 contribution toward a film version of *The Miracle*, on the condition that Hearst's sweetheart Marion Davies play the virgin Madonna. (Twelve years later, he'd try again, with Max's son Gottfried, doubling the amount.)

Max saw no reason to recast the Madonna role played so well in London a dozen years before by Maria Carmi, then-wife of *The Miracle*'s author, Karl Vollmoeller. But Gest's promotional instinct called for new blood and exploitable glamour. He was confident that English rose Diana Olivia Winifred Maud Manners Cooper was the ticket. "Lady Di," thirty-one, had grown up in Belvoir Castle, the product of a hermetic, aristocratic world shattered by the First World War. Much to her family's disappointment (they'd hoped for a match with the Prince of Wales), she'd married Alfred Duff Cooper, soon to be elected to Parliament, later ambassador to France.

Diana knew little about Reinhardt other than that he was "a genius." Gest offered her $1,500 a week. Blue blood or not, she needed the money. Max eventually gave his approval, but not before making Diana audition with a pair of chiffon drawers tied on her head.

•

JUNE 1923. THE Barnsdall entourage (Aline, five-year-old Sugartop and her nurse, Aline's secretary, and two dogs) was joined at the docks by the Geddes clan (haggard husband, giddy wife, six-year-old Joan, howling, six-month-old Barbara) plus seven crates of *The Miracle* designs. Five days later, three chauffeured motorcars awaited them in Cherbourg. First stop, Mont Saint-Michel, so Norman could study the cathedral firsthand. It was, he would write, "my personal rendezvous with history."

Once settled into their Saint-Malo hotel, Norman and Aline got down to business. She planned to purchase Olive Hill, thirty-six acres with a sloping view of citrus and olive groves and the Pacific. Hollyhock House would grace the rise, and Wright's theater would incorporate some of Norman's ideas.

All was amicable until the old problem of Norman's presence in Los Angeles came up.

Aline, who planned to endow her project, her life's work, with millions of her own dollars, offered him one more year to learn what he felt he needed to in New York, but after that she wanted an exclusive commitment, in return for which she promised $25,000 a year and a place to live—for life. The offer was generous, the opportunity, he knew, a great one. He waffled. She finally exploded. The next morning, she and Sugartop were gone, off on a long drive along the coast. His two weeks in Saint-Malo over, Norman left for Salzburg, as planned, after which Aline returned to enjoy the Barnsdall-Geddes family vacation without him. He would never see or hear from the heiress again.

•

THE CASTLE REINHARDT called home was set into the Tyrolean Alps at the edge of a mile-long private lake. "Leopold's Crown" was an ornate confection, its walls hung with enormous mirrors and paintings, its ceilings encrusted with allegorical reliefs. As a boy, Mozart had performed here. In 1740, Archbishop Leopold von Firmian had commissioned the Schloss to celebrate his having evicted all Protestants, Jews, actors, and gypsies from his duchy. The great irony was that Leopoldskron's current owner, an actor and a Jew, had spent much of the past twelve years presenting Roman Catholic pageants to tens of thousands.

Since Reinhardt had acquired it, the Schloss had seen the launch of numerous careers, the brewing of various intrigues, and the birth and shattering of diverse love affairs. Guests had included Lillian Gish, Toscanini, Noel Coward, Winston Churchill, Kurt Weill, Hedy Lamarr, Louis B. Mayer, Otto Kahn, and President Roosevelt's mother.

Norman was put up in a guest cottage with another young American, Ernest de Weerth, a Reinhardt assistant. He was given full access to a horse and carriage, a coachman, and a flush young maid who appeared each morning to draw him a bath. Reinhardt's secretary, Fraulein Adler, instructed him in German.

Dinner was served in the grand dining room at 10:00 p.m., usually attended by a mix of poets, playwrights, actors, Nobel Prize winners, diplomats, and duchesses, sometimes followed by candlelit chamber music or a play, with the Marble Hall's enormous fireplace serving as a stage. Reinhardt's workday began at midnight amid the crystal chandeliers and hand-carved boiserie of his library. To Norman's immense relief, his host was delighted with the contents of the crates.

When the subject of a crew came up, Reinhardt suggested Ryszard Ordynski for the stage manager slot. It seemed an obvious choice. Max's former lieutenant had already mounted *The Miracle* in London, Paris,

Schloss Leopoldskron, where Mozart performed as a boy.
(Courtesy: Hotel Schloss Leopoldron.)

and Vienna, he and Norman had a track record, and Ordy was still in New York.

Norman hesitated. Perhaps because the "Polish Reinhardt" had known him when he was wet behind the ears, or because he feared Herr Professor would insist on imposing his own ideas. Ordy's *Everyman*, seven years back, had jeopardized Aline's entire project. No, what Reinhardt's spectacle needed was eight or ten *assistant* stage managers, a job, Norman reasoned, Ordynski could only find insulting.

As for *The Miracle's* principal male role, a kind of evil, agile, seductive faun, Norman's first thought was Nijinsky. Max didn't need convincing; he considered the dancer one of the greatest performers of his day. Vaslav was only thirty-four, in the prime of life, living in a Swiss asylum about 200 miles away.

He'd been diagnosed with "paralysis of the brain," but what did *that* mean? Schizophrenia? The word didn't even exist ten years before. Perhaps the dancer was just exhausted, or depressed. Permissions and travel arrangements were made. But when Max and Norman finally sat down with him, it soon became clear that Diaghilev's protege—having endured "experimental" medicines and dozens of insulin-induced shock treatments[87]—had no idea what they were talking about.

Every day for the next five weeks, Norman and Max hunkered down from midnight until well past dawn going over Norman's charts and drawings. "Every detail has been carried out with positive genius," Reinhardt wrote to Morris Gest. "No one will recognize the rich but banal Century Theatre. Everyone will be held spellbound."

Once back in New York, Norman was anxious to find out how things stood. Max was due to arrive within a week and Morris Gest's contract called for the structural part of the scenery to be in place by then. But Gest, preoccupied with last-minute arrangements for the Moscow Arts Theatre and Eleanora Duse's tour, had yet to obtain a lease on the Century.

It was past the middle of October. The opening was set for Christmas. Max had left in May with strict instructions.

"You *what?*"

"They want too much for the rent."

"You mean to tell me that thousands of dollars and six months of fever-pitch work have been invested in a production *precisely fitted* to a theater *you haven't even leased yet?*" Norman was stunned.

"If the Century's been booked, you won't be able to save enough rent *in a year* to cover what it will cost to do all that work over again. Jesus, Morrie! Do you realize what you're going to pay in overtime? The scenery covers New York's largest stage *and* its auditorium, which is twice as big as the stage, the equivalent of a city block. It's going to take a hundred men working day-and-night shifts for three weeks just to get the stuff *installed.* Reinhardt sails the day after tomorrow! He's a businessman. He'll go directly to Kahn."

Before the day was out, a lease had been procured, but it would be six weeks before they could move in. In the meantime, they needed a rehearsal space large enough to accommodate the basic cathedral layout. With hundreds of mob scene actors and no dialogue to cue in entrances, exits, and movement, a layout was essential. Norman secured an option on the old Metropolitan movie studio and, after some Gest-worthy subterfuge, got Morrie to sign the lease, made out in his name.

It took two days for Gest to realize what he'd put his name to. "Four thousand for a *rehearsal* space?!"

Norman had had enough. "It's *Kahn's* money and *you're* the one spending it. I estimated scenery expenses at $86,000, which was *well* under our \$100,000 budget.[88] This is all the result of you sitting on your hands for the past four months! And trust me, I can prove it." Gest's delays ultimately increased scenery expenses by more than half.

As Norman was leaving the room, Gest ventured a joke. "Up the block," he said, "is an undertaker. Order three coffins, will you? Have them initialed M.G., N.G., and M.R."

•

IN LATE OCTOBER, Max Reinhardt boarded the *Aquitania*, bound for New York, still without his Nun, despite having auditioned hundreds of young women for the coveted role.

And then he found her, a fellow passenger.

Lady Diana Cooper was a pretty, pampered society darling. This girl was something else entirely. Tall, coltish, and athletic-looking, broad-shouldered and long-legged. A full, sensual mouth, flawless skin, thick blond hair, and enormous pale eyes. Her beauty was leonine, astonishing.

Kommer's investigative skills soon filled in the blanks. Rosamond Pinchot was a member of a privileged and illustrious clan. Her father, Amos, was a prominent progressive politician. Gifford, Amos's brother, had been Pennsylvania's governor, twice, and as the country's first secretary of the interior had had a million-plus-acre national forest named after him. Rosamond and her mother were headed home from a Paris shopping trip in anticipation of Rosamond's coming out party. They'd had tickets for Gertrude's favorite Italian liner but at the last minute were diverted to the British ship; a similar switch had happened to Reinhardt. It was meant to be.

A theater career was not a proper path for a refined young woman of means. On the contrary, it verged on the scandalous. (Unlike Lady Di, the Pinchots weren't hurting for funds.) And why her? Rosamond hadn't even been able to make the dramatic club at Miss Chapin's.[89] If she was known for anything, it was her habit of riding bareback in the moonlight or swimming naked at Grey Towers, her family's 3,600-acre Pennsylvania refuge. But by the time the *Aquitania* landed, the nineteen-year-old had done considerable soul searching. Whatever it was she thought she might do, was *expected* to do, with her life, this was an opportunity she'd be mad to ignore.

The final decision rested with Gest. But given Rosamond's lineage (Social Register on both sides, multimillionaires, philanthropists, Yale connections), and her extraordinary looks coupled with exquisite grace, there was no reason for doubt. The ladies Pinchot and Cooper were a publicist's dream.

Before Rosamond's parents could decide whether Max's offer was even worth considering as a one-time "experience," Gest's machine was cranking at full throttle. "Reinhardt Selects Girl with No Stage Experience the Moment She Passes Him on Ship; Calls Providence Guide" was splattered across *The New York Times'* front page. Meanwhile, the press was busy reporting *The Miracle's* statistics: 2,000 costumes by 100 seamstresses, 5,800 pounds of lampblack, 2 tons of stage snow, 1 million feet of wood, more than 900 draftsmen and workmen, 40 electricians, dozens of stagehands, 22 assistant directors, and a cast of 700.

When truckloads of parts—cloisters, capitals, arches, parapets, altars, nail-studded great doors, turrets with winding steps, ironwork grills, candelabras—began arriving, Morris Gest panicked. Accustomed to great quantities of canvas, he saw the terrifying collection as a shambles that could never be put right.

The cathedral columns, which would be raised and lowered on cables, were fifty feet high and six feet in diameter. "Didn't you tell me these weigh a ton apiece?" he asked their designer.

"That's right, Morrie. No exaggeration."

"There are twenty of them!" He waved his cane in the direction of the gridiron high above the stage. "Do you think that was built to hold twenty tons of scenery, plus all the rest? It's all going to come down on our heads! One of us ought to be run out of here."

If only, Norman thought. Five years earlier, he'd been stalking the promoter, hoping against hope for an interview. Now he couldn't wait to be rid of him.

One moment Gest would be enthralled as Norman's stained glass windows were put in place, the next he'd burst into curses because the architecture had been painted dead black (ergo, the 5,800 pounds of lampblack). The explanation—that the structure was a frame for the windows, that the flat black, painted over a rough layer of plaster, created a rich receiving surface for the light, that the dark architecture would be silhouetted by that colored brilliance *and* be a foil for the candlelight—fell on deaf ears. Gest disrupted rehearsals without ever seeing one through, his steel-tipped cane incessantly tapping on the newly laid "stone" floors, playing unwanted middleman between the construction foreman and the carpenters.

"What I wouldn't give," the foreman confided, his face strained with the effort of back-to-back ten-hour shifts, "to get that man out of here for a few days."

"What if I could get you twenty-four hours?" Norman asked.

"Sure."

"What's it worth to you?"

"Hundred bucks."

His pockets filled with nails and bolts, Norman took the elevator up twelve stories to the gridiron, the vertiginous perch where years before, at the Garrick in Detroit, he'd stretched out to watch the patterns play out below. The actors' voices were inaudible, but the sound of Gest's cane was not, and his slouch fedora was easy to locate. Looking down through the matrix of vertical cables and horizontal pipes to make sure no one stood immediately below, Norman dropped a dozen small nails. A few struck metal on their way down. Gest could be seen looking up, talking excitedly. Norman dropped a dozen more over another part of the stage. Terrified, Gest ran far over to the side.

"Where is that son-a-of-bitch Geddes?" he yelled. "I told him hoisting all that scenery would pull the roof down!"

Next came a handful of large bolts, which made considerably more noise, clattering from pipe to pipe past the hanging scenery. Before they reached the floor, Gest was running from the building. "If the roof falls in on people, they'll hold me responsible!" he howled.

Morrie stayed away for four entire days. Norman took the opportunity to send him a note:

> *The three coffins have been ordered. I've arranged for a funeral service during dress rehearsal in New York's soon-to-be-most-talked-about church.*

On January 5, 1924, with the cast suffering everything from bad colds and nose bleeds to chronic lack of sleep, the first of five dress rehearsals took place. Norman introduced his costumes. There were 470 different designs, many of them challenging (the heaviest weighed 300 pounds) but all important psychological aspects of the play. Handled incorrectly, he told the cast, they would look ridiculous. Moving in them with any kind of grace (not to mention *getting* into them) would require practice.

The Madonna's garb was sprayed stiff with concrete to look and feel like stone. The "miracle" of the play was her transformation from statue to warm flesh. The costume was part of the column on which she sat stock-still for forty-five minutes on a high, built-in stool, with only her hands and face, painted to appear stone-like, visible. As there was no curtain, she would take her place there half an hour before the play began. At a critical moment near the end of the performance, the Piper Trickster would distract the audience as a crew member behind the column pulled a thin iron bar, causing the concrete costume to open on hinges. Two seconds later, the Nun would stifle a scream as the Madonna stepped down to the altar. With the proper lighting and music, it was a breathtaking effect.

Diana thought her stone coat was "rather wonderful." What she couldn't abide was the foundling Christ Child destined for her arms, an "abomination" wired to glow "like a spectral fetus."[90] She switched it for a simple doll, but the "fetus" had a way of reappearing—until a mysterious explosion secured its banishment.

With a week to go, Max began sprinting, unstoppable, pulling all the disparate pieces together, and everyone, down to the smallest players, could feel it. Something extraordinary was taking place. Like a piano teacher who insists on the relentless practicing of scales, Reinhardt had pushed the company to the point where they were almost bored with repetition; now he gave them something new, sketching in the fine lines, adding tautness, nuance. His face and gestures came alive in a new way. Rehearsals went on for hours, but even those who *could* leave the theater didn't.

Then, contrary to all established theatrical codes of courtesy, Morris Gest announced, through an underling, that opening night—already delayed long past Christmas—would be advanced by five days. In other words, the day after tomorrow.

No explanation was given. Reinhardt's exhaustively worked-out rehearsal schedule was scuttled and Gest was, conveniently, unavailable. The house had been sold out for a month. Otto Kahn refused to intervene. With five dress rehearsals planned, now there was time for only a last run-through.

In the excitement and confusion, Norman completely forgot about the coffins.

He'd arranged to have them carried in the opening procession of the final dress rehearsal. There they were, halfway down the aisle, draped in black, each bearing two initials, borne by solemn carriers (who assumed it was a bit of last-minute staging) amid the priests, acolytes, nuns, and monks. Gest, for whom the whole thing had been staged, wasn't there. Norman spotted Max in a far corner of the auditorium. What could he think, under the circumstances, except that his designer had a monstrously perverse sense of humor? Norman rushed toward him between a row of seats.

"I see dem," Reinhardt said simply.

A mechanized section of the stone floor rose up to the organ's swelling notes. "M.G." was lowered out of sight, followed by "N.G.," then "M.R." descended on top. The stone slab was replaced and the rehearsal continued.

"I'm so sorry. I'd forgotten. It was planned weeks ago."

"Of course," said Reinhardt, attempting a smile. "Don't verry."

•

TWO HOURS BEFORE the much-ballyhooed, twice-rescheduled opening, only Norman and the prop man remained in the theater, making last-minute checks and arranging flowers on the Madonna's altar, where Norman now sat, exhausted. He hadn't left the building, slept, shaved, or showered in three days, catching only an occasional catnap in a theater seat or the couch in the Madonna's tiny dressing room. In six hours, the notices would be on the presses! Suddenly came the familiar sound of Morris Gest's cane.

"You!" he shouted. "You've been ruining my theater for six months! Painting all the scenery black! Spending all my money! You get out! OUT!" Had he heard about the coffins? Norman was too tired to care. Ignored, Gest flew into a rage, angrier than Norman had ever seen him, his cheeks, neck, and ears flushing. He was screaming now, nudging Norman up the aisle, telling the stage foreman he'd be fired if Bel Geddes was *ever* allowed back in the theater.

Six years before, Gest had barred him from the Century after installing a glass dance floor. Now he was being expelled for installing a Gothic cathedral.

Norman walked out into the freezing late-January afternoon and hailed a cab to his friend Kirk Brice's penthouse on Fifty-Fourth Street. A dinner party was in progress, a prelude to his friends' attending the opening. Seeing the expression on Kirk's butler's face, he realized how he must look.

"Don't tell him I'm here," Norman said, "but could you bring me a double brandy in the library?"

Half a bottle later, Norman tiptoed out the front door and headed back to the theater. At the stage door, the stage foreman smiled and waved him in. The sound of bells told him that the performance had only been underway for a few minutes. Walking to the door leading to the front of the house, he ran smack into Morris Gest, who raised his cane and brought it down with serious intent. In the best silent-movie-chase tradition, Norman ducked under his arm, with Gest in hot pursuit, shouting to the ushers to stop him. Around the horseshoe-shaped foyer they ran, as latecomers in white ties, tails, and evening gowns stopped to stare. Norman managed to slip through a door that led to the basement, where he found a large coil of rope and, reasonably sure that Gest had been outwitted, lay down and fell asleep.

In the audience that night were the Whitneys, the Vanderbilts, the Astors, the Lippincotts, U.S. Senator Simon Guggenheim, Mrs. Charles Dana Gibson, Conde Nast, playwright Luigi Pirandello, and the Duke and Duchess Richelieu. Bowing to discretion, W. R. Hearst brought his wife rather than the mistress he still hoped would play the virgin Madonna on the silver screen. Alfred Steiglitz and Georgia O'Keeffe would be among the many thousands who attended in the coming months. "You are a rare spirit," the photographer would write in his gorgeous, florid hand. "Perhaps you do not need to be told that, but still I (we) tell you."[91]

Everything, even the Breugelesque scene in which the Nun nearly gets her head chopped off, went flawlessly.

The standing ovation lasted fifteen minutes; the chancel was banked with flowers. The enormous cast took their curtain calls, then Max Reinhardt appeared. After half a dozen bows, he sent the backstage crew scurrying to find his collaborator. It took a while. Then, unable to shake Norman awake, one of the assistant stage managers doused him with a fire bucket of water. Exhausted and drunk, he scarcely budged. Slung across someone's back, he was carried up the stairs, still dazed when Max led him onstage—disheveled, water-soaked, and blinking—to receive his hard-earned accolades.

Four of the next days' papers, from *The New York Times* on down, ran front-page reviews. Aside from a few critics who complained that using nuns as ushers was going a bit too far, superlatives were exhausted all around.

At some point, either during *The Miracle*'s lengthy New York run or the four-year national tour that followed, a review reached Toronto. Having read it, Cherokee Will de Haw, now nearly sixty, sat down to write his old charge a letter.

There was the night Norman was born, when Will jumped on the family steed without stopping to saddle him. There were the broomsticks they used to carve. Prince, the St. Bernard, and Jocko, the pony. White-washing the fence, picking watercress by the stream. Riding in carriages drawn by bobtailed horses in silver-plated harnesses. "Boy oh boy, if we could live that life over. Nothing to worry about and blue skies all day."

When Norman turned four, Will wrote, his mother gave him a party, with four candles burning in the center of the cake. "And she asked you what they stood for. You said—right away you said—'For light.'"[92]

CHAPTER 7

Hollywood/Paris

1925

The cinema stands today fully matured, on the threshold of the greatest era of artistic accomplishment the world has known since some aboriginal, with blunt stone knife, carved his dreams upon the walls of his cave.

—CARL LAEMMLE, 1923[93]

Thanks to Otto Kahn, Jesse Lasky, president of the Famous Players–Lasky Corporation, offered Bel Geddes $500 a week for ten weeks plus roundtrip transportation to Southern California to observe all phases of motion picture development and production.

He would function as an unofficial member of Cecil B. DeMille's staff and help with composition during shoots. Lasky believed that the designer of "the greatest of all Broadway stage spectacles" and the director of the recently released *Ten Commandments*—which boasted the largest set in motion picture history—would share a common language. Implicit was the promise that at the end of Norman's contract, pending Lasky's approval, he'd be allowed to direct.

In April, the Bel Geddes drove to Arizona. Leaving Bel and the girls in Phoenix, Norman went ahead by train. The day after arriving, he reported to the Lasky-DeMille barn. The receptionist suggested he return the following day. When he did, a different receptionist awaited him.

"Your name, please? Three names? Odd. Who did you say sent you?"

"I've been trying to see Mr. DeMille for three days," he explained on his second return. "I was sent out by Mr. Lasky. There are telegrams in your files."

"Of course, Mr. Geddes. But Mr. DeMille is busy working on a story. We dare not interrupt him."

"But I'm here to work on the story *with* him."

"Oh! *You're* the author! We've kept it a secret from Mr. DeMille's secretary, being that she's the one writing it. Can you please return tomorrow?"

He spent his afternoons wandering the lot. He observed Howard Hawks and Victor Fleming at work, met John Emerson and his petite collaborator, Anita Loos. (Norman would help proofread Loos's sellout story collection, *Gentlemen Prefer Blondes*, published the following November.) He rented an apartment with a tennis court (having taken up the game) and arranged to give half a dozen evening talks about theater development at the Hollywood Community Studio (he'd brought along a few models).

Much had changed since his time with Aline and Ordy. The population had swelled to 100,000 and the citrus groves were disappearing, though coyotes still howled in the distance after dark, and abalone were there for the picking when the tide went out. Instead of the colossal Babylonian columns atop enormous elephants, leftovers from W. D. Griffith's *Intolerance* set, which had loomed over Hollywood and Sunset, there was a real estate sign—HOLLYWOODLAND—on the side of Mount Lee, its gargantuan letters legible from twenty-five miles away. Restaurants shaped like oranges, windmills, and derby hats competed with mansions disguised as Spanish haciendas and movie houses decked out like Egyptian palaces.

Once a novelty, the "flicker" business had morphed into a lucrative dream machine, so lucrative that thugs hired by Edison's Motion Picture Patents Company had been known to travel five days cross-country for the explicit purpose of shooting holes in non-Edison cameras.

Distractions were relatively few. Cocaine and bootleg liquor were abundant. Sunday afternoons beside glistening private pools were replacing beach sojourns. Friday night boxing matches in the American Legion's stadium were a big draw. Charlie Chaplin could sometimes be found there, studying the fighters' techniques; the Marx Brothers had a habit of climbing into the ring and wreaking havoc. The Cocoanut Grove on Wilshire was a popular dance spot, a landscape of towering artificial palms left over from the set of *The Sheik*.

Returning to the studio a fourth time, he was eventually ushered into a dark enclosure, the only illumination a spotlight directed at the floor. As his eyes adjusted, he made out the figure in equestrian clothes standing just beyond the circle of light.

"Is your name Geddes or Bel Geddes?" came the voice.

"I was born Geddes."

"Bel Geddes sounds better. But I'll call you Geddes."

"May I approach?" Norman asked, not without sarcasm.

Cecil Blount DeMille stepped into the spotlight and motioned his visitor to a chair. For the next hour and a half he talked, dropping a stack of twenty-dollar gold pieces, one by one, from his right hand into his left, then back again.

"Ten years ago, I directed a picture that had seventy-nine scenes in it. Today I have scripts with over five hundred. Why?" *Clank.* "Psychology! With long shots, we could only capture physical action. But what about *the soul?!*" *Clank.* "The soul can't be filmed from twenty feet away. But with close-ups and semi-close-ups, we're capturing what was once thought *impossible.* The surging of love. Hate. Fear. How they move up from the heart"—*clank*—"into the muscles of the face, into the light of the eyes!"

"I'm very interested in the camera's capacity for close-ups, Mr. DeMille. In fact, it's one of the main reasons I accepted this—"

"Fine shadings and distinctions. The proper rise and fall of an eyebrow can carry a picture for eight, ten, twelve weeks! The audience can *see* the actors *thinking!* That's how modern photoplays will give the world something of permanent value."

"Yes, I—"

"Rapid action and a good plot matter, for course. *But there's nothing to grip the mind!*" *Clank.* "To help people know more about the complex civilization we're living in!" *Clank.* "Motion pictures can be *more* than mere panoramic devices. They can be a moral and ethical *influence!*"

"I really appreciate you telling me all this," Norman managed. "But aren't there any questions you'd like to ask me?"

"Oh," said DeMille, returning the coins to his pocket, "I know all about you."

The interview was over.

•

NORMAN'S JOB CONSISTED of standing around and observing. The more he did, the more DeMille reminded him of Flo Ziegfeld—humorless about most things, himself in particular. DeMille's daily rituals were a case in point. Though he insisted on punctuality, he was often late. His car, equipped with a distinctive-sounding Klaxon horn, was blown repeatedly as he was driven through the studio gates. All work and conversation stopped as hushed voices spread the word: *C.B. is here . . . C.B. is here.* He exited his car resplendent in jodhpurs and knee-high laced boots, an outfit he wore in all weathers. Approaching his chair, he swept off his coat and loosened his tie, tossing the end over one shoulder, then tore open his shirt collar, buttons flying, to reveal a matching undershirt.

Whether shouting at his actors and crew—Reinhardt *never* raised his voice while directing—or whispering at close range, DeMille was never without his megaphone. During a shoot, he kept himself walled up behind white canvas screens manned by uniformed guards. A chamber music trio was kept on hand (piano, violin, cello), useful for supplying love scene moods or blocking out nearby noise.

One particularly hot morning, three weeks into his "apprenticeship," Norman boarded a motorboat to Catalina Island with his mentor for some filming. Toward the end of the day, DeMille confided he was frustrated with his art director.

"A month ago, I asked him for something new for the garden party set. Any ideas?"

"I think," Norman ventured, "a more interesting party would inspire a more interesting set."

DeMille handed over a few sodden photographs.

"How soon?"

"Right away."

Given an assistant and small crew, Norman headed back to the mainland against a darkening, starless sky. The sea was rough but too warm for rain gear. Schools of flying fish, fifteen inches long, were drawn to the speedboat's searchlight, hissing as their stiff wings cut the air. Two hit Norman so hard he nearly fell overboard. The sight of the delicate, birdlike fish committing unintentional suicide, drawn to the bright light, struck him as an ominous metaphor. When they docked at San Pedro two hours later, battered and exhausted, three dozen fish lay dead in the cockpit. As he helped throw them back into the sea, Norman found himself wondering if Hollywood, too, killed three dozen every two hours.[94]

DeMille was so impressed with Norman's garden set (which he proceeded to dress with cartloads of atrocious knickknacks) that he commissioned designs for a dance cabaret to be built in the center of a lagoon, with high, interlacing arches of neon over the orchestra and a revolving dance floor. And he offered Norman two more sets—a modiste shop and a Purgatory-like afterworld—for his next venture, *Feet of Clay*, a ten-reeler about a young couple who, in DeMille's words, "are turned back from the shadowy borderland of Eternity to finish and rectify their prematurely ended lives on earth."[95] The script included a millionaire mauled by a shark, a fatal fall from a window, and a double suicide, lightened up with beautiful girls in lavish costumes, "passionate lovemaking,"[96] and a jazz band.

He was creating, historian Margaret Thorp would write, "the first educational films" for the decade's newly rich who "wanted to know all about high-powered cars, airplanes, ocean liners, yachts, villas, exotic food, wine, jewels, Paris dresses, perfect servants."[97] Bel Geddes, for his part, thought *Feet of Clay* was a title best suited for a biopic on the perennially jodhpur-clad director himself.

•

ONE AFTERNOON AT the Montmartre Restaurant, Norman spotted Chaplin walking toward him. The actor, just four years Norman's senior, was in the process of shooting *The Gold Rush*, a project that would ultimately require 16 months, 600 "imported" tramps to climb a crew-created mountain pass, truckloads of rock salt "snow," a licorice "boiled shoe," the soon-to-be-famous "Dance of the Dinner Rolls," and endless retakes. (Chaplin, like Bel Geddes, was a perfectionist.) Greeting Norman warmly, he led him over to his table.

The playing field had leveled, somewhat, since their last meeting. Norman's work with Reinhardt had changed that. Now he and Chaplin were both acknowledged professionals who knew people in common— including Frank Crowninshield, Otto Kahn, and Morris Gest, whose "large, kidney eyes" reminded Chaplin of "a coarse edition of Oscar Wilde."[98]

At Charlie's table was a journalist named Jimmy Gruen. Professing a great interest in newcomers' first impressions, Gruen got Norman talking. And talking. The studio system was designed to encourage jealousy and distrust, not cooperation. Shrewdness lorded it over brains, and "the fourteen-year-old theory"—that audiences were adolescents and entertainments should be created accordingly—was the order of the day. Gruen, it turned out, was a new kind of journalist, one who specialized in gossip; Norman's observations appeared in print the next day. "I'm sure I was tactless enough," he later admitted, "but I couldn't have been as vitriolic as Gruen made me sound." C.B. responded with an edict forbidding Geddes from giving interviews *of any kind*. When Norman's apology note went unacknowledged, he went looking for Chaplin. "You're lucky to have learned so early, and so easily," the Little Tramp offered, blue eyes twinkling. "Next time you'll be smarter."

At a farewell lunch thrown in Norman's honor, DeMille offered him a two-year contract as his newest art director, arguing that the experience could only help Bel Geddes reach his goal as a director. *The Miracle* designer turned it down.

During his ten weeks in Los Angeles, he'd often thought of telephoning Aline Barnsdall, despite their painful falling out in Saint-Malo. Before returning to New York, he decided to drive to Olive Hill in person.

The approach was winding and uphill. In the late morning light, Wright's design suggested "a Petit Trianon from Mesopotamia," though he may have been projecting some of his frustration with the grandiose DeMille. Still, "there was no question of it being a thing of exquisite detail."

Though she would surround her luxurious manse with pine groves, eucalyptus, and masses of flowers that mirrored the custom-made carpets and install a Japanese cook (a detail duly noted in her FBI file), Aline had lost her passion for Hollyhock House and would only occupy it sporadically.

Norman's appreciation had expanded over time. Aline had created an unprecedented opportunity for a group of unknown talent who were too young—all novices but for Ordynski—to recognize their great good luck and her "enormous vision and idealism" in financing the entire enterprise with no interest in profit.

The house was closed, the gardener informed him. Miss Barnsdall had moved to San Francisco. He couldn't (or wouldn't) provide a forwarding address. As for Wright's theater, it would never be built. A few stakes, driven in five years before, marked where the foundation would have been.

•

LASKY AGREED THAT Norman deserved to direct a "photoplay." In the meantime, it was suggested he spend some time observing a director Lasky knew he respected. D. W. Griffith (another high school dropout), originator of the long tracking shot years before the invention of camera cranes, was filming *The Sorrows of Satan* in Lasky's studios on Long Island. Norman would be kept on salary, and in an appeal to his ego,

Lasky mentioned that the creator of *Birth of a Nation* and *Intolerance* was in a bind.

Eighty percent of *Sorrows* had been shot, but Griffith was floundering over the climatic sequence: Lucifer's expulsion from Heaven. Expenses were mounting, but given D.W.'s stature, calling in another director to advise him seemed . . . inadvisable. His ten-page scenario and dozen charcoal drawings approved, Norman was set up with a small staff, including ace cameraman Fred Waller (the future inventor of Cinerama). Lasky, meanwhile, told Griffith that Geddes *might* be coming by to observe, and would he mind? "Didn't you see *The Miracle?*" Griffith reputedly replied. "He might teach us more than we can teach him."

The contrast between Cecil B. DeMille and David Llewelyn Wark Griffith was noteworthy. Where one was flamboyant, the other was thoughtful, precise, introspective, and incapable of posing; where one (in Norman's estimation) was imitative, the other was genuinely creative.

Tacking an enormous sheet to the floor, Norman painted a stairway in perspective to a fixed camera; on film, it could be made to appear as if it stretched for two or three miles. Adolphe Menjou as Lucifer (costume by Bel Geddes) was lit from all directions so as to cast no shadow, then the correct backlit shadows were painted on the treads and risers of the steps. The entire sequence was shot in less than a month using stop-motion photography, and for relatively little, then edited to the exact length of Griffith's original footage.

Norman claimed to be flattered when "about half" of it was incorporated—recut, rearranged, and without credit—into Griffith's original footage.

•

EVA LE GALLIENNE—A Broadway star at twenty-one and pioneer of off-Broadway theater—wanted to mount a new production of *Jeanne D'Arc*, based on a script by her lover-of-the-moment, Mercedes de Acosta. Firmin Gemier, director of the state-subsidized Théâtre National Populaire, had offered the use of Paris's Théâtre de l'Odéon, rent-free. It was an audacious plan, presenting an American version of the French heroine's story on French soil.

Could Norman join them as designer and director?

Though they'd never met, Norman was a Le Gallienne admirer, and the prospect of a Paris contract was tempting. Even more, it would be the first time he'd be in complete charge of a production.

It was a three-month commitment, plus five days of sailing each way. Norman would go to Europe alone this time; having Bel and the girls along would double or triple expenses. Beyond that, Bel's father was seriously ill back in Toledo. Norman formed a partnership with stage producer Richard Herndon, and together they got Otto Kahn to invest a substantial sum slated for *Jeanne* and two other plays of Norman's choosing. The plan was to repay him from profits earned when *Jeanne* opened in New York following a sensational Paris debut.

He set to work designing a single *Divine Comedy*–like, multilevel, almost futuristic complex of huge cubes and plinths that "painted with light" and augmented with props and impressionistic costumes, would morph into a dozen different scenes.

He determined that Joan's theme would be played out on two dozen little brass bells. Every mention of "burning at the stake" during the trial scene would be accompanied by a faint roll on six kettledrums (located overhead in the gridiron). As the scene progressed and the mention became more frequent, the intervals between rolls would gradually become shorter. In the prison scene that followed, a continual low drum roll would serve as ominous background. In the street scene, the sound would increase in tempo, almost continuous now, as the crowd mobbed the streets, rooftops, and ledges, so omnipresent that the audience would barely be aware of it. Then, as the flames encircled Joan and she whispered "Jesu," the drums would abruptly stop. Using sound for psychological and emotional effect would later become commonplace in radio and film. In live theater in 1925, it was not.

•

IN THE MIDST of this latest crunch came an invitation from Richmond, Virginia, asking "for the good of little theatre everywhere" that Norman give a trio of talks. His protests were answered with a handsome increase in his usual fee.

Met at the train station by a committee of powder-faced society matrons, their ample bosoms sporting corsages, he was driven to a hall filled with similar brethren. It was a hot, humid afternoon. Halfway through his talk, the assembled began to nod off. Theater planning was clearly not a subject destined to enthrall.

"With your kind permission," he said, abruptly shifting strategy. "Leaving behind the theoretical . . ."

He managed to conjure up some romantic and risqué anecdotes about Rudolph Valentino, Theda Bara, John Barrymore, Diana Cooper, and Rudy Vallee. The change in the audience was "electric." He finished to resounding applause.

That evening, after a second talk to an attentive, overflowing crowd, he was ushered off to a ball, where he spent the evening surrounded by a contingent of vivacious young belles, all chattering in Southern accents. He was the dashing New Yorker. Everyone wanted to talk with him, dance with him, and someone kept refilling his champagne glass.

He woke to insistent ringing and a massive hangover. The president of the University of Virginia was on the phone. "Had it skipped Mr. Geddes's mind" that he was due to address the student body that morning at ten? It was already quarter to eleven. With Charlottesville an hour's drive away, the talk was cancelled. After packing his bag, he went down to the hotel's restaurant for some much-needed coffee.

"Mr. Geddes," said the maitre d', "I've taken the liberty of increasing your reservation to fifteen."

"What an amusing man," Norman managed, cradling his throbbing head.

Then in walked a young Southern belle with her mother in tow, followed by six more belles and *their* mothers. It was all—vaguely—coming back to him. During the course of the evening, already quite drunk, he'd invited one particularly vivacious young lady to join Miss Le Gallienne, Miss De Acosta, and himself on the SS *Mauretania*. Perhaps he might get her a small part in the play. *Ah really ought to ask ma mo-tha*, she'd said, and that's when Norman had suggested they all meet for lunch the next day to discuss it. How could she not share such news with six of her closest friends?

Doing his best to save face, Norman said how splendid it was to see them all again and thanked them for such delightful hospitality. Considering the occasion, he ordered half a case of chilled Rhine wine, and soon enough they were talking and laughing. As for his proposal, how could the girls' mothers ignore such a once-in-a-lifetime opportunity? Needless to say, they'd travel at their own expense.

It was the champagne talking, the Rhine wine, the humidity, the lack of sleep, all liberally laced with antebellum charm. Late that afternoon, the group saw him off at the station. Heading home, Norman did his best to concentrate on the work yet to be done: an elaborate prompt book in which the entire *Jeanne D'Arc* script was index-tabbed and cross-referenced and every piece of stage business was linked to his meticulously detailed sound and lighting charts; the fabrication of a state-of-the-art switchboard designed to run borders, spots, and floodlights; plus an $8,000 eight-color "boomerang" of his own design that could, with great subtlety, change the color and intensity of a light shaft, altering mood as it moved over Joan's head.

On April 1, almost exactly a year after he'd set out for Hollywood with his family, Norman and his principal assistant, Gerstle Mack, exited a taxi at the *Mauretania*'s dock with Norman's drawings, models, and assorted luggage. Bel, the children, and her friend Margaret arrived in a dead heat with Eva, Eva's dog Tosca, Mercedes, and three friends as well-wishers began gathering along with a covey of press photographers and reporters. *Was it true they were going to burn Joan in the nude?*

Suddenly, a jazz band appeared, playing a swing version of Dixie, fast and loud. Hearing his name above the din, Norman turned to find the Richmond girls, all seven, marching abreast with their mothers! They'd brought along their own musicians in gaudy, unmatched outfits. Rushing up, they threw their arms around their "sponsor," kissing his cheeks. *Did you eva* dream *we would really come?* Flashbulbs popped, reporters pushed and shoved, and Norman attempted to introduce the girls to his companions and family. Except that he couldn't remember their names.

Bel's face was a study in hurt astonishment. Then she burst into loud sobbing. Eva and Mercedes stared in wonder. Joan threw her arms around her mother and began weeping, too, setting off baby Barbs. The departure whistle began to blow.

The next day, a full-page photo was splashed across the cover of the *Daily Graphic*, New York's "number one" tabloid. "Wife and Kids Weep Bitterly While Husband Sails to Europe with Seven Southern Belles."

•

THE FIRST BLOW, in what would prove to be an ongoing comedy of errors, was the discovery that the Odeon's stage was a lot smaller than they'd been told, far too small for Norman's set, and the voltage of his custom-made switchboard (which cost a fortune to ship from the States), would, Mercedes observed, "have blown out all the fuses on the Left

Bank."⁹⁹ The only available venue large enough cost $7,000 for thirty days (more than half their budget), plus an obligation to hire the theater's stage crew, box office staff, in-house orchestra, and permanent company of thirty actors, whether or not they were needed.

The task of raising funds fell to Eva, who was busy hiring actors and trying to find time to rehearse. As the martyred Joan, she would appear onstage for almost two straight hours, highlighted in Norman's moving shaft of light; hurled, in full armor, from a twenty-foot height; jostled between various strapping extras; wrapped in heavy chains; thrown to the ground; spat on; and otherwise abused before finally being tied, with some relief, to a stake for burning.

Norman wanted tall actors, six-footers, for the knights. Available Frenchmen being short and stocky, he selected a crew of White Russians, escapees from the Revolution, former schoolteachers, soldiers, and aristocrats who, he admitted, "knew more about courtly manners, posture, carriage, and clothes than I ever would."¹⁰⁰

As the director, his ignorance of French and Russian was exacerbated by his English, which was American, with a decidedly midwestern twang. Gerstle Mack, who helped with sets and lighting, and Thomas Farrar, in charge of costumes, proved invaluable. During rehearsals, Norman bellowed *"Mack!"* or *"Tommy!"* so often, and so loudly, that his actors assumed these were newfangled curse words and adopted them as such for their own use.

•

L'EXPOSITION INTERNATIONALE DES Arts Décoratifs et Industriels Modernes was an unprecedented display embracing everything from French Cubism, Italian Futurism, Russian Constructivism, and German Bauhaus to African and Aztec design. Norman walked the grounds with his *Orange Blossoms* friend Paul Poiret.

The couturier's popularity had skyrocketed thanks to the discovery of King Tut's tomb, the most lavish archaeological find in history. "Orientalism," an ideal fit with his designs, was the rage. For the Exposition, he'd hired three barges to ply up and down the Seine: one devoted to gastronomy, one to his couture collection (with original Raoul Dufy frescos), one for his fragrance line. He commissioned Norman to design a series of private, post-Expo barges and promised to introduce him to Jean Cocteau, a fellow *enfant terrible* (and dropout) with a lifelong love of theater, what he called "the crimson and gold disease."

Of the twenty-one nations represented at the Expo, the United States was conspicuously absent. Invited to send examples of good American design in the modern spirit, U.S. Secretary of Commerce Herbert Hoover had declined, saying that the United States didn't have any. Modernism, he declared, didn't exist in the United States, and on the outside chance that it did, it in no way reflected American culture.

To his credit, Hoover sent a small commission to take a look around and file a formal report. The result was scathing.

The United States was, the report read, a nation living "artistically on warmed-over dishes." The French, in contrast (more importantly, the French *government*), believed that art and industry were not only compatible and mutually beneficial but that "artists led the way and industry followed," which was why France had gained a distinctive advantage in both domestic and foreign trade. Waiting around for industry to take the initiative was a recipe for "irreparable damage." America needed to develop its "own creative genius."[101]

•

FOR THE FIRST time since adolescence, Norman allowed himself the luxury of leisure—something impossible to imagine in New York—the pleasure of sitting in the Ritz Hotel's garden simply watching the world pass by, soaking in the Paris of Hemingway and Joyce, Picasso and Chagall, De Beauvoir and Brassai.

The aura of sex was palpable. Colette was in a liaison with her stepson, half her age. Josephine Baker was shaking her backside like a hummingbird in *La Revue Negre*. Caroline "La Belle" Otero, the most notorious of the so-called *grandes horizontales*, remained formidable at fifty-seven. Thanks to a list of lovers that included a Vanderbilt and several crowned heads, she'd accrued $20 million in jewels, cash, and real estate without ever having set foot in a classroom.

There were late-night dinners in Russian bistros (Gypsy singing, dancing Cossacks) and the occasional decadently rich repast at Foyot's. There was the Moulin Rouge and Joe Zelli's, known for its stable of gorgeous hostesses and charming gigolos. (Just as he'd incorporated the Palais Royale into one of his *Lady Be Good* sets, Norman would include a Zelli's-inspired set into *Fifty Million Frenchmen*.) As for Prohibition, *n'exist pas*.

On one such evening, a tall, good-looking young American introduced himself. At twenty-eight, Dudley Murphy seemed to know everyone worth knowing on both sides of the Atlantic. He'd recently worked as cameraman and codirector of *Ballet Mechanique*, one of the first avant-garde "art" movies, a pastiche of oscillating discs, repetitive close-ups, and visual puns, with a cacophonous score of sirens and airplane propellers.

As it happened, he was invited to the same private ball that night as Norman, Eva, and Mercedes. "See you both there," Norman said, as the four of them left the restaurant. Dudley had mentioned a wife (his second, as it turned out).[102]

"Well, no." Dudley hesitated. "Katharine won't be coming. Doesn't have the right clothes. You know women."

"I have some experience in the costume department," Norman offered, emboldened by several glasses of wine. "If she's not too far away, would you mind if I went over and took a look? Maybe snap something together?"

The attic studio was occupied by Katharine, her mother, and a very young baby. Norman explained his fairy godmother mission and after considerable persuasion was allowed inside to inspect Mrs. Murphy's wardrobe. After pulling apart several out-of-date frocks and rearranging some pieces, Katharine was sewn into them, and off they went.

Eva and Mercedes deemed Norman's handiwork a great success, Dudley seemed to take it as a good joke, and Norman was happy to have found a charming, if temporary, dance partner. Dancing had become something of a passion, one that Bel, ever self-conscious, submitted to rather than enjoyed.

It was lonely living with Eva and Mercedes, who had each other, and there was an immediate emotional connection with (admittedly gorgeous) Katharine Hawley Murphy. By the end of the evening, she opened up to Norman. She'd married in the wake of a traumatic miscarriage, at her mother's insistence. Their son had been born in January. Being left alone while her husband painted the town was nothing new.

Within a few days, Katharine had moved into an anonymous hotel, hired a nurse for her baby, been given a small part in *Jeanne* under an assumed name, and become Norman's lover. Dudley, she warned, had a decidedly nasty streak. "Consider me warned," he said.[103]

A week or two later, a drunk, belligerent Dudley tried, without success, to get up to Bel Geddes's hotel room; thinking him all hot air and bluff, Norman declined the concierge's suggestion to get the police involved. Then late one night he managed to get upstairs and somehow pick the door lock. Alerted by the noise, Norman crawled in the dark toward the figure silhouetted against the hallway light, rose to his knees, landed his best punch, then pulled out one of the intruder's legs, sending him backward into the hall, a revolver peeking out from his pocket.

•

THE PARIS EDITION of *The Chicago Review*, and stateside newspapers as far afield as Walla Walla, Grand Forks, and Duluth, announced *Jeanne D'Arc*'s June 12 opening. Reviewers arrived from New York, London, Copenhagen, and Berlin.

The audience included the U.S. ambassador and his wife, the Minister of Beaux Arts, and a representative of President Doumergue, each in a flag-draped box; Elsie de Wolfe, Elsa Maxwell, Arthur Rubinstein, Mrs. Vincent Astor, and Mrs. Oliver Harriman; Princesse Edmonde de Polignac, Conde Nast, the Cole Porters, Algonquinite Dorothy Parker, and Mrs. Archer Jones of Richmond in the company of the Richmond girls (minus three who'd been given "crowd scene" parts).

All ran smoothly, though there was at least one moment of unintentional comic relief. During the scene in which Joan is captured, one of the tall Russian knights became so caught up in the action (lights, noise, weapons) that he refused to give in as the script required. Finally, Eva, as Joan, summoned up the very little Russian she knew and yelled, "Please die! I implore you to die!"[104]

But in the end, no amount of spectacle or repeated curtain calls could compensate for Mercedes's mediocre script. Eva's acting was praised, but Norman was the star of the hour, his innovative staging and painterly use of light called everything from astonishing and ingenious to triumphant.

Given the financial losses, and Eva and Mercedes's personal "break-up," the play would not open in New York as planned. Still, given all the difficulties they'd overcome, Norman was returning home imbued with confidence.

But before setting sail, three notable things occurred.

The first was an attempted assassination. Exiting a local theater with

Katharine one evening, Norman was whacked on the head from behind by Dudley's ever-present cane; his rival had been lying in wait. A wrestling match ensued. Several newspapermen, there to cover the scheduled performance, happened to have cameras. Early next morning, Norman managed to find an American correspondent to go around town speaking with various managing editors to ensure that no photo or story would be published, especially in an edition that might reach New York.

The second was the arrival of a handwritten letter from Jean Cocteau:

Monsieur
Le mort de mon pauvre Erik Satie m'a fort brouillé. On l'enterrait ce matin et, depuis hier, j'avais ma tete vide.
  Pardonnez moi. Je desirais tant vous voir et m'entendre avec vous.[105]

Cocteau's longtime friend and collaborator had died of alcoholism in a sad little room strewn with bowler hats, a pair of grand pianos (found on top of each other), musical compositions thought lost or totally unknown, and 100 umbrellas.

The third was a farewell-to-Paris celebration that would go down in memory as the Night of the Twenty Doubles.

What began with martinis and champagne at various venues morphed into a drinking wager at Zelli's. Norman ordered twenty double *fines* (the equivalent of forty shots of brandy) be lined up on the bar. He didn't remember much after the eighth. After ten, his friends tried hard to intercede. Sometime after a dozen, he became absolutely rigid before falling on his face.

While Tommy went looking for a taxi, Norman disappeared, breaking through a screen in an open apartment window and settling on the floor. It took a while to find him. Once inside a second taxi, he crashed his right elbow through the glass, trying to open the window. It took four hours to locate a doctor—it was Easter morning. The hangover lasted almost a week. The scar on the underside of his arm was a permanent souvenir.

CHAPTER 8

Colossal in Scale, Appalling in Complexity

1926–1935

In esoteric circles, gamester Geddes is acclaimed Manhattan's greatest. Auction bridge and poker are dismal to him, and so with the fervor and precision of a half-mad mathematician, he creates games colossal in scale, appalling in complexity.
—*TIME* MAGAZINE, MARCH 4, 1929

Back from Paris, Norman was offered the directorship of Yale University's nascent School of Drama, but seeing it as a full-time commitment, he turned it down. Instead, he started up his scenic design classes again, wrote a scenario for Paramount (never made), cowrote a pantomime for Reinhardt (never produced), designed a Factory Concert Theatre adjacent to Bethlehem Steel and the promised barges for Poiret, and wrote articles for *Encyclopedia Britannica*. Of his designs for *Arabesque* at Manhattan's National Theatre (political refugee Bela Lugosi was in the cast), *The New Republic* wrote, "Good or bad, it's a work of genius."[106] It closed after three weeks in the red. Meanwhile, he'd been asked to design a second Gershwin musical, *Strike Up the Band*, set to open on Broadway in the fall.

But what he really wanted was a concept worth committing to celluloid. He liked the idea of "steel" as a protagonist—steel as both hero and demon, from sewing needles and scalpels to giant ships, the thing that affects thousands of lives. He eventually abandoned it, though not before reading dozens of books on the subject, tracking down Charles MacArthur (who'd reported on the 1919 Gary, Indiana, steel strike), and getting Sherwood Anderson to work up an eighty-page treatment. In longhand.

Then late one evening, walking past a pawn shop, he noticed a large brass pendulum on an antique clock, which reminded him of "The Pit and the Pendulum." Returning home, he reread the tale. By morning, he'd completed a scenario.

Though Edgar Allan Poe had died half a century before cinema was invented, "the movies knew a kindred spirit when they saw one."[107] Norman wouldn't be the first to stir up the unhappy writer's ashes. But his would be the first shot solely from the *victim's* perspective; audience members would feel themselves in the protagonist's place, strapped beneath the swinging blade, experiencing his every reaction without ever seeing his face, because his face was *their* face. Special lenses would create the illusion of thoughts dissolving into each other across the screen.

He imagined sound throughout—the visceral sound he'd devised for *The Divine Comedy* and *Jeanne D'Arc*. Prerecorded sound effects—the "tumultuous" beating of the protagonist's heart, the "dreamy, indeterminate hum" of inquisitorial voices, "the hissing vigor" of the pendulum as it arced ever closer. He'd experimented with prerecorded discs synchronized to the action in stage productions, but the technique had never been used in a film.

Casting was the easy part. For the hero? Leslie Howard, who promptly volunteered upon reading the script. Norman's "dream cast" for the five black-robed Inquisitors (no acting ability required) was young Broadway wunderkind Jed Harris, Alex Woollcott, cartoonist Don Marquis, Algonquinite Heywood Broun, and Morris Gest (his animosity having been placated by *The Miracle*'s considerable profits), all of whom said yes.

Determined to make his movie independent of Lasky or any other studio, Norman contacted his friend and former Bronx neighbor Howard Dietz, now head of advertising and publicity at Metro-Goldwyn-Mayer. Dietz put him in touch with art photographer Paul Strand, who was freelancing as a cinematographer; Strand agreed on the condition that the job be completed within two weeks. Norman estimated that the shoot would require three days and 900 feet of footage. Adolph Stuber, vice president in charge of sales for Eastman Kodak, donated the film stock.

As a final stroke of luck, Norman met and hired a brash, personable, young Princeton dropout by the name of Leland Hayward; together, the two formed Pendulum Enterprises. As general manager, Hayward would get 25 percent interest of any future profits. He immediately proved his worth by setting up a shooting schedule, detailed down to the minute.

For the first time Norman was genuinely elated about the movie business. His script had been perused by a dozen "severely critical" people, all of whom had enthusiastically endorsed it. Everything was coming to-gether beautifully.

Then, in an instant, it all fell apart.

COMING DOWN THE steps from his workroom one afternoon in March to have lunch with Bel, Norman found her "tight-lipped and grim." Visibly trembling, she announced she was leaving and taking the children with her. She'd purchased a piece of land in New Jersey near her friend Marga-ret's farm and planned to build a house on it.

Dumbfounded, he managed to find his voice. Was she really that un-happy?

I know your father's been ill, he offered, and the kids can be a hand-ful. Shouldn't they at least *talk* about this? He didn't deny being wrapped up in his work, but she had known that about him from the beginning. Look what they'd achieved, he said, flailing his arms. A four-story brown-stone in the world's greatest city!

It was hard to argue with someone who refused to argue back. As for Katharine Hawley, he'd assumed what Bel didn't know wouldn't hurt her. Given all the temptations of the theater world, he'd been remarkably loyal.

And hadn't he always tried to include her? My work is social—you can't have theater without *people*! But she preferred to stay home, or with her suburban friends who work for the telephone company, for Christ's sake . . .

Bel winced at the blasphemy but said nothing.

The problem, he suggested, was that she had very old-fashioned ideas about marriage. A modern wife was interested in her husband's work, supported it, helped make it a success.

And what about biology? A man wants to know that his partner *desires* him.

Bel sat quietly, waiting for him to finish. Then she said, simply, "How do you want to divide the furniture?"

"Jesus, Bel. Can't we—"

"The furniture, Norman."

"Just take it."

She asked him to leave the house for a while. When he returned four hours later, suitably inebriated, everything that wasn't technically his—clothing, toys, kitchenware—was gone. Which meant that Bel had arranged things with the movers days, maybe weeks, before and had known the previous evening that it would be their last.

Unwilling to remain in a house made doubly empty by the absence of familial voices—Joan, the studious ten-year-old, dark like her mother; Barbs, four, fair-haired, and already incorrigible—or anything to sit on, he moved into a hotel.

When Leland Hayward phoned to remind him where to be and at what time, Norman knew he couldn't go through with it. The shoot, he said, would have to be indefinitely postponed. That meant losing Paul Strand. Within a week, Leland left for California.[108]

•

ONE DEFINITION OF luck is when preparation meets opportunity. Norman had had both, but the window was irrevocably lost. When Al Jolson's black-faced *Jazz Singer* premiered in October 1927, it did more than put audible words into actors' mouths. It introduced the idea of descriptive music, "pseudo-jazz crescendi" and "mock-syncopated sforzandi," that marked every twist and turn of the plot,[109] rendering Norman's visceral sound concept moot.

Weeks of misery followed. Only now did it occur to him that he hadn't taken responsibility for anything. That he'd lectured his loyal companion of a decade, tried to browbeat her into submission. Because she'd never said anything—or had he been too preoccupied to notice?—he'd assumed she was happy with their whirlwind life. *His* whirlwind. Perhaps if they hadn't started a family so quickly, if they'd had time, if *she'd* had time, before making the leap from bride to mother, her world might have stretched, as his had. New York frightened her. Her father had always frightened her. Maybe he did, too? That she was lonely beneath her shyness had never really crossed his mind. She'd uprooted herself for him, done most everything for him. In return, he'd said exactly the wrong things, no doubt brilliantly confirming the rightness of her decision.

Enough was enough. Heywood Broun—at six feet five and 250 pounds, the most lovable, slovenly, easygoing, and intellectually serious of the Algonquin group—finally dragged the brooding, boyish genius out for an afternoon's distraction at the Belmont Stakes. The pageantry and theater of it all, culminating with the victor posed like a Caesar under a blanket of white carnations, was enthralling. The seeds of a plan were planted. He would funnel his self-recriminations into an elaborate new entertainment.

Despite a reputation for gregariousness, Bel Geddes was, at heart, a private man who preferred socializing with people he already knew. More than half a century before Pac-Man, Nintendo, or cross-platform, trans-media game consoles, he'd begun creating a series of elaborate interactive diversions that would gather friends and associates around him.

Though he wasn't athletic, the "hard competition" of sports held a certain fascination. One of his first small-scale amusements, designed when he and Bel were still living in the Bronx, was a five-by-three-foot football field, with players kicking, passing the ball, running, and blocking and tackling. He followed up with a mechanical baseball game. Players pitched, hit, and ran the bases.

More ambitious had been an indoor golf game, an elaboration of the one he'd conjured up on his future father-in-law's lawn. Stuck in bed with the flu, he suspended a drafting board on wooden horses and built up a realistic, nine-hole relief course of wooden blocks, wire screen, and paste-soaked newspaper, complete with sand traps, an overlay of green velvet, and hand-fashioned velvet shrubbery. For clubs, inch-high spring-loaded cylinders; for golf balls, ball bearings painted white. Par was set at seventy-two for eighteen holes. It took over the dining room table, becoming so popular and so disruptive to any semblance of family life that sportswriter Grantland Rice helped get it moved to the lounge of the Links Club for tournaments.

•

THE NUTSHELL JOCKEY Club—homage to a man he never spoke of (*My father won the cup for the champion saddles of Michigan*)—would usurp all Norman's previous games. Following his afternoon at Belmont, he immediately got to work, commandeering several of his employees to help. In a matter of months, the Nutshell was ready for a trial run.

"The only two-mile track in the heart of Manhattan, located on the beautiful estate of Norman Geddes," boasted the official program, the "Bel" in Bel Geddes now conspicuously absent. Crafted from unevenly dyed and waxed felt, the twenty-inch-wide track was built to scale atop a twenty-eight-foot-long straightaway table. Races ran from four furloughs to the full track length. There was green velvet turf, white fencing along the sides, and sixteen reserved boxes behind a locked brass rail for stable owners, who got to view the proceedings at eye level. Behind the boxes was a triple tier of spectator bleachers, room for 100—it quickly proved inadequate—and a judges' stand opposite the finishing post. The setup took over the entire basement of Norman's brownstone.

The game was entirely electrical and featured cast-bronze horses, each three inches long, all realistically painted. Emerging from the stable, they lined up twenty abreast behind a thin steel gate, a device Norman claimed predated, by four years, the introduction of electric starting gates at nationwide tracks. Each contender had an individual motor controlled by an individual rheostat set at a winding speed based on the horse's past performance (zero to 1,000). They ran on copper rails, pulled by nearly invisible silk threads connected to unseen pulleys.

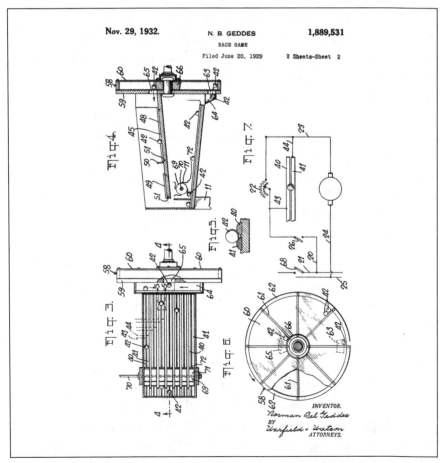

Nutshell Jockey Club patent drawings, 1929.
(Courtesy: U.S. Patent Office)

Randomness was provided by the Chance Machine, a mechanism that worked against the individual motors. A second series of motors shuffled fourteen large ball bearings across the copper rails; completing the electrical circuit, they doubled a horse's speed by one-half furlong. "Thus," observed *Time* magazine, "any horse might suddenly frisk ahead, outdistancing rivals with a higher starting speed, only to 'stumble' in the middle of the race or 'blow up' at the finish."[110] They could even jump hedges and ditches. So exacting was the Chance Machine that it gave averages comparable to those on any recognized track. The entire Nutshell apparatus was reported in the press as having cost $4,000 (in excess of $55,000 today), minus the considerable hands-on labor. All the machinery was lit and locked under glass, just beyond the finishing line.

Every Thursday afternoon, for three consecutive springs and summers, the dials were tested and set. Every Saturday evening eight races were run, beginning at 9:00 p.m. and shutting down (at least officially) at 1:00 a.m. Races were announced over an amplifier, and an authentic recording of racetrack bedlam, from shouting to pounding hooves, lent atmosphere.

An onsite printing press cranked out term definitions, billboard cards, weekly event programs, and forms for purchasing a stable and for buying, selling, or trading additional horses. All transactions were made through the Racing Secretary; other officials included Stewards and a Clerk of the Course. A Betting Board listed which horses were running, their past performances, and their odds to win or place. A Notice Board listed the purses, ranging from two to fifteen dollars. With the exclusion of breeding, weather conditions, and crooked jockeys, it was a live horse race in all the particulars.

There were lots of rules. Individual bets were limited to one dollar, with no percentage for the house. A "good horse" could win as much as ten dollars in a single event. Anyone could purchase a stable (five horses minimum) and actively participate. Before the game was two months old, nearly 100 people owned one; the number eventually doubled.

Norman's Chinese houseboy-cum-butler ("a blackamoor in jockey silks," reported *Time* magazine) served as chief usher at every game, resplendent in a tall white hat emblazoned with INFORMATION in large red letters. He also blew the starting bugle and struck the finishing gong. Any horse that fell, stopped, or lost its place in line, regardless of the circumstances, was disqualified, along with any horse that fell or stopped as a result. Norman remained above the fray, operating the switchboard.

Emotions ran high, fueled by prodigious amounts of alcohol. Rumor had it that on one particular evening $3,000 changed hands in a friendly wager. It was strictly invitation only. Amelia Earhart and Cole Porter were known to stop by. Regulars included Frank Crowninshield and theater critic Kenneth Macgowan (both golf game veterans), publishers Harold Ross and Horace Liveright, Rudolph Kommer (he owned a string of twenty steeds), cartoonist Rube Goldberg, Bel Geddes's protégés Henry Dreyfuss and Aline Bernstein, and most of the Algonquin crew. Alexander Woollcott christened all twenty in his stable after characters in Dickens's novels. Naturalist William Beebe's Arcturus (named after the steamship that carried his recent expedition to the Galapagos) quickly proved himself, winning thirteen of his first eighteen starts and placing in the others.

Newlywed Rosamond Pinchot arrived with husband William Gaston. It seemed an ideal match. "Big Bill" was a Harvard grad, a former World War I pilot, and a fellow blue blood (his father had been governor of Massachusetts, his grandfather Boston's mayor); the family's law firm dated back to 1844. Rosamond was an accomplished horsewoman, tennis player, and golfer; Bill had played football and rowed crew. Both were charismatic and blessed with extraordinary good looks. And he knew how to make her laugh.

After cartoonist Peter Arno lost his favorite, Parade, to an unknown, he auctioned off his entire stable to Clare Boothe for thirty dollars. Recently seen riding atop a different kind of animal—a live ostrich in the *Ziegfeld Follies*—Boothe renamed her expanded stable the "Clare Luce Stud." (She was keeping company with publishing magnate Henry Luce.) A week later, Arno's fury reached monumental pitch when Parade, along with two of his old stablemates, won twenty dollars.

The game was also a magnet for Hollywood notables who happened to be in town. Charlie Chaplin, Douglas Fairbanks, Sr., King Vidor, Leslie Howard, Adolphe Menjou, Mr. and Mrs. Basil Rathbone, Ethel Barrymore, Edgar Selwyn (cofounder of Goldwyn Pictures), and Natacha Rambova (the former Mrs. Rudolph Valentino) all crowded into Norman's basement.

One evening when the Classic for two-year-olds was running, one of Raoul Fleischmann's horses was the favorite. The courtly *New Yorker* cofounder got so caught up in the excitement that he forgot his family and luggage were waiting for him at the dock, scheduled to sail for Europe that same night.

"It was a tense race and a close finish but Raoul's horse won," Bel Geddes would recall decades later. "Not until five minutes before sailing did someone remember and shout: 'Raoul, your ship!' into his ear." Bolting from the house without hat, coat, or winnings, he arrived at the pier as the ship was floating out to sea, its railings draped with wife and children hysterically sobbing. "Raoul won his second race of the evening when an amiable tug captain put him aboard, mid-channel."[III]

Horse auctions were held after the evening's last race. The bidding was always fast, furious, and loud, and long after Norman went off to bed, which was rarely before 3:00 a.m., taxis would be screeching up to the door, dispelling the inebriated and the tardy—purportedly several hundred on one particular night—and infuriating the neighbors. By the middle of its second season, the Nutshell Jockey Club was attracting gate crashers and a disturbing criminal element. Norman appealed to Mayor Jimmy Walker, one of the Nutshell's enthusiasts, who assigned a patrolman to keep "the wrong people" out

In the end, the game died of "extreme popularity." A gambling syndicate twice offered the "blond little Leonardo"[112] $1,000 per night to operate it at Saratoga Springs, New York, for a season. Harry Payne Whitney, whose real-life stables had already garnered numerous Triple Crowns, also expressed interest. Financier "Barny" Baruch promised to provide a Manhattan venue. Norman turned them all down.

By the end of the Twenties, his mind was on other things, not the least of which was a major career shift. The "other things" would soon include the first of two major fairs; an elegant, members-only venue to replace the Algonquin's now-defunct Round Table; and a theater set that would leave Broadway audiences literally gasping when the curtain went up. Just as the Jazz Age was ending and the Great Crash was about to steep the country in a debilitating, protracted Depression, Norman was entering the most creative and lucrative decade of his life.

And yet . . . another at-home entertainment was brewing, one that would make the Nutshell seem modest in comparison.

•

THE WAR GAME was an idea that dated back to the so-called Great War. Bel Geddes had viewed that slaughter and destruction as a "hopelessly futile" exercise enacted under the guise of patriotism by ambitious, willful men, criminals in positions of power. An anti-enlistment pacifist (he claimed to have "languished in jail for two days" for speaking out against the draft),[113] he was nevertheless fascinated by military history and tactical maneuvers. Now, with a second European conflagration brewing, he decided it was time to clear away the mothballs and develop the game in full.[114]

Not for the last time, he had a finger on the zeitgeist. The Thirties ushered in a growing interest in board games (the hugely successful Monopoly allowed would-be entrepreneurs to make a killing that the real economic landscape all but prohibited), more obvious forms of gambling (Bingo, Irish Sweepstakes), and participatory sports (dance marathons, roller derbies, six-day bicycle races)—activities that offered escapism, camaraderie, and community at a time when large-scale cooperation seemed largely absent.

Though quieter and slower-paced than the Nutshell, the War Game was far more complex. The rules binder, cranked out on Norman's faithful printing press, was as thick as a phone book. Not one to do things by half measures, he supplemented his research with a personal library of 1,200 to 1,500 books on the Civil War, the Russo-Japanese War, and World War I (all of which he read, half of them more than once), along with declassified military records (some translated at his request), subscriptions to the *Quartermaster's Review*, and a journal devoted to cavalry.

What had begun modestly, years before, with Rand McNally topographical survey maps pinned to the wall and two chess sets pushed together now expanded into an alternate universe soon to be frequented by flesh-and-blood five-star generals, retired naval commanders, and international chess champions. This particular toy reputedly ended up costing $13,000 out of pocket,[115] a whopping $9,000 over the Nutshell.

The board was twenty feet long and four feet wide, its surface a varnished relief map of the coastline of two imaginary countries, Redegar (the red team) and Yelozand (yellow). The battle line between them was as long, by scale, as the distance from Verdun to Paris, with a sea area as big as the Adriatic.

All told, the landmass represented 20,000 square kilometers—approximately the size of the Western Front. There were a dozen layers of land above sea level (cut from cork), each an eighth of an inch thick; the water had three measured depths. One journalist reported that 9,000 cities and towns, not to mention every mountain, valley, bridge, river, harbor, and railroad, were delineated with fictitious names.[116]

Pins of various colors, sizes, and shapes (12,000 to 25,000—accounts varied) stood in for forty-plus types of military units. Each pin had a different kind of move, and each occupied the approximate amount of space on the map that its equivalent force would occupy in the field or at sea.

Military equipment included motorized supply carriers and artillery, PT-like patrol boats, minesweepers, submarines, tanks and armored cars, machine guns, and landing barges. Norman had the complete Navies of all five leading powers constructed in the model shop of his office. The battleships were built to exact scale (1 inch = 100 feet), complete with brass hulls, armaments, and planes. As time went on, there were also destroyers, aircraft carriers, capital ships and cruisers, tankers, tenders, barges, depot ships, and merchant vessels, ranging from the *Queen Mary* to a Chinese junk. At some point, the Coast Guard Artillery Association made Bel Geddes an honorary member.

Each country had the same railroad mileage, imports, and resources, from shipbuilding to sheep. (If a country's iron, coal, munitions, and agriculture industries were all overtaken, it had to surrender.) Military units and ships moved at speeds contingent upon the terrain they were covering; their artillery's range (and destructive power) increased a kilometer for every contour level occupied above its target. Those who found themselves several positions down could rally by sending up aircraft. Three dice were thrown, in a Geddes-designed case, to determine how much destruction an attacking force might do when it got within range of an objective. Each side had a general staff (equipped with maps, charts, typewriters), field commanders, and assorted advisors. As with the Nutshell, scrupulous records were kept in huge loose-leaf binders.

While one side maneuvered, the other watched and conferred among the ranks. A Geddes-designed electric clock, circa 1919, "before there were any commercially available," was jury-rigged to tap a ball at ten-minute intervals. Three short warning taps marked twenty-five minutes, and a sharp gong announced the half-hour mark—the other side's turn.

Field commanders could move units and cause them to fire, but only once. Thirty minutes, representing twenty-four hours in real time, was rarely enough to move more than half one's force; much advanced planning and discrimination was required. Stenographers kept score. At the end of the evening, captured enemy pins were placed in coffin-like boxes and handed over to a special secretary, who tabulated the dead (one of every sixty-seven casualties, just as in real wars) and staked them out in

WEEKLY REPORT OF _____ ARMY											
WEEK	LOSSES IN UNITS						ADVANCED		GAINS		
(5)	KILLED & WOUNDED BY CORPS					SHIPS SUNK	SECTOR LENGTH DEPTH		OCCUPIED CITY · FORT	POINTS GEOGRAPHIC	UNITS CAPTURED PRISONER SHIPS TRAINS
	I	II	III	IV	N						
(29) AM				8	3		Trent 25	3			
PM				18	4	X			Warminster		
(30) AM				15	6					Ilfracombe	
PM				10	11	1					
(31) AM	3			13	5	2				Alton	
PM	2			3	8	1					
(32) AM			2	8	5	X					
PM				9	2						
(33) AM				16	3						
PM	X	X	X	9		X					
(34) AM	X	X	10	4			Dale 10	5	Dale stn Start		
PM		0	2	15			14	6			
(35) AM		0	20	7	1		22	3			
PM		6	3	20	17		26	4	Leigh		
TOTAL						18					

DIVISIONAL ASSIGNMENT	DEAD TOTAL TO DATE		SHIPS SUNK TO DATE		OUR FRONT LINE	ENEMY CAPTURED		SHIPS CAPTURED	
CORPS I	HEADQUARTERS	0	DREADNOUGHTS	1	Coleshill	HEADQUARTERS		DREADNOUGHTS	
	INFANTRY	26	BATTLESHIPS	1	Shrewsbury	INFANTRY		BATTLESHIPS	
CORPS II	CAVALRY	15	BATTLE CRUISERS	1	Scout E. Roots	CAVALRY		BATTLE CRUISERS	
	LIGHT ART.	7	LIGHT CRUISERS	1	Leigh	LIGHT ART.		LIGHT CRUISERS	
CORPS III	TANK ART.	2	DESTROYERS	3	Dale	TANK ART.		DESTROYERS	
	HEAVY ART.	0	SUBMARINES	2	Tredegar	HEAVY ART.		SUBMARINES	
	PERMANENT ART.	0	MINE LAYER	0	Brombone	PERMANENT ART.		MINE LAYER	
CORPS IV	ENGINEERS	1	MINE SWEEPER	0	Cardiff Bridge	ENGINEERS		MINE SWEEPER	
	SUPPLY TRAIN	1	COAST PATROL	1	Gravesend	SUPPLY TRAIN		COAST PATROL	
RESERVE	AIRCRAFT	10	SUPPLY TENDER	1	Dover N.E.	AIRCRAFT		SUPPLY TENDER	
	FLAGSHIPS	0	TRANSPORT	0		FLAGSHIPS		TRANSPORT	
	NAVY CREW	14				NAVY CREW		RAILROAD TRAINS	
	TOTAL		TOTAL	3		TOTAL		TOTAL	

A complex scorecard for a complicated game.
(Courtesy: Harry Ransom Center, The University of Texas at Austin)

graveyards. Wounded troops were returnable in three weeks. Other rules applied to prisoners.

The most remarkable thing was that despite there being as many as twenty-eight belligerents—fourteen on each side, simultaneously engaged—there was no need for an umpire.

The game ran every Tuesday night throughout the winter months. The battles commenced, or more often continued, from 8:00 p.m. until midnight, with no time-outs. Sunday was "staff day." Each side was allowed ten minutes to study its opponent's formation on the board (secured beneath a locked plate-glass cover to prevent tampering) in anticipation of the following week's play. This privilege came, however, at the cost of five sacrificed airplanes.

Whereas several horse races had run on any given night, a "war" could last indefinitely. One conflict dragged on for six months, another for three years. Players took the proceedings seriously. Faces reddened, lips tightened, fists clenched. So much Prohibition scotch, rye, and gin was consumed at Bel Geddes's expense that he was finally reduced to serving water.

Unlike the Nutshell, it was mostly a man's game; some wives came and knitted through the proceedings. The stalwarts forfeited everything from vacations to business commitments, returning week after week, month after month. There was a broker, an architect, several engineers, an underwear manufacturer, a rare book dealer, and a shipping magnate. Playwright Maxwell Anderson was a devotee. Theater critic Bruce Bliven doubled as a referee and war correspondent, madly punching away at a typewriter set up between the opposing sides.

Other regulars included a member of Woodrow Wilson's Peace Commission at Versailles, a U.S. Navy Rear Admiral, a Brigadier General of the British General Staff, an Italian cavalry captain, and the former Chief of New York City Detectives. Annapolis football coach Tom Hamilton credited the game for inspiring one of his primary technical strategies. "Had the Kaiser had one of these to work over," remarked a young German player, "there wouldn't have been a war!"

One morning, a pair of government agents appeared at the door, sent from Washington, D.C., to ascertain if the designer was a spy. Norman commenced a tour, explaining the game's finer points. His guests managed to escape partway through.

In the game's early days, Norman chose to play, often neglecting his professional work in order to read Foch and Joffrey and plan out tactics; later, he preferred to watch, pacing up and down "like Jove overseeing the assault of Troy."[117] He saw his creation as chess on a heroic scale, chess carried to infinity, a concept seemingly confirmed by the frequent appearance of five-time international chess champion Edward Lasker.

One evening, Alexander Alexandrovich Alekhine knocked at Geddes's East Thirty-Seventh Street door; he'd read about the game in a White Russian newspaper published in Paris. Alekhine was widely considered chess's greatest and most inventive player. (Lasker, who'd competed professionally against the Russian, compared Alekhine's chess obsession to a morphine addiction.) He was intrigued enough to return for four consecutive weeks before leaving town.

As commander in chief of the Red Force, Lasker subsequently lost an entire field army, his whole command plus a quarter of his side's total force; thus failed, he was up for court martial. The future author of *Chess for Fun and Chess for Blood*, and an expert at the strategically challenging Japanese board game of Go, he threw himself onto Norman's sofa and sobbed like a baby.

As with the horse race game, rumblings about merchandising came to naught. When Norman finally put the War Game into storage, it was partly, he would claim, because its complexities required "a school for the recruits."

Though he couldn't have known it then, the miniaturization and three-dimensional landscaping of this particular diversion laid the groundwork, literally and figuratively, for what would become his most enduring legacy, the most iconic World's Fair exhibit of all time, as well as for government military projects, some of them top secret, created during the real-life war to come.

CHAPTER 9

Goods into Roses

1920s

He's a big Rabelaisian bull in their pasture. They know he started the whole thing.
— EARL NEWSOM, BEL GEDDES FRIEND & FORMER COLLEAGUE[118]

America is . . . the future . . . a land of desire for all those who are weary of the historical lumber-room of old Europe.
— FREDERICK HEGEL, CIRCA 1820

The real turning point that led to Bel Geddes's controversial career switch occurred in the autumn of 1926 when, already under contract to design *The Patriot*, he was invited to spend the weekend at the Long Island home of Ray Graham, president of the Graham-Paige Motor Company, a small family-owned factory. To his delight, his fellow weekend guest turned out to be Gene Tunney. The boxer had recently defeated Jack Dempsey in Philadelphia for the world heavyweight title before 130,000 fans (originally scheduled for Chicago, the venue was changed for fear that Al Capone, a big Dempsey fan, might "get involved").

At some point, Graham asked if Norman had ever thought of focusing his talents on the refinement of everyday objects. "Something sent the blood rushing through me," he would recall decades later. "I didn't glow; I burned all over."

The designer, boxer, and executive talked late into the night. By the time they parted company, Graham had offered $50,000 for a series of automobile designs, all expenses paid, the work to be carried out over the course of a year at Graham's Detroit plant. If all went well, there'd be an option "at a figure considerably higher."

On the train ride back to Manhattan, Norman watched a dozen reapers mowing down a wheat field, almost in a single swoop; later, instead of the two or three silos of his youth, he saw a row of twenty. Industry was the future.

•

AFTER CONFERRING WITH a handful of businessmen and with encourage-ment from Tunney, who would become a lifelong friend, Norman made a counteroffer. He would accept Graham's fee but retain his independence, working out of his own office, traveling to Detroit when needed to confer with Graham's engineers. And he'd pay his own expenses for drawing and creating models. When Graham protested that the latter would cost him a fortune—cars were designed using full-scale wooden models built on actual chassis—Norman said he planned to work one-quarter size, us-ing modeling clay (a medium he "understood") over wooden block forms, with the equivalent chassis mass provided by the plant. A technique he often used for theater commissions, it had never been applied to car de-sign. The plan was to follow those up with full-sized clay models which he could easily improve and refine. Using quarter-, then full-size clay models would later become standard practice.

Within a few months, he'd managed to negotiate contracts with Toledo Scale Company, Franklin Simon Department Store, J. Walter Thompson, and the Simmons Bed Company. In order to take up Graham's challenge, he would need to hire a nucleus of specialists, which would cost more than Graham's commission, so there needed to be similar jobs going on that those specialists could help with.

The plan was for five increasingly innovative automobile designs—progressive radicalization—to be introduced over a period of five years. Norman numbered his concepts in reverse, from least radical to most. For Car #5, he introduced a teardrop-like shape and rear engine (pref-erably water-cooled) that eliminated the long transmission shaft, allow-ing for a lower center of gravity (decreasing the danger of overturns on curves), and better insulation against heat, noise, and fumes and simpli-fied repairs and service. He designed a louver-cooled radiator and seating for eight. Wraparound windows offered panoramic views and eliminated sightline obstruction from the hood. *Landscape as spectacle.* It was a con-cept he would return to again and again.

For the "ultimate" Car #1, he anticipated the radiator placed behind a rounded grille (allowing the engine hood to integrate with the car body) and a windshield that slid vertically *into* the body. Driving lights, set low and between the mudguards, would turn with the front wheels; the fuel tank and trunk compartment would be built in, rather than added on; and the back license plate and back driving lights would be incorporated into the fender to help reduce air resistance. Graham's engineers estimated that it would be able to travel fifty-eight miles per hour with the same power that existing cars required to go forty-five miles per hour. And for a touch of dash, white "trim" was added to the tires. According to at least one contemporary source, "whitewall" tires were a Bel Geddes innovation.[119]

The first test body was scheduled to be built and put through its paces at a maximum cost of $8,000 at the end of October 1929. Unfortunately, the stock market crash refused to wait a week, and few felt experimental in its wake. Nonetheless, Norman credited Graham with being more influential than anyone for setting him on his new path.

•

THE EPIPHANY CAME, as it had with *The Pit and the Pendulum*, while window-shopping one evening. Norman was struck by the drabness of the displays, "as cheerless and uninviting as a mausoleum."[120]

Fifth Avenue's department stores, *Forbes* would note in hindsight, were no different from rural shops that stuffed a single forlorn window with a miscellany of rubber boots, pastries, fertilizer, and bicycle pumps, under the assumption that there was something to appeal to everyone.[121] Even Saks Fifth Avenue, the neighborhood's high arbiter of style, favored potted palms and old-fashioned strip lighting.

It suddenly seemed obvious that here was another form of theater. *Store windows were stages, the products on display were actors, and prospective customers were the audience.* Why mount a production unlikely to engage? Why market twentieth-century goods with a nineteenth-century approach?

He drafted a proposal for displays that would highlight and glamorize merchandise, the opposite of "showcasey" cramming-in, using standardized, interchangeable display units made of unorthodox materials like glass (frosted, etched, mirrored), stainless steel, aluminum, Bakelite, and Lucite, plus freestanding lettering (a first) cut from metal and wood. Everything would be arranged to create a sense of unity—sleek, refined, and "radically plain" to offset coordinated color ensembles and contrasting textures. Thousand-watt focus lamps (his famous "spots"), mounted on adjustable swivels with interchangeable lenses and soft-edge irises, would create emphasis and variety.

Mr. Gimbel felt the plan would render the bulk of his display equipment obsolete. The owners of the Franklin Simon department store offered a two-year contract.

In anticipation of the 1928 yuletide season, Norman created an abstract Christmas tree of chromium-plated tubes and triangular glass shelves; Worth's famous Dans la Nuit perfume lined the "branches." Lit from the front by a pale red light, the "tree" cast brilliant green shadows. Another window contained only three items—an aluminum bust topped with a designer turban, its neck wrapped with a vermilion and chartreuse scarf, and a matching handbag. A background of large triangular shapes focused the viewer's attention; spotlights cast dramatic blue shadows.

The displays were simple, theatrical, narrative, almost dreamlike, the windows nearly empty.

Unveiled just before Christmas, they began attracting crowds, which in turn lured curious pedestrians, even from across the street. On one occasion, the ranks swelled to the point that police were called in to keep traffic moving. Five weeks later, Norman installed a similar design with different merchandise to the same result. Months went by. Then, seemingly overnight, the entire street changed. Window "dressing" had become window "design."

Nearly a quarter century later, Raymond Loewy would claim that, circa 1919, he'd designed a minimalist window with "violent" shadows for Macy's—a single mannequin in a black evening gown illuminated by spotlights, with a "luscious" mink dropped nonchalantly at her feet. "Simplicity blended with a dash of French logic" that the horrified store owners had misunderstood.[122]

Over the next two years, Norman's displays would continue to draw crowds (one featured custom-made glass mannequins lit from within, revealing the delicate workmanship of lingerie), helping to shape the expectations of a growing consumer society. After touring Fifth Avenue, one visiting German display man remarked that Americans had learned "how to transform goods into roses."[123]

•

THE UNITED STATES had emerged relatively unscathed from World War I. Now the machines were too idle, ran the argument. They needed to be "reloaded" for peacetime production. Karl Marx would have approved.

So began an era of unprecedented convenience—tea bags and instant coffee, toothpaste in collapsible tubes, precanned vegetables and prerolled cigarettes, toilet paper in rolls, soap bars that fit comfortably (the brilliance of small ideas) in the hand. Throwaway "safety" blades replaced razors honed on leather strops—the difference, claimed Gillette, between taking an elevator and walking the stairs. Thin, throwaway latex condoms replaced thick rubber ones used until they fell apart. Sanitary napkins (mass-produced thanks to a huge World War I cellucotton surplus) were followed by tampons, designed by a Connecticut doctor (to the eternal gratitude of his wife, who was, the story goes, a professional dancer). Band-Aids were the brainchild of a man whose wife was clumsy with kitchen knives.

Another purely American invention was a jiggly, shimmery, prepackaged dessert that could be molded into endless forms. Jell-O was created from the gelatin patents of New York industrialist Peter Cooper, founder of Cooper Union college and, not coincidentally, owner of a glue factory that processed slaughterhouse parts (collagen, derived from slaughterhouse waste, being a main Jell-O ingredient).

In the rush to "reload" the machines, products were often invented first, their purpose concocted later. Initially marketed for irrigating wounds, Listerine did a turn as a floor cleaner, an aftershave, a dentifrice, a deodorant, an antidepressant, a dandruff remedy, a contraceptive, and a cure for venereal disease before being reinvented, yet again, as a bulwark against halitosis, a term exhumed from a British medical journal. Backed by an aggressive ad campaign alerting the public to this debilitating "social disease," sales (finally) skyrocketed.

With mass production came mass psychology. The human mind can be regimented "like the military regiments the body," wrote propaganda guru Edward Bernays. "We can effect change . . . just as the motorist can regulate the speed of his car by manipulating the flow of gasoline."[124] (Bernays's uncle was Sigmund Freud.) Creating an appetite and keeping it "hungry," convincing the public that it needed those goods-turned-to-roses as much as oxygen, was the job of advertising, still in its infancy but about to leap into a no-holds-barred adolescence.

Everything was accelerating.

A second Industrial Revolution was underway, one that Detroit would contribute to with dreams of mobility and autonomy, Hollywood with narcotic images of love, wealth, and intrigue. Artists and writers were deconstructing time into slices and jags. Even death was adapting to the new century, with heart attacks (a quicker exit) displacing tuberculosis. There were X-rays and Freud's messy inner landscapes. Suddenly things were not—had never been—how they seemed. Millions of young men had, only yesterday, been lost in a horrific world war for reasons that remained unclear. Life was absurd. Innocence was a myth. There was no such thing as a guarantee.

Everything was modern. Just *now*. *Only* now.

"New" was the last word, the cat's pajamas, the frog's eyebrows, the bee's knees, the goat's whiskers, the duck's quack. "New" was the demise of the homemade and the ascent of the store-bought. "New" meant you were a "consumer," not just a buyer—"in the know," "in the swim," on "the cutting edge."

"Wants are almost insatiable," concluded a 1929 government report created under the auspices of then–Secretary of Commerce (*there is no American design*) Herbert Hoover.[125] "Economically we have a boundless field before us . . . new wants will make way endlessly for newer wants, as fast as they are satisfied." "Where Babbitt, as a boy, had aspired to the Presidency," wrote Sinclair Lewis in his famous satire, "his son, Ed, aspired to a Packard Twin-Six."

To ensure that people would succumb to Eve's shiny apple not once but repeatedly, General Motors' president Alfred Sloan had initiated the Installment Plan. By 1925, the vast majority of America's cars, furniture, appliances, pianos, phonographs, and radios were being purchased "on time." Then in 1927, Sloan introduced the "yearly style change," a challenge to Henry Ford's puritanical and ubiquitous Model T. Ford had refused all suggestions of updating or change, though his "any color as long as it's black" approach had a very practical side: black paint dried faster than other colors, thus cutting production time while reducing inventory and supplies.

With help from DuPont's new faster-drying lacquers (the result of chemists "fooling around" with nitrocellulose left over from World War I), GM introduced a stylish new Chevy with hubcaps, a better paint job, and less milk-wagon-like corners. Stubborn but no fool, Henry Ford rallied with a "Rich Windsor Maroon" Model T, then discontinued his bestseller altogether for the Model A, making used cars a menace comparable, as Robert Benchley put it, to a biblical plague of locusts.[126]

Charles Kettering, head of GM research, referred to the annual style change as "the organized creation of dissatisfaction"—a strategy designed to make the public feel out of step, even *diminished*, if they didn't go along. Intrinsic to that strategy was planned obsolescence, also known as "death dating." By creating "time bombs"—products made far less well than they could be—even the recalcitrant would be obliged to trade in the old for the newer or newest.

Thirty years later, when Vance Packard's *The Hidden Persuaders* alerted the public to the fact that advertising, installment plans, and updates were designed to make them "buy things they can't afford with money they don't have"—by tapping into their subconscious, using brainwashing techniques developed by the military, even digging into the busy minds of children—it would come as a shock.

The Bel Geddes & Co. logo.

In 1927, the same year GM instituted its style challenge to Ford, Bel Geddes set up shop as a designer for industry. Three competitors soon followed suit. Together, they became known as the Big Four. All hailed from the arts. Only people with theater and advertising backgrounds understood the connection between design and sales, George Nelson would later write. Architects "were busy turning out gentlemanly forgeries of European buildings," and engineers, "when detached from their slide rules, tend to become self-consciously arty in an oddly spinsterish fashion."[127]

Henry

Henry Dreyfuss was, like Bel Geddes, a high school dropout with a talent for drawing. Scion of two generations of costume suppliers, he came by his interest in theater naturally. At sixteen, he signed up for one of Norman's design courses, which led to a three-year apprenticeship.

He was the quartet's youngest member by a decade and the only native New Yorker. His first office consisted of a borrowed card table, two folding chairs, and a twenty-five-cent philodendron that, twenty-five years later, would be ceiling high, luxuriant. Demonstrating Geddes-like chutzpah, he got himself hired at the Strand, where he created a stylized peacock whose tail unfurled to reveal a jazz band, and designed sets for several plays, a roof cabaret for Maxim's, and a shatterproof, early-plastics chandelier.

Meanwhile, he stayed busy ferreting out small commercial jobs—designs for neckties, garters, a fountain pen, a shaving brush, an egg-candling machine, a sonar fish finder, a woman's urinal, a fly swatter, a package redesign for Higgins Vegetable Glue, a legless grand piano. His search for a secretary proved providential. Doris Marks, a Vassar grad from a wealthy family (the story goes that she arrived at the interview in a chauffeur-driven Pierce Arrow), had an excellent head for business, leaving her boss free for creative challenges. She quickly rose from secretary to partner to wife.

Raymond

Raymond Fernand Loewy was the son of a bourgeois French mother and Jewish-Viennese father. Drafted into the French army, he decorated his WWI dugout with discarded chairs, carpet scraps, flowered wallpaper, and tufted pillows; planted geraniums; and sewed himself a new pair of breeches, deeming military issue badly cut.[128] By his own account, he rose from private to captain and was awarded the Croix de Guerre.[129]

In 1919, just as Bel Geddes was mounting his enormous ship at the Chicago Opera House, Loewy followed one of his brothers to Manhattan, where he fell into a career as a fashion illustrator. His first paycheck was spent on a custom-designed suit, his second on a tuxedo.[130] After several years of dreaming up diaper ads and elevator operator uniforms, he decided to initiate a one-man crusade against what he deemed America's shockingly bad taste.

His first industry job, in 1929, was a redesign for a mimeograph machine. After covering it with 100 pounds of modeling clay, he emerged with the Gestetner Duplicator Model 66 (claiming the clay technique, which Bel Geddes had been using for years, as his own). His first office followed, a studio on the forty-fourth floor of a Fifth Avenue skyscraper, filled with sumptuous furniture and a repeating phonograph that softly played dance music throughout the workday.[131]

Walter

Walter Dorwin Teague was the oldest of the quartet by a decade. An Indiana minister's son, he arrived in New York City at nineteen with aspirations to follow in the footsteps of Maxfield Parrish. He eked out a wage lettering signs and drawing for mail-order catalogs while attending night classes at the Art Students League, eventually establishing himself as an illustrator specializing in "Teague borders"—Victorian trellises, cornucopias, festoons, filigrees, and phoenixes used to frame ads for floor varnish, silverware, cigarettes, shirt collars, and more.

At forty-two, he was financially secure, married with three children, and bored. A trip to Europe introduced him to the Bauhaus.

His first major industry commission, for Eastman Kodak, would lead to a thirty-year relationship creating decorative camera casings with names like Bantam Special and Beau Brownie. Another major client would be Texaco, whose iconic white-tiled gas stations with green horizontal stripes were his design. W. Dorwin Teague, Jr., a fellow industrial designer, described his father as "practical and dogmatic, but [with] no interest whatsoever in aesthetics."[132]

•

MANUFACTURERS WERE QUICK to follow GM's lead and to recognize that, with the possible exception of automobiles, "Mrs. Housewife" and "Madame Homemaker" were the nation's primary consumers. "When women refuse to wear ostrich feathers, South American farms languish," observed Doris E. Fleischman, public relations pioneer and wife of the influential Bernays. When women bought home appliances, thousands of men were employed to make them.[133]

"Other countries are old. We are youth personified!" proclaimed evangelistic home economist Mrs. Christine Frederick, whose books and magazine columns influenced housewives nationwide. "Progressive obsolescence," she explained, was "the very knife-blade" severing modern, superior America from the shackles of "shoddy Europe." Why? Because quality goods impede progress![134]

"The business man cannot tolerate durability," echoed economist Stuart Chase, "because of the brake it puts upon sales . . . Imagine a modern department store deliberately seeking to sell a vacuum cleaner good for a generation, as sewing machines once were!"[135] "That way is defeat," wrote

Egmont Arens, "a turning back to medievalism."[136] Let Europe have its "past," a Goodyear business manager wrote dismissively—their art, their savoir faire. "We'll take care of the future."[137]

Self-indulgence, not self-restraint! The Founding Fathers, Mrs. Frederick informed her millions of readers, had crossed the Atlantic to escape suffocating virtues like delayed gratification. One's most private longings were fulfillable through commerce; even the most isolated rural farmer could order everything from tractors to Parisian-inspired dresses from Sears Roebuck catalogs. Not coincidentally, part of Mrs. Frederick's income was derived from paid endorsements.

•

WHEN ALBERT EINSTEIN, the original scientist celebrity, first visited the States, he was greeted with a frenzy not unlike The Beatles' arrival more than a generation later, though few had the remotest idea what E=mc² meant. Even before his bags were packed, propositions flooded in, offers of tens of thousands of dollars for testimonials for disinfectants, haberdashery. Einstein found the offers insulting, if not corrupt.

Bed czar Zalmon G. Simmons, Jr., had no such qualms.

"H. G. Wells Disagrees With Napoleon on Sleep" ran a Simmons co. headline in the recently founded *New Yorker*. Bonaparte might have gotten along on six hours, but the author suffered from "threadbare nerves" without eight.[138] The genius of wireless inventor Marconi and the fresh complexion of George Bernard Shaw were also attributed to the rejuvenating, hand-tied coils of a Simmons Beautyrest mattress. One ad, never released, suggested shipping a Beautyrest to Stalin "in an effort to relieve international tension."

Seated in Zalmon's office, brash thirty-five-year-old Bel Geddes condemned the frames that held the mattresses—steel tubing disguised with fake wood grain and augmented with Victorian "spinach." Why not exploit materials for their virtues instead of pretending they're something else? "Always happy to double our sales," the fiftyish Zalmon replied, "but we're already selling more than all our competitors nationwide." Customers included the *Titanic* and the *Lusitania*. There are, he added, "only so many folks going to sleep at night." Besides, beds didn't sell on looks but because they were comfortable and "built to wear."

"Look, Mr. Simmons," Norman countered, "you're still a young man. Not too old to know that you'd be more attracted to a beautiful girl than to one *who would just last longer*." Besides, he added, his manufacturing costs were far higher than what an "honest" design required.

It wasn't an easy sell. But a week later, the two traveled to Kenosha together to observe experiments with a new plastic, one that offered not only the strength of steel but the option of transparency or color. There was every indication of success, they were told, but the technology wasn't there yet. Norman returned home with a $50,000 commission.

Norman's beds, introduced in 1928, featured radically simple angles and highly polished or black-enameled steel recast from discarded railroad rails (an early example of recycling). Minimalist chromium trim, reminiscent of De Stijl paintings, anticipated the 1940s rage. "I thanked her and sat in a chromium and leather chair that was a lot more comfortable than it looked," Raymond Chandler would have Philip Marlowe observe in *The Lady in the Lake*.

Norman followed up with Simmons's first "metallic" bedroom suite—chair, dresser, night table, dressing table with a tall rectangular mirror, and an upholstered armchair. Variations, introduced in the mid-Thirties, would include butter-yellow lacquer trimmed with chromium, Viridian green lacquer with brass, and a severely simple burnished dresser with vermilion-lined drawers. "Nothing gloomy or tomb-y," observed *Pencil Points Magazine*. "Everything practical, colorful, controlled—in that best Norman Bel Geddes manner."[139]

The first available suite was shipped to Rome for the Duchessa di Sermoneta, to be installed against the ancient frescoed walls of Palazzo Orsini castle. (Her bloodline, *Vogue* was quick to point out, dated back to the beautiful Vittoria Colonna, beloved of Michelangelo.) "So modern," exclaimed the Duchessa, "and yet so restrained."[140]

•

THE TOLEDO SCALE Company's grocery scale, circa 1897, was the most widely used and imitated device of its kind, but its cast-iron body, to the chagrin of salesmen who had to transport it, weighed 163 pounds. In 1928, an aggressive new company president offered Bel Geddes a commission to redesign it, along with nine other models.

He'd always approached theater design systematically, even if he'd had to invent the system himself. First came the preliminaries: a precise determination of what was expected, a study of existing factory methods and equipment, and of the competition, consultations with experts on materials. Surveys and public opinion polls had never been conducted from a *design* standpoint. Norman decided to change that.

He sent canvassers out to metropolitan markets and rural stores to talk with clerks and customers. As a result, the scale's pendulum was moved to one side and its cylinder cantilevered so purchasers could see *what* was being weighed and *how much* it weighed. He also suggested setting the scale *into* the counter, flush with the wrapping surface.

The first choice for the body was aluminum, but he settled for far less expensive sheet metal, a decision that altered the entire production process. More important, in the long run, was finding a substitute for the traditional porcelain enamel coating, which led to the idea of making the scales entirely from plastic (if beds, why not scales?). Plastics were "way out in front," he told Toledo's president, Hubert Bennett, even less expensive and lighter than sheet metal. There might even be a way to mold an entire scale from a single large piece. Perhaps even a *transparent* piece, so customers could see the mechanism inside. Not to mention that plastic was a renewable and variable resource, like manna.

Bennett was intrigued enough to set up a research and development fund, the Toledo Synthetic Products Company.

One thing led to another, and before long Bel Geddes and Bennett were discussing the idea of creating an entire factory complex. Four months into Norman's initial contract, Bennett agreed on terms for a "preliminary study of the problem," despite the fact that architect Harvey Wiley Corbett had already submitted blueprints. Bennett told his board of directors that Norman's design would "cost lots of money and be entirely different, even weird looking."[141]

Bennett was rewarded, wrote *Fortune*, "with such an exciting set of plans as has rarely ever been seen outside the most advanced industrial areas of Germany and Holland"—a workplace where "even dirty labors may be performed" in "suave" buildings that "will adorn, not sully, the countryside."[142]

The expansion from grocery scale to factory would often be cited as yet another example of Bel Geddes's impracticality and over-weaning ambition. Thirty years later, when Norman was no longer around to take the blame, *Fortune* would write that the "oddest aspect" of the profession was the propensity of designers to expand "a simple request" for a product face-lift "into the redesign of the corporation that manufactured it."[143]

An eleven-story administration building, a porcelain factory, a flexible-use laboratory space, a central heating plant, a truck dock, and a parking lot would cover one-quarter of the eighty-acre property. There would be a four-story machine shop with diffused natural light streaming in, cathedral-like, from a circle of high, wraparound windows (with overhanging cornices to eliminate glare and eye strain). All radiators and pipes would be concealed with no overhead belts or pulleys in sight, and all phone, gas, and electricity lines would be carried through occasional columns and between the floors, everything designed for easy cleaning and minimal dust.

The remaining sixty acres would be landscaped, with an entrance through a broad avenue of Lombardy poplars and a series of reflecting pools that did double duty as fire protection. Employees would have a wooded picnic area, an athletic field complete with grandstand, a baseball diamond, eight tennis courts, and a swimming pool with showers and lockers. There was also space for an executive airport.

Factories had always been built as economically as possible. As a result, they were dark, depressing, not necessarily efficient, and often unhealthy. Could a first-rate product be produced in a third-rate plant? "Suave geometrical forms with walls composed almost entirely of glass," noted *Fortune*. Sleek, graceful, monumental, thrilling.[144] The contractor's estimate came to $2.88 per square foot. The average factory cost $3.00 per square foot to build.[145]

Bennett was "very anxious" to get things started,[146] but the Crash and the Depression brought indefinite postponements. Norman was paid in full (despite the 90 percent listed in his contract); later, when the laboratory building was, in fact, built, Albert Kahn's drawings remained faithful to Norman's vision.

•

IN TANDEM WITH Bel Geddes's career switch was another seismic change. By the summer of 1928, he had, for the first time as an adult, fallen deeply in love.

Two weeks after Norman had decided he was too busy to continue his scenic design classes, three determined young ladies had arrived at his office to enroll. Two of them, Bryn Mawr archeology majors just back from an Egyptian dig, had been persuaded, after attending a performance of *Jeanne D'Arc* in Paris on their way home, to switch career plans.

"I've always been a softie for sincere youthful eagerness," Bel Geddes recalled years later. One of the three was Frances Resor Waite, a well-traveled brunette with a regal nose and pedigree (her great-grandfather had been Chief Justice of the Supreme Court). She juggled Norman's class with courses in architectural drafting, followed by outside jobs designing costumes and props. By the fall of 1927, she was on his payroll. In the course of things, Norman discovered a willing dance partner who also introduced him to sailing. It was Frances, too, who gained him access to J. Walter Thompson, by then the nation's top ad agency. Company president Stanley Resor was her uncle.

The agency offered a $100,000 commission for an assembly room in the recently completed Graybar building. The catch was that it had to accommodate board of directors' meetings, lectures, and small conferences, informal teas and luncheons, cinema showings and ad layout exhibits; had to be divisible to accommodate two separate functions simultaneously; required storage space for up to 200 comfortable chairs; and still had to be sufficiently unusual to create an impression.

It was Bel Geddes's first architectural project since the Palais Royale. Collaborating with the engineers who'd designed the Graybar, he began by eliminating four interior and three exterior girders.

Half a dozen windows were joined to form two-story-high floor-to-ceiling views. Intense blue-green draperies stretched up like columns to an opaque glass ceiling, which housed the room's only lights. Dimmer controlled, they could be preset to any time and to increase or decrease to specific intensities over a specified number of seconds or minutes—something particularly useful on winter afternoons, when the sky darkened early.

The same vivid fabric was used to cover the low, soft-edged Bel Geddes–designed cube-shaped chairs with expansive arms for note-taking. Pale gray walls were relieved by strips of brass, black Vitrolite glass, and burnished copper that continued around the room in a series of rectangles.

There were no paintings, murals, tapestries, or other distractions, other than an adjustable brass rack for hanging advertising sheets.

Radiators and ventilators were concealed behind decorative bronze grilles, allowing in fresh air but without weather and traffic noise. Telephones could be plugged in anywhere along the wall.

The room (eighty feet long, thirty feet wide, twenty-five feet high) was entered from above. A balcony connected to a broad, beautifully proportioned staircase that offered a sweeping view of the space below. For conferences, there was an ebony table with brass foundations and a lectern that housed a loudspeaker. Curtains concealed an automated motion picture screen.

"Restfully colored and supremely comfortable," noted *Forbes*, without "the senseless complexity of geometrical . . . forms which characterize the so-called 'modernistic' vogue."[147] The *New York Times Magazine, Theatre Arts Monthly,* and *American Architect* all ran illustrated features on Bel Geddes's "Office of Tomorrow."

In typical fashion, the designer had managed to juxtapose his conference room work with the creation of a mobile marionette theater for the Yale University Puppeteers.

•

WITHIN MONTHS OF joining Bel Geddes's staff, Frances was promoted to executive in charge of the Simmons account (at seventy-five dollars a week) and was overseeing costumes for the Broadway production of *Fifty Million Frenchmen.* Come summer, she embarked on a month-long family vacation to Paris.

June 29, 1928:

Frances darling,
Ever since you left I've fidgeted around like an old woman . . . I just feel like hell and there is no use pretending otherwise . . .

July 3, 1928:

For the last year, we have been so close together . . . and gotten so adjusted to each other that I didn't realize there was anything unusual about it until you left . . .
I am terribly in love with you.

Frances's parents weren't altogether happy about the liaison; her suitor was still legally married and nearly twelve years their daughter's senior.

July 8, 1928:

LONGING FOR YOU WORST TORTURE EVER WENT THROUGH . . .

July 23, 1928:

NEED YOU TERRIBLY IMPOSSIBLE TO WORK . . .

July 24, 1928:

It is inconceivable that you are so out of reach . . . never for a moment thought you couldn't go away for a month without it messing me up like this . . .

•

BY 1929, GERMANY had successfully launched the world's largest heavier-than-air craft, the DO-X. Weighing in at 48 tons, it was 131 feet long and 33 feet high, with a wingspan of 157 feet, and had three decks, a smoking room and wet bar, a dining salon, sleeping berths, an all-electric galley, and room for 100 passengers and crew. That same year, Bel Geddes trumped it, at least in theory. In collaboration with German aeronautical engineer Otto Koller, he created Airliner #4, which bore a striking resemblance to a B-52 Stealth Bomber circa 2003.

It boasted more than three times DO-X's wingspan (528 feet), triple its height (9 decks), and five times its capacity (606 passengers plus crew). Given its twenty 1,900-horsepower motors (plus half a dozen in reserve), 37 propellers, and in-flight refueling over Newfoundland, Norman estimated it could travel from Chicago to London in forty-two hours. A passenger liner sailing the shorter distance between New York and London took four and a half days.

Comfort, like safety, was key. There were four tennis courts, a gym, twin solariums, a spacious deck games area, a veranda café, a nursery, a barber shop and hairdresser's salon, single and double staterooms, some with private baths, and even a lengthy promenade deck. The

A PEEP INTO THE FUTURE

BY ~ The Spectacular
NORMAN BEL GEDDES
former Detroit Artist

Norman Bel Geddes

Design of aerial restaurant by Mr. Geddes in 1931

Proposed air liner designed by Mr. Geddes, 1929, to carry 451 persons.

Motor car for 9 persons, designed by Mr. Geddes in 1931

Takes a Peep Into Future

Airliner #4 gets pride of place, along with the Aerial Restaurant, Motorcar #) 8 and Locomotive #1. January 1, 1933.

(Courtesy: Detroit News, Inc.)

main dining room accommodated its own orchestra. The requisite crew would include masseuses, manicurists, a gymnast, a librarian, headwaiters, wine stewards, and the orchestra's musicians.

"It is my firm belief," Bel Geddes wrote, "that this is in no sense a mad or foolish idea but sound in every particular."

[It's worth noting that when Bel Geddes was seventeen, a hugely popular sci-fi series, *The New Tom Swift Jr. Adventures*, began appearing. It featured a protagonist who, never having gotten past high school, invented—by instinct and blind experimentation—everything from a photo-telephone to an electric rifle. Making a frequent appearance in these tales was an atomically powered airplane the size of a Boeing 747.]

Airliner #4 could, its creator insisted, be built, equipped, and furnished for $9 million dollars. With three Atlantic crossings a week, it would give ocean liners (which cost $60 million each) a run for their money. Ticket prices would be comparable; top dollar would buy a double stateroom with private bath. Tallying up all the costs (crew wages, fuel, insurance, depreciation), it would pay for itself in three years.[148]

Like many of his more ambitious commercial concepts (e.g., a Floating Airport, designed for proximity to Wall Street), it was never built. But eighteen years later, when Howard Hughes debuted his massive birchwood flyer, the *Spruce Goose*—with a cruising speed of 250 miles per hour to Airliner #4's 150 miles per hour but with a considerably shorter wingspan and minus Norman's masseuses, solarium, and orchestra—rumor had it Airliner #4 had served as inspiration. True or not, it did strongly influence Alexander Korda's 1936 *Things to Come*, an early masterwork of sci-fi cinema.

CHAPTER 10

Skyscrapers/Streamlining

1931–1933

We admire the swordfish, the seagull, the greyhound, the Arab stallion and the
Durham bull because they seem made to do their particular work . . . Their beauty is
inherent.

—NORMAN BEL GEDDES[149]

I n a world that had yet to experience TV or tape recorders, where horse-
drawn ice wagons were commonplace and few knew what "the sound
barrier" was or that it could be "broken," Buck Rogers was practically a
national hero. Traveling 25 trillion miles beyond the solar system and
into a twenty-fifth century, Buck introduced comic strip readers to magnificent "metalloglass" cities where people commuted via jet-propelled
belts.

In nineteenth-century America, the future had been horizontal—
Westward ho!—with earthbound folk heroes like Davy Crockett and Paul
Bunyan. In the new century, it was decidedly vertical—*Up, up, and away!*
"The face of the old world takes on a new makeup," observed furniture
designer Paul Frankl,[150] peering down from an airplane seat 5,000 feet
above the Caribbean, a makeup that confirmed the "rightness" of those
crazy Cubists—Braque, Picasso, Cezanne, Juan Gris. Even radio was
"on the air," transmitted through a seemingly limitless sky in which (or
above which, who knew?) the Almighty dwelt. *The sky's the limit!* was the
new catchphrase.

Enter skyscrapers.

The flappers of the Twenties, with their slender sheaths and ropes of
pearls, segued neatly into the impossibly tall buildings of the early Thirties. In October 1929, the economy had snapped shut with the velocity of
a guillotine. With optimism suddenly in short supply, skyscrapers were
secular cathedrals of hope, symbols of an indomitable spirit, and architects were their avatars.

Manhattan's silhouette was being transformed by them—a technology made possible by the advent of steel and reinforced concrete, elevators, and high-pressure water pumps. "Who can look on the majestic skyline of New York in sunshine or shadow and not be moved?" claimed an early paean. "If we never build another building . . . posterity will have to accord us the creation of a great new architectural style fixed beyond all changing."[151]

They were an indigenous art form,[152] as uniquely American as detective novels and jazz. They inspired skyscraper bookcases and skyscraper suicides, a Bel Geddes's chrome-plated cocktail set,[153] and Ayn Rand's love letter to Frank Lloyd Wright, *The Fountainhead.*

Norman Bel Geddes's Skyscraper Cocktail Set for Revere Copper & Brass Co., 1937
(Courtesy: Brooklyn Museum)

Developers vied to construct "the world's tallest" in the world's greatest city. The pyramids at Giza, Rome's Pantheon, the Washington Monument, and the Eiffel Tower were all well and good, but the challenge was to create an *inhabited* sky-topper. City planner Harvey Wiley Corbett predicted exterior moving stairs, allowing people to alight on any floor. The cacophony of riveters punctuated the air. The phrase "the American Dream" was about to enter the vernacular.

Prior to Wall Street's implosion, it had been decided that Chicago would host a World's Fair. Officially, it would celebrate the four-hundredth anniversary of Columbus's voyage to the New World and the resultant corporate-driven nation, busy moving toward a shiny future on the wings of science and technology. Unofficially, it was designed to counteract the city's reputation as a mecca for gangsters ("Chicago typewriter" was slang for a Tommy gun; a "Chicago overcoat" was a coffin) and, equally worrisome, the public's fascination with them.

"Brilliant bad boy" Raymond Mathewson Hood[154] was selected to recruit fellow architects to collaborate on the site's layout and content. Hood was known for a generosity of spirit, rare in his profession, that may have stemmed from the fact that he'd been past forty, penniless, and in debt when he got his first important commission, Chicago's Tribune Tower.

At sixty-two, Frank Lloyd Wright remained in obscurity. Hood made a point of including him on his list of potential contributors, despite the fact that the rest of his commission believed (not without cause) that the ego-maniacal elder statesman was incapable of working cooperatively. Wright would spend a spare moment coming up with ideas but ultimately demur.

Bel Geddes was another name on Hood's list. Like Norman's Franklin Simon windows, only more so, Hood's American Radiator Building, with its black-brick facade and floodlit golden top ornaments, glowing torch-like against the night sky, had caused Fifth Avenue traffic jams. It was a unique opportunity for "an Imaginative Specialist," as the press had taken to calling Bel Geddes, to elaborate on his theories about what theaters, as physical structures, should and should not be. By late February 1929, Hood's committee had authorized half of Norman's twenty-five proposals for further development.

"Geddes' vitality and versatility seem sufficient to permeate even so vast a project as Chicago's," *Time* magazine announced in March, adding that "his sandy, habitually tousled hair and careless attire" were more indicative of "a summer camper or a man tinkering in a workshop" than a brilliant jack-of-all-trades. The "blond little Leonardo . . . is donating his services."[155] He was, however, paid $10,043 (approximately $138,130 today) to expand his blueprints and create models of six radically innova-tive theater venues—Temple of Music, Open-Air Cabaret, Water Pageant Theater, et cetera—each large enough to accommodate thousands, one of which was specifically designed for staging his *Divine Comedy*.

"To our way of thinking," wrote *The Boston Transcript*, "Norman Bel Geddes is . . . an indispensable visionary."[156] *Theatre Arts Monthly* ran a seventeen-page illustrated feature. "It was impossible to know," noted *The New York Times*, if Bel Geddes's designs "were castles in Spain or theaters on the moon."[157]

The *Ladies' Home Journal*, with its comforting Norman Rockwell cov-ers, strove to inject optimism into the deepening economic crisis for its 2.5 million readers. Would Bel Geddes play soothsayer?

"Ten Years From Now" was given pride of place in the January 1931 issue. Norman's sixty-plus predictions would prove considerably more grounded than those of Aleister Crowley and other popular visionaries of the day. In particular:

- Synthetic materials will supplement wool and cotton in clothing.
- Arc welding will replace riveting.
- Airplanes, complete with sleeping compartments and dining salons, will attain a speed of six miles per minute, and by 1950, thousands will be in daily use.
- The world's literature will be available at ten cents a copy. (Pocket Books paperbacks would appear in 1938.)

Others, their technology held up by the war, took longer:

- Crops will be artificially stimulated.
- Mechanical devices, controlled by photoelectric cells, will open doors, serve meals, and remove dirty dishes and clothes to appropriate parts of the building.
- Talking motion pictures will achieve a third-dimensional quality (3-D films, Blu-Ray).
- Talking pictures will replace talking professors.
- Events of national interest will be available by television at the same moment they occur (cable news).
- A new fuel of vastly increased power but infinitesimal bulk will supplant gasoline (electric and hybrid cars).
- Paper will be replaced by a material that doesn't require the slow growth of trees.
- Improved machinery will free workers from the drudgery of purely mechanical human tasks.
- Rainfall will be controlled scientifically, ocean power will be harnessed, and exploration of the sea bottom and interplanetary space will make accurate weather prediction possible.

A few were, ultimately, too optimistic:

- With a Commercial League of Nations regulating international commerce, there will be no slumps or booms.
- The workweek will consist of four six-hour days.

- Every roof will be a garden.
- No epidemics. No incurable diseases.

With us always, he concluded, will be cruelty and intolerance, generosity and unselfishness, workers and drones, and women's ever-shifting hemlines.

THE CHICAGO FAIR organizers asked Bel Geddes to develop a program of forward-thinking entertainments that could be performed in his proposed theaters.

Perfect. He'd created them for just that.

He sent a letter to some thirty persons "of international standing" (Picasso, Stravinsky), describing his method and asking for feedback. A second set of letters sought out "distinctly experimental" plays (Cocteau, O'Neill, Pirandello) and "forward-looking" directors (Reinhardt, Stanislavsky) to realize them.

One name was conspicuously absent from all three lists.

After three years at the Met, where he directed fourteen productions, including the U.S. debut of Tchaikovsky's *Eugene Onegin*, Richard Ordynski had canceled his contract. Given Poland's increasingly precarious position, he'd decided to head home. "Ordynski was a big man over there before the war, director of their finest theatre, the Imperial in Warsaw," Eugene O'Neill wrote a friend, "and probably is going back to resume where he left off."[158]

Norman had blocked Ordy from coming on board *The Miracle* crew six years before; it was impossible that "the Polish Reinhardt" remained unaware of the designer's highly publicized collaboration with his former employer. The query for directors and playwrights was mailed out on October 23, 1929. On December 13, in faraway Warsaw, Richard Ryszard Ordynski sat down with pen and paper.

My dear Norman, he wrote, *I was surprised, no more, hurt, not for reasons of stupid pride or fame but deeper reasons.*

I am glad to help others understand your achievements. Years ago, I wrote about you here in Poland. We have together so often discussed purely as artists our dreams . . . We thought always to work together and make a common front against bluff and all-doing greatnesses in your New York. You must remember our talks of The Divine Comedy *and others . . . I have been,*

*only yesterday, asked to a meeting here, where your plan of the Chicago Fair
was discussed.*

Floryan Sobieniowski, the Polish theaters' London representative, was
among those who'd read Bel Geddes's letter.

> *They want to send me. I do not want to come that way. I still think
> that I will be on your list . . . please know that you have in me your
> best friend and can always count on me . . . I am in best form, hungry
> to do good things.*

The missive went unanswered, perhaps for the same reason Norman
had discouraged Max from hiring a trusted lieutenant: Ordynski might
insist on his *own* way of doing things. Except that in this case, Norman
was *soliciting* ideas. Was it the Pole's boutonnieres, striped trousers, and
hand kissing? A vestige of loyalty to Aline, whom Ordy had abandoned?
But all that was sixteen years ago, ancient history, before Odry had helped
him, brand-new to Manhattan, navigate the backstage politics of the prestigious Metropolitan Opera House. Before he'd bent Ordy's ear fleshing
out his *Divine Comedy.*

Norman had his secretary type up a duplicate of the letter for his files.

•

DESPITE THE CONSIDERABLE ink expended in the press on his theaters, it
was his whimsical Fair restaurant designs that truly captured the public's
imagination.

His Aerial Restaurant, a 278-foot rotating tower of steel, aluminum,
and glass, was the first of its kind, predating Seattle's Space Needle "Sky
City" Restaurant by decades.

Each of three cantilevered, counterbalanced levels housed a restaurant (total capacity 1,500), each with ceiling-to-floor glass walls and open-
air terraces that doubled as observation decks. The higher up one went,
the more expensive the food. The top level offered a "near-airplane perspective." The decks made a full revolution every hour, offering diners
a complete panoramic view of the Fair site, the city, and Lake Michigan.
Elevators would accommodate 300. The kitchens were stowed safely on
ground level in the tower's base, facilitated by four dumbwaiters.

An extraordinary example, observed *The New York Times*, of "just how far the twentieth century has got . . . It's nothing more or less than an application of F. L. Wright's principle of a house built like a tree."[159]

Among its other attributes, the Aerial Restaurant solved the problem of there being no attractive spot on the Fair's grounds to situate a large restaurant; its base was small enough to be built within the courtyard of another building, and its mass spread out high enough so as not to interfere with other proposed two- and three-story buildings. There were plans to make it a permanent Chicago fixture.

Bel Geddes's Aquarium Restaurant (with corresponding seafood menu) was inspired by Nutshell regular William Beebe, who'd recently descended thousands of feet beneath the Atlantic inside his Bathysphere ball—the first time a biologist had been able to observe deep-sea animals in situ.

The Aerial Restaurant. Popular Mechanics, July 1930.
(Courtesy: Popular Mechanics.)

The least feasible but most intriguing of all Norman's proposals, the Aquarium Restaurant was designed as a dam built across a lagoon. Water would fall in sheets on an overhanging glass roof, creating a transparent liquid curtain six inches from the windows. At night, the curtain would be illuminated by underwater lights, creating "a pleasant, eerie effect." There was seating for 650 to 800. An outdoor café and open-air roof terrace, capacity 120, would have water flowing under its glass floor.

Diners would enter from a dock, then descend a circuitous maze to sixteen feet below the water's surface. Making their Jules Verne–like descent, they'd be surrounded by sunlit, brilliantly hued fish—the interior walls, floor, and ceiling were all glass—the view gradually shifting to larger deep-sea creatures and accompanying aquatic plants, the water (lit from within) darkening from amber to deeper and deeper green, making visitors participants, not simply observers.

There was a third food venue, too, an Island Dance Restaurant with seating for 600, lit by huge, interlaced neon tubes that reflected in the surrounding water—an expansion of the set he'd designed for DeMille's *Feet of Clay.*

Visiting Bel Geddes's brownstone home office on East Thirty-Seventh Street, birthplace of this "far-flung recklessness," one journalist described the organized chaos: Twenty-five assistants and technical experts at work. Buzzers summoning secretaries. Telephones ringing. Wood and metalworking machines humming. Typewriters clacking. Clerks running up and down stairs with blueprints and specs.[160]

•

BEL GEDDES WAS appointed by unanimous vote as the Fair's Consultant for Illuminations on all matters concerning exterior lighting, charged with looking into special apparatuses "and to begin scheming certain general ideas." In short—dramatizing the grounds. Who better than the master magician who painted with light? The position came with a fifty-dollar-a-day honorarium, plus expenses for work done in his office, one-hundred-dollar-a day for work that required his presence in Chicago. He was also offered a fifty-fifty partnership as a member of the planning group collaborating on Ray Hood's upcoming Radio City Music Hall project, nineteen buildings covering twenty-two acres.[161]

Fair president Rufus Dawes envisioned a Rainbow City "as modern . . . as a straight-eight motor" and "as colorful as the vanished palaces of Carthage."[162] Joseph Urban was hired as director of exterior color, the daytime equivalent of Norman's after-dark illumination. He custom-ordered twenty-five intensely bright shades of oil and casein paints—hot pink, blood red, scarlet, deep orange, purple, and green, two yellows, periwinkle blue, plus grays, whites, silver, and gold—more than 25,000 gallons—and employed them in striking combinations, swipes of pure color 40 feet high and 400 feet long. It took 350 workers more than 6 months to complete the job.

Even as a spate of enthusiastic responses to Norman's queries was coming in, thousands of letters from struggling architects deluged Ray Hood's office, expressing outrage at the appointment of a self-taught, un-certified practitioner.

In March, Norman traveled to Albany to petition the Board of Regents to waive the education requirements for certification, in light of his many building designs: the Palais Royale, the landmark conference room for J. Walter Thompson, not to mention his Fair projects, which were being praised by outstanding men in the field. Illinois officials had noted that his blueprints were the first set of Fair plans submitted without errors.[163]

Called "the party of the decade," Fete Moderne featured architects (including Ray Hood) "dressed" as their buildings. Bel Geddes was very likely there. January 23, 1931.
(Courtesy: Office of Metropolitan History. NYC. Photographer unknown.)

Norman was also, at that moment, at work on a vast Ukrainian State Theater to be built in Kharkov; he was one of only six Americans paid by the Soviets to enter an international competition (his design would place second to Walter Gropius's eighth).[164]

Petition denied, he traveled to Albany a second time to meet with then-governor Franklin Roosevelt, hoping to gain accreditation in his home state. Frank Lloyd Wright, Norman's lawyer, and Basil O'Connor, Roosevelt's former legal partner, came along.

Hadn't Haussmann, who lacked architectural credentials, permanently reconfigured much of Paris? And what about Robert Moses, another non-architect? Look what he was up to! But Norman had two strikes against him: lack of an architectural degree compounded by lack of a high school diploma. Roosevelt's hands were tied. "I couldn't have passed the exam, anyway," Bel Geddes would admit. "I hadn't had a year in the drafting room."[165]

Returning home, he was forced to resign from both his Consultant of Illuminations post and Hood's Rockefeller Center group. In future, all Bel Geddes & Co. contracts would stipulate that neither Norman nor any of his staff were architects. Though rumors would circulate, then and later, that his Fair proposals were axed for impracticality, Rufus Dawes would insist, in a letter to *Forbes*, that, given cutbacks, there had simply been "insufficient funds."

•

IN APRIL, NORMAN'S talent was referenced in a *New Yorker* poem about a luxurious meal made unaffordable by the Depression:

> *Rosy salmon and lima beans,*
> *Minestrone in great tureens,*
> *Lobsters seated on curling lettuce*
> *With color effects by Norman Bel Geddes.*
> *Crabmeat salad and breast of hen—*
> *Will I ever behold such sights again?*[166]

In November, Little, Brown and Company released the designer's first and most important book. Given his recent legal bouts, *My Theories on the Future of Architecture* had been renamed *Horizons*. Both unrealized and completed projects, and the thinking behind them, were show-cased—everything from motorcars, buses, railways, aircraft, and floating airports to stoves, theaters, amusement centers, restaurants, retail window displays, and factories—in one neat package.

Images ranged from bison cave paintings circa 50,000 B.C. and Michelangelo's layout of St. Peter's Basilica to grain elevators and dirigible hangars. The Rose Window at Reims was juxtaposed with a star-shaped Lycoming Airplane Motor. There were photos by Margaret Bourke-White, Edward Steichen, and a despairing Imogen Cunningham. "Anyone who will go to the pains, after all the years of fruitless struggle, to try to prove the artistic worth of photography deserves encouragement," the forty-eight-year-old Cunningham wrote him, "but being in one of my periods of great discouragement, I cannot give it to you."[167]

A bestseller, *Horizons* anticipated books by the rest of the Big Four by five years. Within a month, the New York Public Library chose it as one of the year's Top Fifty, and a jury of writers, publishers, and critics (Woollcott among them) included it among the 200 contemporary volumes to be presented to President Roosevelt, to supplement the White House's permanent library. *The Times* mentioned it in the same breath as Aldous Huxley's *Brave New World*.[168] If *Horizons* isn't an exact picture of tomorrow, wrote the *San Francisco Chronicle*, "it is as near to that as any mere mortal could come."[169]

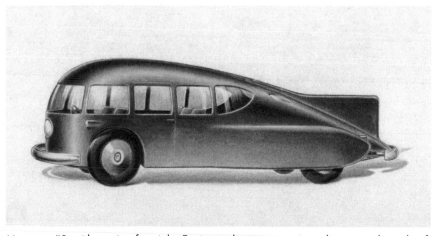

Motorcar #8, with seating for eight. Engine at the rear, opera windows on either side of the tail fin. 1931.
(Courtesy: Harry Ransom Center, The University of Texas at Austin, from Horizons, 1932)

There were, of course, detractors.

His locomotive resembles "an irate earwig minus its legs," observed the *Los Angeles Times*, his motorcar "an inebriated beetle."[170] A Canadian reviewer had a field day comparing Norman's ocean liner to local catfish, his theater designs to plum puddings, and his Aerial Restaurant to a toadstool.[171] "Anything falling short of the hygienic standards of a hospital dormitory," complained a British critic, was now "looked upon with suspicion."[172]

Frank Lloyd Wright also had doubts. "He has notions concerning the future of practically everything, but pretty nearly everything begins with the Greeks, the Egyptians and Cezanne and ends up with Norman Bel Geddes," he wrote in the *Saturday Review of Literature*.[173] (That same year,

"Gentlemen, I am convinced that our next new biscuit should be styled by Norman Bel Geddes."

"Gentlemen, I am convinced that our next new biscuit should be styled by Norman Bel Geddes." The New Yorker, December 10, 1932.
(Courtesy: CartoonBank.com. By Kemp Starrett.)

in his autobiography, Wright called Bel Geddes an exploiter of already-exploited "skyscraperism," a stupidity that bored when it didn't alarm.[174]) Two years later, a mutual acquaintance relayed a comment Wright made: he thought he'd "lost a friend" as a consequence of his review.

"I was flabbergasted to say the least," Norman wrote, unaware that Wright had penned a review at all. "If you jumped on it, you were not alone and why should I, who am no writer, take my writing so seriously as to be concerned over a book review by you, who are no reviewer? A review by you [could not be as significant] as the constructive criticism you've given me many a time.

"Not only haven't you lost a friend [but] you couldn't lose this one no matter what extremes you went to."[175]

•

IN JULY, AMELIA EARHART phoned. Did Norman and Frances have bathing suits?

Aviation was very much in the news. Handsome, one eyed Wiley Post had just circled the globe, solo, in seven days, nineteen hours, breaking his previous record; fifty thousand turned out to greet him. The aviatrix understood the value of marketing (Earhart "air age" luggage, and a women's activewear line, both sported her signature transected by a disappearing plane). Hoping to promote "the social possibilities" of flight, she was arranging "a thoroughly modern day's amusement," a picnic deluxe in Atlantic City. Gene Tunney would be there, along with WWI flying ace "Eddie" Vernon Rickenbacker, novelist Fanny Hurst, and *Fortune* managing editor Ralph Ingersoll. The guest of honor was Scottish aviator Captain James A. Mollison, still bandaged after a nonstop Atlantic hop that ended in a crack-up, followed by a Broadway ticker-tape parade.

At 10:00 a.m. on August 2, 1933, the Bel Geddeses (Norman dapper in a sporty white suit) boarded a giant twin-motored Curtis-Condor. Earhart, busy playing hostess, left the actual flying to others. Two planeloads of reporters followed close behind.

Tunney was mobbed by autograph hounds; Earhart was presented with keys to the city. She, in turn, praised the "splendid" Floyd Bennett Airport (located, she confided to Norman, in a "forsaken hole").[176] Guests were driven down the boardwalk in cars—a first—with a motorcycle escort; after lunch, they enjoyed a roped-off section of the beach and a tented private bar.

•

INTENT ON EXPANDING their customer base, Pan Am retired their pilots' romantic WWI–era attire (leather bomber jackets, silk neck scarves) for a more military look (double-breasted blazers with braid loops at the shoulder), and in late 1933, Bel Geddes was brought in to create interiors for a trio of Glen Martin Model-130 Clippers, upgrading the mail carriers to accommodate human cargo. *Horizons* had twenty-two pages dedicated to innovative plane and airport design. Pan Am's higher-ups had studied them and were convinced enough to offer a $20,000 commission.[177]

Existing commercial planes offered dismal lighting, dismal snacks, glare-flooded windows, no soundproofing, poor ventilation, and bolted-down wicker chairs. For their Clippers, the first commercial planes to connect San Francisco to Manila, among other international routes, Pan Am wanted "the last word" in airline design, including passenger sleeping facilities, but with minimal structural changes.[178]

To start, Norman's team lowered the existing floors for comfortable clearance. As a bulwark against noise and to hide cables, the walls and ceilings were covered with cool gray fabric. A lounge was created with daytime sofas that converted into lean-back chairs, the seats overstuffed with eiderdown. Nighttime sofas converted into four roomy upper beds, larger than Pullman berths; the seats pulled out and turned over to form four lower berths; and all could be installed or disassembled in forty minutes. Upholstered furniture got zippered slipcovers for easy cleaning and required inspections. Leather hassocks concealed life preservers. Bed linens coordinated with the interior's color scheme. Curtains broke the lounge area into two spacious dressing rooms. Lavatories offered the luxury of hot and cold running water.

With weight a priority, anodized aluminum was used for windowsills, hardware, lighting fixtures, and ventilation ducts. A quartet of windows offered unobstructed views; indirect lighting allowed for reading. Controlled air-conditioning ran through concealed ceiling ducts; felt-covered card tables kept games and writing materials from slipping. There was a bookcase, a desk, and plenty of room to walk around.

The first real airplane kitchen boasted an electric stove, refrigerator, and garbage receptacle. A wheeled cabinet locked into position for serving meals; there were table linens (instead of paper), custom silverware, and vacuum-bottomed Beetleware plates. Norman's team went so far as to test various wines to determine the most motion-resistant varieties.

The overall result was "a fantastic deluxe dream,"[179] closer to top-ticket ocean liner standards than what would pass for first class air travel more than half a century later.

On November 22, 1935, the redesigned Clippers soared over the Golden Gate Bridge en route to the Orient. Within five years, the number of nonbusiness passengers would double,[180] much of the credit going to the upgrades. But for all the difficult and innovative work—solving problems of space, sound insulation, sleeping, comfort, food prep, and more—Norman and his team were mostly credited in the press for "a decorative job." His office had, he noted a decade later, gone "way beyond anything [Pan Am] ever dreamt of," creating a standard that, a decade later, was still in play. [181]

•

IT SEEMED INEVITABLE that Norman and Margaret Bourke-White would cross paths.

One of the great chroniclers of the machine age and a brilliant news photographer, Bourke-White was the first to capture, even romanticize, the striking visual power of industry. ("Dynamos were more beautiful to me than pearls," she famously wrote.) It made sense that cameras, which were machines, should provide the means for capturing the beauty of other machines.

Like Norman, "Maggie" was enamored with light, inventing soft focus and other techniques to dramatize her subjects. Her work was grandly theatrical—close-ups, multiples, artful composition. Like Norman, she'd lived and worked in the Midwest and exhibited the classic midwestern qualities of resourcefulness and determination. In June 1935, Eddie Rickenbacker took her for a 140-mile-an-hour spin on the Indianapolis Raceway, the first time a woman had ever set foot on the track. It wasn't the first "men only" barricade she'd crossed. At twenty-four, she'd shot photos from the catwalks of Cleveland's Otis Steel Mill; tradition forbid females even entering the premises. Next came hydraulic dams. In 1943, she'd become the first woman allowed on an Air Force bombing mission. Like Norman, she'd made a name for herself while still in her twenties and, like him, modified that name (adding a hyphenated "Bourke," her mother's patronymic) in pursuit of a distinctive identity.

In early 1929, Henry Luce had hired her as the official photographer for *Fortune*, a new monthly he was adding to his roster. A radical break from typical business tomes, this one featured hand-stitched, oversized pages printed on expensive paper, colorful, contemporary art covers, and an exorbitant one-dollar price tag (the Sunday *New York Times* cost a nickel). Her first, decidedly unladylike, assignment had been to document a Chicago stockyard and processing plant. Another early directive was a trip, the first of several, to the Soviet Union to document the country's industrial transformation (the first Western photographer ever allowed to do so). In the spring of 1930, she shot photos for a lengthy profile on the theater-designer-turned-industry-upstart.[182]

Soon after, she phoned to say that Sergei Eisenstein had arrived from Moscow, en route to Hollywood to write and direct for Paramount. And he wanted to meet Norman Bel Geddes.

He was happy to oblige. Five years before, he'd attended a private screening of *Battleship Potemkin* (its circulation was highly restricted). As a former art director, he recognized the influence of Russian film—the carefully orchestrated tempos and unusual camera angles, the juxtaposition and crosscutting, shots within shots—on everything from still photography to automobile and perfume ads. He'd even prevailed upon Otto Kahn to arrange a public showing (Kahn did, minus the squirming maggots sequence).

Maggie and Sergei arrived for lunch at the Murray Hill brownstone. The Russian, who'd spent his youth in legitimate theater, was fascinated by Norman's *Divine Comedy* model. Norman walked his guests through the production scene by scene, as he'd once done for Reinhardt. "I'm flattered to think," he later wrote, "that the close similarity between [Eisenstein's] method [of preparing a script] and the one I devised for . . . *The Divine Comedy* wasn't entirely accidental."[183]

Maggie left at midnight, pleading other business. Scheduled for a series of talks, Eisenstein thought it best to get back to his hotel. *Why not stay here?* With Bel gone, there was lots of room. Sergei's luggage was retrieved from the Savoy.

Sergei Mikhailovich Eisenstein, a dreamer from half a world away, was thirty-two to Norman's thirty-seven. With their broad foreheads, thick mops of sandy hair (Eisenstein's receding), and a shared intensity laced with mischief, they might have been mistaken for brothers. Despite various commitments, they were inseparable for the better part of two weeks, taking in films, exploring Harlem, walking the waterfront. Eisenstein seemed interested in everything, from the homeless who were erecting makeshift shacks in the recently drained Central Park reservoir to the way a shaft of light lit the corner of a building.

He was writing a screenplay based on Theodore Dreiser's *American Tragedy*. A visit to Sing Sing, where the novel's hero meets his end, was arranged. Double-checking that the infamous chair was disconnected, he sat on it. At a luncheon in his honor at Otto Kahn's palazzo, Sergei found himself, as Norman had during *his* first visit, discomfited by so much wealth—the Rembrandt, the arrangements of exotic flowers, the blindingly white tablecloth and endless silverware, the liveried butler who insisted on standing in attendance behind his chair.

Summoned to the telephone, he returned to find the other guests had finished their appetizers. Soon he'd be in Hollywood, wearing white flannels and playing tennis with Norman's old pal Charlie Chaplin. But for now he had to imagine a way to swallow the "malicious" and spiky object on his plate, a foodstuff he'd never encountered. *How do millionaires eat artichokes?*[184] Decades later, Norman would still remember "the whole gathering . . . watching with folded arms to see how the Russian barbarian [would] get out of this tricky situation."[185]

•

THE CONCEPT OF streamlining, if not the word, had been around since at least the sixteenth century when Leonardo da Vinci, compiling data on the flight of birds, got to thinking about air resistance. Isaac Newton tackled the subject in the seventeenth, and in the nineteenth Sir George Cayley determined—with the help of a trout—that "true forms of least resistance" were ovoid with a tapering end. It was, simply put, a method of minimizing the resistance through which an object passed, the result being more speed and distance, less noise and expenditure of fuel. Bel Geddes's contemporary Arthur Cheney credited Norman with coining the term "parasite drag," the sum of all air resistance influences.[186]

With the publication of *Horizons*, "streamlining" suddenly exploded into the public arena, used indiscriminately to describe things that moved (railroads, steamships, airplanes) and things that didn't (lipsticks, politicians' speeches, women's undergarments.) A pencil sharpener "stupidly modeled after the teardrop," complained Henry Dreyfuss, "couldn't get away if it tried because it was screwed down."[187] The sharpener in question was a much-lauded Raymond Loewy prototype.[188]

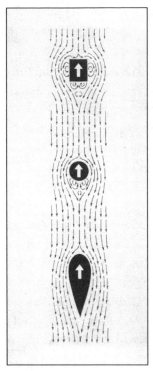

The principle of
streamlining.
*(Courtesy: Harry Ransom Center,
The University of Texas at Austin,
from* Horizons *by Norman Bel
Geddes, 1932)*

Streamlining was everywhere, replacing the exuberance knocked flat by the Crash, waving goodbye to cumbersome Victorian trappings (oil lamps, coal stoves, iceboxes) and hello to electricity (cheerful, instantaneous, clean). It was Einstein's elegant theories and Hemingway's stripped-down prose. It was Fred and Ginger gliding effortlessly across a gleaming, Bakelite-coated dance floor and Busby Berkeley's hypnotizing geometrics. Streamlining, like the future, was sleek, powerful, efficient and SHINY, like chrome, aluminum, Lucite, glass, burnished steel—materials that spoke a vernacular rivaling alchemy. It was Vitrolite building facades and mirrored vanity tables, satin lapels and mother-of-pearl cigarette lighters. It was clinging, sleek-as-seals, cut-on-a-bias gowns with plunging backs, the gleam of Big Band trumpets and trombones. Streamlining was also about precision, exemplified by factory assembly lines and quick-stepping, high-kicking Radio City Rockettes.

It was romanticism married to pragmatism, the first all-American aesthetic. Fresh and affordable. One didn't have to be a second-generation Harvard grad, dress in cashmere, or frequent a weekend country place to own a piece of it.

He didn't invent streamlining, noted a contemporary, "but he works in it as easily as though he had."[189] *Perfection is finally attained not when there is no longer anything to add,* Antoine Saint-Exupéry would later write, *but when there is no longer anything to take away.*[190]

Norman was streamlining's combination poster boy and village explainer. It was a role he would come to regret.[191]

CHAPTER 11

The Democratization of Design

1930s

My stove is automatic,
You don't have to burn wood or coal.
I just strike your match, baby, and stick it right in the hole.
—BAWDY BLUES SONG MADE FAMOUS BY BESSIE SMITH[192]

It had taken twenty years for electricity to revolutionize people's thinking. In 1899, Ringling Brothers Circus displayed an incandescent bulb in a dark room—flicked on, flicked off, flicked on again—a wonder deemed memorable enough to share the marquee with Ringling's aerialists, bareback somersaulters, and Barretta, the Boneless Wonder. In 1915, many still thought of electricity as a dangerous (and unnervingly invisible) force. It wasn't until the Twenties that electric lamps began edging themselves into people's homes. General Electric poetized them as "unroll[ing] a path of light" in every room—in contrast to "dim bulbs," flapper lingo for bores.

By 1932, ads promoting the virtues of electric vacuum cleaners, refrigerators, dishwashers, washing machines, and other automated devices were ubiquitous, though most middle-class households didn't own any. It wasn't just the expense. Seventy-nine percent of U.S. homes still lacked electricity, not to mention hot running water (absent in 60,000 buildings in New York City alone). Only 32 percent of farm households had *any* kind of running water.[193]

National advertising was created, a survey revealed, by "extraordinarily affluent" white males who employed servants and rarely attended church or public amusements.[194] Former ad man James Rorty described his co-workers as Bakelite-boned "dead men" feeding the "gargoyle's mouth . . . of a $2-billion industry." The gargoyle's audience was largely female and relatively poor, with a vocabulary of about 1,200 words (the average high school grad used 13,000), and that wasn't counting the 3 million or so who were illiterate.[195]

"But all of them know that household appliances have broken three centuries of hardship," noted a 1938 bestseller with the provocative title *Slaves by the Billion*. Those who were lucky enough to own them were "queens in a sense which might have turned Victoria of England, Maria Theresa of Austria or Catherine of Russia green with envy." Not to mention Martha Washington, "in spite of all her slaves."[196]

Gone was the sheer drudgery of nineteenth-century housekeeping— soft-coal cookstoves, drinking water gathered from outdoor pumps (in all seasons), washboards, and heavy iron skillets, not to mention the time required to make simple fare like baked beans (eighteen hours) or oatmeal (overnight). A modern housewife, wrote Earnest Calkins, "could not do in a week what my mother did everyday of her toil-bound life."[197] Not everyone was happy about the increasing availability of labor-saving domestic devices. Many politicians and clergy, arguably the same contingent who rallied behind corsets and against tampons, birth control and suffragettes, found the idea of a female populace with "free" time to fill a frightening, even dangerous, scenario.

The upstart fuel of the future didn't go unnoticed by the nation's gas companies, even if electricity remained, for the moment, more expensive, and not necessarily more efficient. "Politicians, social workers, novelists and journalists continue to paint rosy pictures of a future when all the houses will be so well equipped with labour-saving devices that we shall just sit around pressing buttons while Robots do everything," mocked a British gas provider in a fight against the inevitable.[198]

By 1935, nearly 90 percent of United States urban dwellings, at least, had been "wired." GE promised that the average servant-less housewife could now maintain modern standards without becoming a "domestic drudge," backing up its promise with a promotional film disguised as a love story. In its cast was a struggling young actress who, three years later, under the pen name Hedda Hopper, would find her true calling as one of Hollywood's most powerful gossip columnists.[199]

"The artist came in the back door, along with the delivery boy and the plumber, the electrician and the iceman," wrote "industry insider" Mrs. Christine Frederick. "His sculpture took the form of a new bathtub, the shape of a ketchup bottle, a gas range, an enamel sink . . ."[200] But the idea that beauty, utility, and profit were mutually beneficial, and that everything from buildings to beds could reflect both their purpose and their era, were concepts that had made little headway. Artists were viewed by businessmen as long-haired, impractical layabouts. In February 1930, Bel Geddes felt obliged, once again, to address his critics. "The trouble" was that most nineteenth-century manufacturers and inventors "made their designs with attention almost exclusively riveted to the work the product was to perform. Decorative touches might be added afterward, but the thing was never originally designed to be both useful and beautiful."[201] Design, in most people's vocabularies, means "veneer," Steve Jobs would tell *Fortune Magazine* three quarters of a century later.[202]

A company flirting with bankruptcy stepped up to the plate. Standard Gas Equipment (SGE), manufacturer of "ranges" (combination stove-ovens), approached Bel Geddes with a commission. Due to rising competition, domestic sales had been falling for several years, they explained, offering him $1,500 for sketches of a redesign, deliverable in two weeks.

He wasn't, Norman told company president W. Frank Roberts, in the business of drawing pretty pictures. He would take the job if his people could come up with something *better* that could be manufactured for *less*. His counteroffer was a yearlong deadline, $1,000 a month for expenses, and a $50,000 advance against royalties.

The change from wood- and coal-burning stoves to gas models had improved the housewife's lot but had had little impact on the appliance's Victorian appearance. As *Forbes*, already a respected financial bellwether, observed, the industry's chief concern was "sales analysis rather than stove analysis."[203] Despite the introduction of the Magic Chef in 1929 (designed by Frank Parsons of Parsons School of Design), women still had to contend with greasy shelves and interiors, drip pans, spillage on burners and thanks to a vogue for Chippendale-like cabriolet legs, grease and dirt invariably collected underneath. More frustrating was temperature-and-timing guesswork, resulting in scorched roasts and charred biscuits, plus the aching backs and singed hair that came with stooping to inspect the oven's contents.

Bel Geddes compiled a dozen fundamental survey questions. *How often do you bake or roast each week? Does your kitchen stove heat excessively? How much does it cost to maintain? What don't you like about it? Would you prefer stainless steel or an enameled surface over cast iron? What color combinations do you prefer? Would you like a waffle iron, a toaster, an electric percolator built into it? A safety valve?*

Refining his Simmons's research technique, he hired a man to train a crew of surveyors to go door to door in both large cities and small rural towns across the country, speaking with homemakers, shopkeepers, and suppliers. He preferred "personable men and women who are not in a hurry," and he preferred them to travel by train. Dining and club cars were a valuable resource for extracting "the average mass viewpoint," assuming the interviewer had "a pleasing personality [and] something resembling a sense of humor."[204]

Answers were never written down in the interviewee's presence but noted immediately afterward on Bel Geddes–designed forms. Mailed daily to his office, the data was sorted and compiled using a complex, Bel Geddes–designed card index system. Only then did preliminary drawings begin. Visualizing the actual object would be the last, and possibly the easiest, step in the process.

"Has the artist changed his spots and become 'practical' overnight?" asked a treatise called *Consumer Engineering*. "If so, is he still an artist?"[205]

•

"THE CONSUMER'S WORLD unfolds before an alert watcher like a reel of film," Henry Dreyfuss would write some two decades later.[206] But this picking-the-average-citizen's-brain approach was unusual in the late 1920s and early 1930s. Real-life, "outside" input was considered meritless. The public was seen as an immense, revenue-generating maw waiting to be fed. What did *they* know? It was advertising's job to convince people that they wanted what manufacturers came up with.

Over the next six months, the frustrations and desires of 1,200 housewives were catalogued. Collated into usable form, the file ran to 300 pages.

While the canvassing was in progress, an engineer was assigned to observe SGE's plant, a team made studies of competitive stove models, and Norman turned his attention to children's blocks.

SGE had an inventory of 100 models, each with a different broiler, oven, cooktop and other basic features. Norman had scale-model color-coded wooden blocks made, one for each component: blue for ovens, green for broilers, et cetera. "You couldn't get into his office for stumbling over them," *Fortune* reported.[207] Which is exactly what W. Frank Roberts did when he came by to take a look. Then the designer laid out a different set. Adopting the interchangeable module concept he'd used for *The Patriot* and Franklin Simon's windows, he'd condensed the hundreds of components into a standardized sixteen (one oven, three cooktops, three broiler sizes, five utility drawer sizes, four bases), all of which satisfied the demands of SGE's previous stoves. These sixteen could be arranged in twelve combinations, depending on a household's needs, reducing the manufacturer's inventory to four basic models.

The contrast between what was and what could be was "inescapable."[208]

Cast iron, a nineteenth-century "wonder material," had resulted in domestic stoves and ovens weighing hundreds of pounds. Bel Geddes proposed an independent, skeletal steel frame hung with lightweight sheet metal via Bel Geddes–designed hooked clips, a "curtain wall" method borrowed from skyscraper construction. Like the interchangeable modules, it was something Bel Geddes had employed for *The Patriot*, with its hung-from-wires suspended flats.

Another issue was the glossy enamel coating used on cast iron. It often suffered from stress cracks and spalling during assembly (the rigid joints were bolted) and shipping.

Like Simmons, Toledo Scale, and many manufacturers to come, Standard Gas was hesitant to scrap existing inventory and invest in expensive new equipment, materials, and assembly-line training. That said, both the weight and enamel problem were eliminated by Norman's flexible frame and hook-hung sheets, or "skins." His skyscraper method also reduced the cost of materials, labor, and shipping, ultimately saving the company thousands (Depression-era thousands), which allowed for a lower retail price, thus underselling the competition with a superior product.

For color, Bel Geddes chose light reflecting, vitreous white, a bold contrast to dingy black and the marbleized finishes of other domestic appliances. The first all-white stove, it was a harbinger of soon to be ubiquitous all-white kitchens.

Hand in hand with modern appliances was the notion of modern domiciles to put them in. In April 1931, *The Ladies Home Journal* ran an elaborate spread with diagrams and photos of a three-dimensional model, detailing Norman's plans for the "House of Tomorrow." It featured adjustable interior lighting—diffuse or localized (for reading or crafts), dim or strong, all controlled at the push of a button. Living rooms, he argued, should be positioned at the back, to allow for privacy and access to the outdoors. Also included was a two-car "turntable" garage (reminiscent of the turntable landing strip in his Rotary Airport), which eliminated the inconvenience of backing out.

THE ORIOLE GAS Range debuted four months before the Chicago Fair. It seemed to have it all: "Flashomatic" burners that turned on instantly and removed easily for cleaning; accurate heat-control dials; non-tip grates; and a spring-balanced enameled top that hinged down to cover the burners when they weren't in use (the gas automatically shut off when the lid closed), creating a working surface, a boon in small kitchens. Some compared its modular construction to a Piet Mondrian painting.

The oven was oversized, insulated, and heated evenly, with flush, "seal-tite" doors and a timer clock that turned it on and off "at any predetermined minute." The broiler was set high to eliminate stooping; a fully extendable pan allowed the cook to see "what was what" and turn a steak without singeing her hair or burning her hands. Burner castings were protected from spillage by easy-to-clean "aeration plates." There was a warming compartment; storage for pots and utensils; rounded, non-dirt-catching corners; and a flush-to-the-floor base (no cleaning underneath), all subtly enhanced by chromium-plated hardware and polished Bakelite valve handles. Available for two dollars down and easy monthly payments.

Not surprisingly, Bel Geddes's Oriole met with mixed reactions from rivals.

In a lengthy *Fortune* feature ghostwritten by twenty-six-year-old George Nelson, Bel Geddes was called "the P. T. Barnum of industrial design" and a "bomb-thrower," "the man who has cost American industry a billion dollars." Included was a chart citing Bel Geddes with thirty employees, Dreyfuss with five, Dorwin Teague four, and Loewy one. Their top fees were listed as: $100,000, $25,000, $24,000, and $60,000, respectively.[209]

SGE's foundry was let go, sold off at a profit, its assembly line reconfigured for greater efficiency. According to Bel Geddes's files, the Oriole sold so well that, within two years of its debut, SGE pooled its patents with its competitors' for a fee eight times higher than the stove's development costs. In 1935, the Philadelphia Gas Works Co. released a similar Bel Geddes design, the Acorn, adding a Bel Geddes's monogram.

If, as Henry Dreyfuss would privately confide, Bel Geddes "didn't care

Norman Bel Geddes's Acorn Stove, near-twin to his Oriole, 1932.
(Courtesy: Harry Ransom Center, The University of Texas at Austin, photo by Richard Garrison.)

about cost or pleasing the public," that he was unwilling to compromise, that he was "a great designer but not a great industrial designer,"[210] the Oriole—informed by both consumer desire and the manufacturer's need for volume and profit—was the exception. Four years after its debut, the "aristocratic" Oriole was being referred to as "a legend."[211] Dignified and immaculate, it would remain an industry standard for decades.

•

NEXT TO AUTOMOBILES, no piece of retail machinery proclaimed American prosperity like refrigerators. Beyond issues of spoilage, these futuristic monoliths made it easier for families to economize by buying in bulk and saving "leftovers." The trade-off for convenience and modernity (versus labor-intensive iceboxes) was a cocktail of dangerously toxic chemicals—sulfur dioxide, methyl chloride, and ammonia gases—used as refrigerants.

When non-toxic, non-flammable Freon came along, Sears Roebuck decided to enter the refrigerator sweepstakes and called on Henry Dreyfuss for help. Dreyfuss had already transformed their Gyrator washing machine into the Kenmore Toperator, which sold a record 20,000 units.[212] Too busy designing the iconic Model 300 telephone for AT&T, Henry suggested that Sears take on Raymond Loewy instead.

"Those years of pioneering"—1931 through 1934—"taxed my perseverance to the end," Loewy would write. Weeks and months of dismal treks to the Midwest, grimy streetcars, bad food, bleak hotels. He was a nuisance, a "frog" (Frenchman) "reeking with a foreign accent,"[213] who had the nerve to tell captains of industry how to run their own ships. In New York, meanwhile, he insisted on living large. His first bankruptcy, which came with the Crash, was followed by a "second wave,"[214] helped along by the poor reception of his 1932 "aerodynamic" Hupmobile redesign, his first car commission.

By his account, it took two years of repeated visits to Chicago, at his own expense, to convince a reluctant Sears Roebuck that product appearance mattered. The result was a $2,500 commission to dress up the Coldspot refrigerator, "a dust trap with spindly legs . . . an ill-proportioned vertical shoebox." Collaborating with engineer Henry Price, Loewy relocated the motor from top to bottom, thickened the legs and installed one-piece, rustproof aluminum shelving. The 1935 Loewy Coldspot featured a sleek, white enamel front interrupted only by a nameplate, "like a fine piece of jewelry." The door opened on unobtrusive hinges. Chrome ribbing on the freezer compartment and storage drawers complemented the machine's overall verticality.

Its most innovative selling point was a "feather touch" latch, as substantial and attractive, wrote Loewy, as the door handle of an expensive automobile. Ostensibly a long vertical bar, it allowed a housewife, her hands full, to press the door open with an elbow. It also connected "by remote control" to a foot-operated pedal near the floor."[215] The job, he claimed, cost him nearly three times the amount of his commission, out of pocket, in order to do it "as I wanted it done."[216]

The Coldspot was the first true jewel in Loewy's "father of industrial design" crown. The Gestetner mimeograph machine, his much lauded breakthrough, was neither populist nor "sexy," and there's no evidence that either the quality of the copies or the company's sales statistics improved. Throughout his career, Loewy would claim that his (alternately "spectacular" or "sensational") transformation of the Coldspot into something with the sleek styling of an automobile put industrial design on the map, increasing annual sales more than 400 fold in two short years.

The new profession was a collaborative enterprise, many hands and skill sets working together. "Due credit" was at the discretion of the man whose name graced the company letterhead. Bel Geddes had a reputation for being "scrupulous about giving full credit to others."[217] Loewy held fast to his autonomy.

The Sears Coldspot commission came about thanks to Henry Dreyfuss, not two years of repeated visits to Chicago courting intractable executives.[218] Loewy's claim that the Coldspot put industrial design on the map ignores the unavoidable precedent of the Oriole.

As for sales statistics, "It is fascinating to compare the energetically articulated myth with the less-well-publicized reality," notes design authority Stephen Bayley, who, as a young instructor and initial Loewy admirer, met twice with the Frenchman.[219] Every magazine and newspaper article on the Coldspot's transformation tirelessly repeated Loewy's own estimate of an annual sales jumping from 15,000 to 65,000 machines, then skyrocketing to 275,000 per annum—"a record," he claimed, "a thing unheard of in its field."[220] No confirming evidence of those statistics has ever surfaced.

It's difficult to parse out the details of this or subsequent Loewy achievements, as the designer instructed his office to routinely destroy its files, including those documenting his most seminal and successful work. These "clean-up" days continued until the mid-1950s.[221] Henry Dreyfuss, in contrast, maintained microfilm copies (400 reels' worth) of office records dating back to 1929. Bel Geddes held on to hard copies, carbon copies, and Dictaphone recordings of just about everything. The same is true, in one form or another, of Walter Dorwin Teague, George Nelson, John Vassos, Charles and Ray Eames, Russell Wright, Eero Saarinen, Eliot Noyes, and other Loewy contemporaries.

Does it matter, ultimately, who established the first office dedicated to the melding of art and industry (discounting Peter Berens in Germany, it was Bel Geddes, followed by Dreyfuss) or designed the first "modern" refrigerator, or streamlined train, or Coca-Cola bottle (story to come) in an age when so much was being tackled for the first time, when design evolution was contingent on technology's evolution and overlaps were almost inevitable?

It mattered to Loewy.

Aware that the public was beginning to equate up-to-date styling with modern technology, Servel corporation approached Bel Geddes about redesigning their "air-cooled" Electrolux refrigerator. Once again, Norman's surveyors ventured out, train tickets in hand.

"He starts as if nothing had ever been designed before," wrote Sheldon Cheney in his study on artists infiltrating industry. One might find his assistants blowing smoke into a refrigerator to study its air circulation, or hanging it full of thermometers.[222]

The Bel Geddes–restyled Electrolux debuted in 1934, sporting the same pristine, white enamel exterior as the Oriole. Thanks to a brand-new technology that allowed steel to be stretched into simple, U-shaped shells, the traditional steel-wrapped-around-an-inner-box construction was eliminated, along with sharp, right-angled corners. Like the Oriole, it was lighter weight, cheaper to manufacture, and skyscraper "curtain-wall" inspired. And despite Loewy's claim that his 1935 Coldspot "Super Six" (as in six cubic feet) was the first with concealed mechanisms,[223] Bel Geddes's Electrolux was there first.

Its door, on patented concealed hinges, closed flush with the exterior cabinet, whose vertical bulk was ameliorated by several carefully placed horizontal indentations. Other features included "electromagnetic shielding," a double-sealed gasket for minimized leakage, chrome-finished interior storage compartments, shelves on the inside of the door, a bottle-storage liner "for beer and ginger ale," and, in the freezer, cube-ejecting trigger-release ice trays.

The only ornament on the otherwise smooth exterior was a large, circular recessed nameplate. This subtle piece of "jewelry" (to use Loewy's term) was the Electrolux's most innovative feature. When pressed with a hand, elbow, or shoulder, a pivot-mounted rotating snap-action bolt automatically opened the door. No longer would a housewife have to put down whatever she was holding, be it a grocery bag or a roast, to get inside. As one sales director famously put it, "Fifty percent of our business is preserving women, not fruit."[224]

Norman Bel Geddes's Electrolux for Servel, 1934.
(Courtesy: Harry Ransom Center, The University of Texas at Austin, photo by Richard Garrison.)

Six years later, during an interview, Bel Geddes agreed to answer certain questions about his "nameplate" versus Mr. Loewy's "feather-touch," on the condition that his remarks not appear in the published piece.

> *We were doing the Electrolux box. This was a year ahead of the Loewy operation [on the Coldspot] and Loewy offered a great deal more money to a boy in our office who was not on this job but who could nose around in the drafting room, of course, if he was that kind of fellow. All during the lunch hour, you could walk around in there. Anyhow, this man—who is still with Loewy—went down there with this idea and Loewy got it on his box before ours came out. Only he hadn't been smart enough to get it patented, and the Electrolux people for whom we did it, had. Poor Mr. Loewy had to take it off his box . . . so it wasn't completely stolen.*[225]

U.S. patents #95,817 (Refrigerator Cabinet) and #2,172,467 (Refrigerator Latch), the latter with six technical drawings, bear his story out. Both were filed in Bel Geddes's name as assigner to Servel. Loewy's

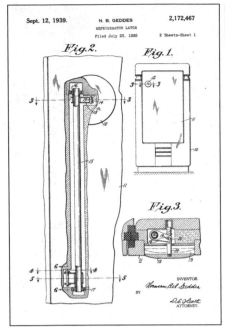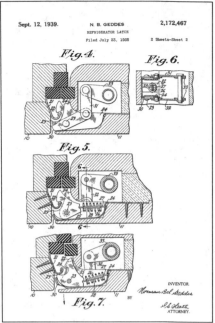

Latch patent drawings for the Electrolux.
(Courtesy: U.S. Patent Office)

Coldspot was released with a simple vertical bar, an external handle to be pulled, with no corresponding internal mechanism.

With the refrigerator wars in full battle mode, magazine ads upped the ante, from beautifully dressed women gazing with rapturous pride at their neatly arranged shelves to refrigerators posed above miniaturized towns—immense, immaculate sentinels emitting celestial beams of light. Movies, even comic strips, helped spread the word about a glorious new standard of living, exemplified by Blondie's new Frigidaire, a counterpart to Dagwood's latest Chevy.

Loewy would create two more Coldspots for Sears before switching his allegiance to Frigidaire. Bel Geddes would go on to design refrigerators for GE, Frigidaire, and Nash-Kelvinator, introducing features like transparent, adjustable shelves and "germicidal ultraviolet ray lamps." By 1937, the number of ever-larger and more ingenious refrigerators assembled in the United States would reach 2.5 million. Gleaming, super-powered treasure chests, they'd come to symbolize the country's aspirations for itself beyond the Depression's harsh grind.

•

THE ONLY SEAFARING vessel Norman had ever set foot on was the *Belgenland*, en route to Cherbourg with Aline. Undeterred, he and his staff designed an immense, ovoid "ocean liner of the future" as an office "exercise," 1,088 feet in length, with 70,000 tons displacement, its aerodynamic form a precursor to jumbo jets.

Employees tested underwater forms in a smoke tunnel and in a crude wind tunnel on the building's roof. Though data regarding ship superstructures was meager, overall data on objects moving through air showed that "streamlining" had the potential to significantly reduce air resistance; in this case, Norman estimated a 14 percent advantage. The smokestacks were oval inside; outside, they disappeared into the ship's mass. Dozens and dozens of windows offered wraparound views. Twenty-four "unsinkable" lifeboats (each for 150 passengers, with radios and two weeks' rations) hung on the exposed deck; if and when they were needed, a section of the deck would hinge out, forming a railed gangway, and the boats would automatically drop into position for loading.

A sliding or roll-up "skin"—operated at the push of a button—would enclose deck areas during rough weather, keeping them usable at all times and eliminating the need for increased engine power. (Experiments with models indicated that speed on ships like the *Mauretania* reduced 15 percent in 60-mile-per-hour gales.) The skin "goes up and down, somewhat like a window shade," observed one journalist. "Closed, the ship looks like a big cigar."[226] Similarly, motorized, heat-insulated sliding panels could cover the sundeck and stern lounge area. The streamlined form would also, Norman believed, serve to decrease

Streamlined ocean liner, 1932.
(Courtesy: Harry Ransom Center, The University of Texas at Austin)

sthe impact of enemy munitions, assuming that armor-grade steel could be shaped into the curves of his design.

There was state-of-the-art fire detection and extinguishing apparatus throughout. More radical still, the entire ship was air-conditioned, and all the staterooms soundproofed.

As with Airliner #4, the liner was predicated on luxury— accommodations for 2,000 first-class passengers and a crew of 900, a tennis court, a large deck games area, a swimming pool, and, on the top level, a sun-exposed sand beach and an open-to-the-stars dance floor. There was a library, a fireplace, a writing room, a gymnasium, a card room, a cinema theater, a smoking room, and a nightclub. The main lounge and music room rose up two levels. Given all that, it was still, the designer insisted, more economical to build and operate "under existing conditions" than the fastest liner in service and would cut transatlantic travel by twenty-two hours.[227]

In 1935, the *Encyclopedia Britannica* would include a photo of Bel Geddes's 1932 ocean liner model as a stellar example of machine-made art.[228] "It is a vessel so far removed yet not without its own beauty of line that our work-a-day brain reels before it," observed *The New York Sun.* "In our enthusiasm, we forget to mention, also, that the after-smoke-stack is an airplane hangar."[229]

Toward the end of the decade, Norman would receive a visit from Ernest de Weerth, his cottage-mate at Leopoldskron, later "assistant in charge of costumes" on *The Miracle.*

Accompanied by his mistress (an alleged countess), De Weerth was in New York to negotiate the purchase of the blueprints, sketches, and rights to Norman's ocean liner for Benito Mussolini. (Il Duce may have been feeling flush after his recent victory in Abyssinia.) Norman "very foolishly" rejected an offer of $200,000 (some $3.5 million today), confident that American shipbuilders would be interested.

•

BY THE SUMMER of 1934, thanks to Norman's relentless hustling ("There isn't a corporation in the country that hasn't gone to lunch with him"[230]), his staff had grown to fifty and his firm was earning fees as high as $290,000, with a net profit of $94,000 (in excess of $1.5 million).[231]

By 1936, the Big Four were appearing in the press with regularity. *Forbes* described them as "the chief agitators of the revolution now taking place," though their roles remained "much misunderstood . . . alternately puffed up to the bursting point by over-zealous admirers and deflated by skeptics."[232]

The "dapper and diplomatic" Teague had taken on Corning Glass, Ford, railroad day coaches, mimeograph and X-ray machines, and a new tractor that, in the name of research, he drove for hours. The designer, he told *Forbes*, does not pluck his ideas "out of thin air or out of his own soul, whatever that may be."

Forbes described Dreyfuss, still in his early thirties, as serious, bespectacled, and practical, "with little time to spend dreaming about houses and automobiles of the future." Before his career was over,

Henry Dreyfuss. A reputation for always wearing brown
(Courtesy: Life magazine/Getty Images. Photo by Bob Landry)

he, too, would drive a tractor—and ride in a submarine, spread manure, pick corn, operate a turret lathe, wear a hearing aid for a day, and design a lollypop-packing machine and the first anatomically shaped toilet seat.[233] Loewy claimed to be turning out more than $100 million worth of product annually—filling stations, trains, Greyhound buses (he slenderized the logo's mascot), a line of trucks.

Despite the fact that Bel Geddes was busy with pragmatic accounts— the first gasoline pump with an automatic sales price register, the first all-metal seltzer siphon (the patented, chrome-and-enamel "Soda King"), and a microphone for CBS—he tended to be portrayed as the one who *did* pluck ideas from his soul and dream of futuristic landscapes. Was he "a visionary who has done the profession great harm" or "a prophet born before his time"?[234]

⋅

IN 1936, RAYMOND LOEWY was given permission to make suggestions for improvements to Pennsylvania Railroad's 6,000-horsepower electric GG-1 locomotive. There was a growing sense that if the railroads wished to compete with the fledgling airlines, they needed to pay more attention to comfort and "looks." His suggestion to replace the train's thousands of rivets with welding "caused a sensation," he claimed, "and a shock."[235]

Though Bel Geddes would never boast a manufactured iron horse on his résumé, he'd spent considerable time studying them. His lawyer's wife, Beryl Austrian, remembered seeing him sketch a streamlined locomotive design on the tablecloth at a dinner party in 1928, having just returned to Manhattan on a slow Long Island train.[236] Images of his patented Locomotive #1—sleek, tapered, designed to be sheathed in lightweight, rivet-less aluminum—had been widely disseminated even before the publication of *Horizons* (1932), which devoted an entire chapter to railways.

The design took into consideration issues of safety, stability, cost-effectiveness, temperature control, the configuration and placement of the engine, and, last but not least, passenger comfort.

Six months after *Horizons'* 1932 release, Union Pacific announced plans to build the high-speed M-10,000 Streamliner. "The Geddes train of 1928 is still a mark to shoot at," the company's reps told *BusinessWeek*.[237]

By 1936, when every major railroad was either rebuilding or had already converted one or more of its passenger trains into "streamliners," Pennsylvania Railroad engineers would have been well aware of the benefits of rivet-less, or at least sunken rivet, facades.

In 1935, with industrial-artist-designed objects flooding the landscape "by the thousands, by the tons, by the gross,"[238] Norman joined forces with a tall, bespectacled, Groton-, Harvard-, and Beaux-Arts-trained architect to form Norman Bel Geddes George Howe & Co., Inc., with Norman in charge of production, Earl Newsom covering finances and public relations, and Worthen Paxton, or Pax, as a third partner and head designer. Having George Howe on board increased the possibility of important architectural commissions. He and Bel Geddes were old friends. It's unclear, though, why the conservative forty-nine-year-old chose to leave a senior partnership in Philadelphia for a renegade practice, investing $25,000 in exchange for a quarter of the corporation's stock.[239]

"I believe this step . . . is one of the most important I have ever taken," Norman wrote Maggie White.[240] They would take on everything from skyscrapers to inkwells. A handsome oversized brochure listed John D. Rockefeller, Jr., Kermit Roosevelt, and the U.S. Coast Guard as clients.

Four days after the partnership papers were signed—four years after

Bel Geddes had filed petitions to allow him to remain on the Chicago Fair's Architectural Commission—George Howe received a summons from New York's attorney general, demanding the corporation be disbanded within twenty-four hours. Unbeknownst to either principal, there was a law forbidding a partnership between a registered architect and a non-registered one.

There was a loophole. Anyone who'd worked six years prior to the law was exempt. In the Twenties, Bel Geddes had designed several theaters, including Aline's on Olive Hill ("on a fifty-fifty basis" with Frank Lloyd Wright),[241] the Palais Royale, and J. Walter Thompson's ambitious, multi-use Assembly Room, but he had no proof. As a precaution, his name didn't appear on the blueprints.

CHAPTER 12

What Geniuses Worry About

1930s

His diversions are as tremendous as his orgies of creative endeavor.
— "ARE YOU AFRAID OF THE UNEXPECTED?" *AMERICA MAGAZINE*, JULY 1931

Unlimited imagination is a typical American trait.
— RAYMOND LOEWY

Terminally ill with cancer, Bel finally agreed to a divorce. The cause on her petition read "adultery." Three months before the Geddes-less Chicago Fair opened its gates, Norman and Frances married.

The Oriole and Acorn commissions had, by twenty-first-century standards, made him a millionaire.[242] A multimillionaire, considering his earnings on *The Miracle*. But between salaries for Norman Bel Geddes & Co. employees, War Game maintenance, his increasingly ambitious "home movie" collection (cameras, film, a boy kept on salary to splice, catalog, and assist on weekend shoots), out-of-pocket investments in theater productions,[243] a penchant for "business" evenings at the Cotton Club, Jack & Charlie's "21," El Morocco, the Stork Club and Cafe Pierre, and nights out dancing (Harlem, Central Park Casino), not to mention alimony and child support, Norman's life of "elegant nonchalance," as one colleague put it, was running $350 a month (close to $6000 in today's dollars). In short, he was nearly broke.

He gave up the four-floor brownstone, the War Game (with its astronomical liquor bills), his cook and Chinese houseboy. The new Mr. and Mrs. Bel Geddes moved to a staid, fifteen-story apartment building on East Seventy-Ninth Street, and Norman's office was relocated to a set of "seedy" rooms on East Thirty-Eighth. Possessions that didn't go into storage were severely pared down.

It was a temporary reprieve.

•

IT HAD ALL begun with rose bugs, "no larger than the head of a match," observed during on a Long Island summer weekend with friends, before the move from Murray Hill. Armed with a custom-made 35-millimeter telescopic lens (attaching it to his spring-driven, wind-up, 16-millimeter camera required the invention of a support carriage, "not unlike one on a coast artillery gun") and specially split 35-millimeter film, Norman documented the bugs' mating habits, magnified several thousand times. Emboldened, he set up three lenses and two 1,000-watt lamps on a trio of fly-speck-sized snail eggs and managed to capture the movements of the infinitesimal unborn creatures inside. (Coincidentally, German physicists were busy trying to perfect an "electron" microscope.) An attempt to immortalize the "post-honeymoon gesture" of a praying mantis couple (the female eats her mate) was, alas, foiled, when the bridegroom, "Henry," died in the broiling sun.

Norman's interest in filmmaking dated back to the early 1920s. He'd shot everything from backstage footage, his children's antics, and vacation scenarios to friend Lily Tosch's famous "Hand Dance," but now he'd found his true subject. It wasn't long before word of his telephoto exploits traveled beyond his immediate circle. A society columnist credited him with "painstaking studies" of the aphis and the common louse, among other quarry. "The enthusiastic man is often a prize bore, but Norman Bel Geddes is never a bore, even when he inflicts his hobbies on you."[244]

A unique project of Norman's "insect" phase combined his entomological interests with his love of theater—a production of Homer's "The Siege of Troy" using mock-heroic titles based on Alexander Pope.

The script called for eight sequences, including Helen's rape and Jove's seduction by Juno. Miniature sets were constructed on the roof of a friend's house in East Hampton. All the actors, from principals to soldiers, were played by ants, the opposing armies by red and black ones. The gods and goddesses (winged larger ants) were shown floating on labelled clouds high above Mount Ida. (Eight thousand miles and an ocean away, Ernest Hemingway was busy tracking down lions, rhinos, antelopes, and wildebeests, in the best Teddy Roosevelt tradition.)

He shot 800 feet of 16-millimeter film—"with excellent results." There was a scene of the Greeks' tented encampment with miniature pennants and one of the fully manned fleet approaching the shore. The attack on the fortress called for Paris and Menelaus dragging each other by the leg, Greeks climbing the walls, and Trojans hurling sticks and stones. The Trojan Horse was a large, dead, hollowed-out June bug. And then the denouement, beginning with Menelaus entering Helen's bedchamber, where she is not alone:

> *She sighs for ever on her pensive Bed,*
> *Paris at her side . . .*

Then a close-up:

> *The Fair . . .*
> *Repairs her Smiles, awakens ev'ry Grace,*
> *And calls forth all the Wonders of her Face*

And ending with Menelaus victorious, Paris waving "feebly with his antennae" as he sinks back, dead.

"Moving pictures of Armageddon" was how Norman's friend Murdock Pemberton described it. "He took pictures of thousands of [ants] in trenches he built."[245] There was a problem getting them out of their holes and into small boats, Norman conceded, though "I did get them to fight . . ."[246]

Like many of his more imaginative forays, "Troy" was never completed. (Imagine what he could have done with iMovie software, the transitions and cross-dissolves, the soundtrack.) But the project achieved some notoriety through syndicated cartoonist Edwin Cox's "Private Lives," a high-end Ripley's "Believe It or Not" series focused on public figures. In newspapers across the country, Norman's likeness appeared in a suit and tie, directing his tiny charges with a pencil point and handheld magnifying glass while the cameras rolled. [247]

The caption read: WHAT GENIUSES WORRY ABOUT

Twenty years later, "ant farm" kits ("See the ANTS . . . Building Bridges, Digging Subways, Moving Mountains"), minus Homeric allusions would, like miniature golf, become a national craze.

Within a year of his marriage to Frances, Norman had progressed from insects to amphibians, reptiles, and the occasional carp or minnow, making it a point to know their Latin names. Weekdays were consumed with designing a wind tunnel to test Chrysler's Airflow, a vacuum

(Courtesy: Harry Ransom Center, The University of Texas at Austin, unidentified periodical)

cleaner and refrigerator for Electrolux, the first oceangoing streamlined diesel yacht (a smaller version of his ocean liner), and interiors for the new China Clippers. Weekends found him loading up a car with cameras, tripods, reflectors, homemade specimen boxes, and a burgeoning collection of Abercrombie & Fitch gear for Long Island outings. (Somewhere in the German countryside, another self-taught weekend naturalist, Vladimir Nabokov, was busy tracking down butterfly specimens.)

The East Seventy-Ninth Street living room was now stacked to the ceiling with five enormous, custom-built glass tanks (tropical and fresh-water) and terrariums (dry desert heat to humid woods). The latter in-cluded thermostatic heating, aeration, drainage, sunlamps (with a fan for circulation), and appropriate plantings. Other tanks were specifically for breeding. All told, a thousand-dollar investment, not counting their inhabitants. A false floor was installed to help handle the weight.

Friends en route to Bermuda were asked if their children might be encouraged to catch chameleons, for which Norman offered one dollar apiece, a generous Depression bounty.[248] He had a friend in the Customs Department who could meet them at the boat. But unsatisfied with the lack of diversify, he soon had his secretary tracking down supply houses nationwide that catered to labs and museums. To help ensure the safe arrival of his orders, he designed a Constructivist-like mailing label with red and black lettering, printed on the same press as the Nutshell's racing forms and the War Game's tally sheets.

Typical requests from, in this case, two Florida purveyors, included:

Spotted night snake
Hammond's garter snake
Boyle's king snake
Western ringneck snake
San Diego gopher snake
Red-necked snake
Crowned snake
Blainville's horned toads
Pacific newts
Blue-bellied lizard
Blue-tailed lizard
Legless lizard
Swift lizard
Ground lizard
Worm lizard
Iron worm
Baur's mud turtle
Southern south-shelled turtle
Yellow-bellied terrapin
Red-bellied terrapin[249]

(Courtesy: Harry Ransom Center, The University of Texas at Austin, Design by Bel Geddes)

He shared his reptile passion with Maggie White, who'd had a baby boa constrictor and a tame puff adder as childhood pets and who now kept two small alligators (Mercury and Mars), Florida "souvenirs" from friends, on the sixty-first floor of the Chrysler Building, on the tiny terrace outside her studio.

"I make no pretense at being a scientist," Norman wrote Dr. Gladwyn Kingsley Noble, curator of Manhattan's American Museum of Natural History and founder of its Department of Experimental Biology, "but I enjoy watching the little fellows tremendously." Dr. Noble responded with a gift of several unusual specimens, sent parcel post, with the hope that they wouldn't devour their associates.[250]

To Mr. R. T. Berryhill, Sr., Lakeland, Georgia:

Regarding your recent communication, I do not know what the 4-legged rubber-hided salamanders are. Please send me their scientific classification and let me know what food they eat and if they will devour smaller salamanders.

I appreciate your offer, but regret to say that I am not interested in investing in a frog farm.

Very truly yours,
Norman Bel Geddes[251]

The frog farm inquiry was precipitated by frequent orders of frogs' eggs by the pint. Norman raised them in his apartment, year-round, as food for his snakes. (When one of the snakes had eleven babies, announcement cards were dutifully printed and mailed out.)[252]

"We do not list the blue-tailed skink in our publications," wrote a Chicago supplier, "but knowing Mr. Geddes' interest in specimens of this kind, we are sending it along without cost."[253] Much of this ongoing correspondence was the responsibility of Norman's long-suffering secretary.

The lizards, newts, tortoises, and mollusks were quiet company, but come spring, the same neighbors who'd witnessed Norman's packages piling up in the Georgian building's otherwise respectable lobby were subjected to a cacophony of mating geckos and toads, crickets and katydids. Complaints were made and presumably ignored.

Having read about the difference between fresh and saltwater eels, Norman contrived an experiment.

At the mouth of a stream near Montauk, he collected three dozen inch-long babies and ten gallons of salt water. Back in his Upper East Side apartment, he had a fifty-gallon, aerated supply of pond water on hand. (Fellow New Yorker Clarence Birdseye, already known for his fruit and vegetables, was busy stocking the family bathtub with live fish for his latest "quick freezing" experiments.[254])

"The ten gallons of salt water with the eels were emptied into a tank," Norman reported to William Beebe, still busy setting world records deep beneath the Atlantic. "Once a week, a quart of their salt water was removed and replaced by a quart of pond water. This meant that within ten months . . . all of the salt water had been changed into fresh water. Not one eel died . . .

"It has always been my understanding that eels were bred at sea . . . somewhere off the West Indies . . . Would you have some time to let me know if this is correct? . . . It seems to me that [they] must be bred in *fresh* water, up above where they abound in vast numbers, [before] making their way out to sea . . . This is, of course, an absolute contradiction to the dozens of theories I've read."[255]

Six months after the "adjustment" experiment began, the test subjects had grown to nearly ten inches. Too large for their tank, Norman presented them to the Museum of Natural History. Paying a visit sometime later, he was pleased to find they'd grown to six feet. Dr. Noble offered his congratulations. "I know of no other local naturalist who has tried this experiment."[256] Certainly not within the confines of a midtown Manhattan apartment.

In the meantime, Norman had moved on to larger prey. Fascinated to learn that Central Park Zoo was planning to breed their hippos, a consummation that takes place at night, underwater, Norman obtained permission to document the proceedings, aided by several assistants and great lengths of lighting cable.

The apotheosis of his latest pastime was a pseudo-educational film series, a trilogy spliced together from miles of Long Island weekend footage. "The Amazing Adventures of Rollo, the Boy Naturalist" was an homage, of sorts, to his brief adolescent career as Zedsky, the Boy Magician.

Shot "in actual localities" with "no expense or risk of life spared," and with nods to Buster Keaton, Harold Lloyd, and Chaplin, it starred Melanchthon Pip (Norman) as Rollo. Music from Bel Geddes's considerable record collection was added as a soundtrack.

In "Rollo Amid Pond Life," he's seen foraging about (a floppy sunhat, strainers, sample jars, butterfly net) only to end up, after various misadventures, "dejected, alas empty-handed," read the title card, "after a day of toil." Sitting on his roadster's running board, water gushes from the tops of his waders, soaking him through. In "Rollo After Big Game," he sports a beret, shirt and tie, saddle shoes, and improbable white-cuffed pants to crawl on his knees through high grass in search of spiders and bugs, "a hound on the scent." Falling to the ground, he finds himself face-to-face with an enormous, Bel Geddes–magnified caterpillar, menacingly inching up a plant stem. Gullet-trembling frogs lash out their tongues to catch a meal, tinier-than-a-thumbtack bugs mate, and a praying mantis begins her post-prandial clean-up. ("No lady was ever more meticulous with her toilet.") The closing shot focuses on Rollo's bottom as he crawls away, still intent on his quest. In "Rollo at the Farm," he slides on a cow paddy while maneuvering his tripod and camera but manages to capture two enormous cows in determined flagrante delicto.

•

WHEN THE BEL GEDDESES moved to the upscale River House on East Fifty-Second Street, their burgeoning collection of housemates came along. Two summers later, Norman was still placing mail orders, though on a less ambitious scale. Then, flush with the success of a Bel Geddes–designed and produced Broadway smash,[257] word arrived of an available Park Avenue apartment a block from St. Bartholomew's Church, an address irreconcilable with Norman's brood.

Just as the introduction of Kodachrome (invented by two classical musicians, Leopold Mannes and Leopold Godowsky, Jr., also known as Man and God) was creating a huge boost in the 16-millimeter market and Harold Edgerton was creating dazzling strobe light images (bullets passing through apples, crown-shaped drops of milk), Norman sold off his three telescopic motion picture cameras, along with more than 2,000 fish, reptiles and amphibians (weekly rations now included 200 peeper frogs, 200 to 300 insects, several hundred earthworms, 2,000 mealworms, and the occasional caterpillar), vivariums, terrariums, and equipment to an Upper East Side private school, the Dalton, on East Eighty-Ninth Street.

Good steward that he was, he followed up the sale with detailed instructions to ensure that his brood "will live very happily."

Between camera supplies (he'd bought "everything there was",[258]) film development, and mailing costs, his "splice" boy's salary, and the purchase and feeding of his on-call "actors," Norman's amateur naturalist phase had, he estimated, cost $10,000 to $12,000 a year, in Depression-era currency, to maintain.[259]

Stacks of film tins, housing more than a million feet of footage, followed the Nutshell Jockey Club and War Game into storage.

CHAPTER 13

Too Good to Succeed: The Chrysler Airflow

1933–1937

Genius is 1 percent inspiration and 99 percent perspiration.

—THOMAS EDISON

Design is 25 percent inspiration and 75 percent transportation.

—RAYMOND LOEWY

I n the summer of 1938, when the first issue of *Action Comics* introduced the world to Superman, its cover featured the Man of Steel lifting a steel-framed Chrysler Airflow, what Bel Geddes called "the first
sincere and authentic streamlined car," above his head.[260] It was the
1937 model, down to its rounded, beetle-brow hood and tapered rear, its
grooved speed lines and triangular back "opera" window, its whitewall
tires and condensed, newly horizontal grille.

The following year, when Universal Pictures decided to make a film
version of *The Green Hornet* radio serial, the screenplay called for the
hero to drive a car with "ultramodern lines," something that *looked* fast.
("That thing travels faster than the bullets I send after it," notes a patrol
officer during a chase scene.) But by then the Airflow—a vehicle vastly
superior in speed, safety, and comfort to anything on America's roads—
had been so maligned in the public's imagination, thanks in part to a
competitor's expensive smear campaign, that decades later it would still
be spoken of as the greatest failure in automotive history.

Instead, Universal chose a 1937 Ford Lincoln Zephyr—its speedster
look augmented with stylized lightning bolts painted on the fender skirts
and a "Flight of the Bumblebee" soundtrack—as the 200 miles per hour
Black Beauty, the Green Hornet's signature transport.[261]

•

CHRYSLER'S 1929 COUPE had been inspired, claimed company ad men, by
"the canons of ancient classic art . . . unsurpassed and unchallenged."

Its radiator with cowl molding suggested the repetition motif in a Parthenon frieze, its front elevation replicated the Egyptian lotus leaf pattern. "This patient pursuit of beauty will doubtless prove a revelation to those who have probably accepted Chrysler symmetry and charm as fortunate but more or less accidental."[262] The following year, the new models were said to be "as distinctive and charming" as the Parisian couture of Paquin and Worth.[263] But the focus soon shifted from ancient history and European aesthetics to what was taking shape in the New World's own backyard.

Walter P. Chrysler understood the importance of tenacity and vision. In 1905, he borrowed a considerable amount of money to buy a car that caught his eye for the sole purpose of dismantling it to see how it worked. A few years later, he was General Motors' first vice president, and not long after that, he quit to start a rival company. In 1933, despite a debilitating economy—wages had dropped 60 percent, more than 12 million Americans were unemployed, and business at large was running at a net loss exceeding $5 billion[264]—Chrysler turned a considerable profit, the only company to produce more cars that year than it had in its Parthenon -Egyptian Lotus phase, just prior to the Crash.

And there was encouraging news wafting in from California: A brand-new culture, designed around the gasoline engine, was emerging. "Drive-ins"—from banks and "car-hop" restaurants (with roller-skating waitresses) to outdoor picture shows—were sprouting up like mushrooms after a rain.

Chrysler had, as early as 1932, appropriated some $25,000 for the development, "somewhere in the Canadian woods," of what was known in-house as the RD 124.[265] The plan was to produce lighter, faster cars with better gas mileage, increase riding comfort using balanced weight in lieu of independent springs, and reduce the standard wheelbase to fit perpendicularly into railway containers (doubling shipment capacity), ultimately offering customers a price below anything Ford or GM could manage. In short, it was a reimagining of the promise of the Model T—an affordable, high-quality vehicle for the masses.

"Streamlining" had suited the corporation well enough when it came to the recently completed Chrysler Building, the triumphant art deco stalagmite at the corner of Forty-Third and Lexington. Other auto executives remained reluctant to invest in the substantial retooling costs such a vehicle would require. According to Fred Zeder, one of Chrysler's top engineers, Bel Geddes's *Horizons*—required reading for all company top brass—was "entirely responsible" for giving his employer the courage to proceed.[266]

Before the Airflow, automobiles were based on a horse-and-buggy model, with passengers sitting in a "wagon" behind a motor (its strength measured in horsepower), owners still draping a blanket over the engine when the vehicle was garaged, as if it were a stabled steed—all told, "an ungainly shape" comprised of "unrelated compromises."[267] Now the body was being designed *around the engineering*, with a new format and materials, a body worthy of Chrysler's new high-compression engines, a body conceived, it was said, when engineer Carl Breer spotted a chevron of geese that turned out to be a squadron of fighter planes flying low at low speed.

To say that the Airflow was ahead of its time is to damn it with faint praise. It was the first car with automatic transmission, automatic choke, hydraulic brakes, and automatic overdrive. In an age before air-conditioning, its adjustable, two-piece windshield tilted open from below. The side and vent windows tripled visibility, could be lowered simultaneously with the flick of a lock, and virtually eliminated "wind roar" inside the vehicle. The welded all-steel unibody, forty times more rigid than conventional wooden-frame bodies, increased passenger safety while reducing overall weight.

The taillights and dual headlamps were flush to the body; the rear wheels were enclosed; there were chrome-enhanced, wraparound bumpers, and a dust-proof luggage compartment; and one no longer had to step up from the running board to get inside. The whitewalled inner tube tires, like a double pair of gleaming spats, were something Bel Geddes had envisioned for his Graham-Paige designs in the late Twenties.

The interior featured divan-like adjustable seating. The leather-trimmed cushions set into polished chrome tubular frames created a sophisticated, moderne armchair look.

The eight-cylinder, 130-horsepower engine (powerful for its day) with two-barrel carburetor could run 90 to 100 miles per hour. Moved some twenty inches forward and placed directly over the front axle, it offered nearly fifty-fifty weight distribution; passengers were now cradled *inside* the frame, instead of *on* it. This, in turn, moved the driver's seat and the steering wheel forward, almost directly above the front wheels, resulting in more head- and legroom. The front seat could now, for the first time, comfortably accommodate three; the backseat sat six. The tank held three "emergency" gallons in reserve.

"Independent suspension"[268]—elongated front springs that functioned separately from the rear ones—absorbed "road shock" and eliminated engine tremors, producing a "Floating Ride that has a rhythm like a walk," in sharp contrast to the rocking and pitching passengers had come to expect. And for an additional fifty-five dollars, a custom-tailored, "golden tone" Philco car radio could be added.

When initial feedback indicated that the Airflow might benefit from additional refinements, *Horizons'* author was brought on board. So secret was RD 124 that Chrysler's representative on the project was referred to, in Geddes & Co. memoranda, as "Mr. Q."[269] The chrome "waterfall" grille featured thirty-nine slender vertical bars; Norman replaced them with twenty-one thicker ones, thus strengthening and broadening out the front end while softening its look. He raked the windshield, slanting it back to the sides and top, set the headlights into curved wings over the wheels, and (based on experiments with his own eighth-inch-scale wind tunnel) introduced side grooves along the front that allegedly channeled air around the moving body.[270] The steering wheel's tilt was modified, the spare was brought inside (so as not to interrupt the back's clean sweep), and marbleized vinyl floor mats were replaced with carpeting.

At the same time, Norman took on a second auto-related commission, what he'd later describe as "the first attempt at a slow-leak double-tube tire."[271] The "Firestone Streamline," which never got past the drawing stage, featured tapered sidewalls and a slightly bulging hubcap; the letters of its name were molded into the treads in bold, deco-like capitals. The overall effect was that of a preternaturally handsome (yet to be invented) Frisbee.

The result of six years' work, fifty prototypes, and rigorous testing, the Airflow debuted in January 1934, Chrysler's tenth anniversary, at New York's National Auto Show. A publicity photo showed it posed, in profile, beside a Union Pacific Streamliner locomotive. Before the show was over, thousands of orders had been placed.

"A New Kind of Car that Literally Bores a Hole Through the Air!" read a typical headline.

"There is the sleekness of a racing yacht's cabin," read one ad. "A suggestion of a modern penthouse apartment in the rich upholstery fabrics and gleaming Chromium trim."

"A car that turns gravel into asphalt . . . and makes asphalt seem smooth as glass," read another. "A car that cleaves the air like a bird."

One copywriter simply paraphrased *Horizons*: "You only have to look at a dolphin, a gull or a greyhound to appreciate the rightness of [its] tapering, flowing contour."[272]

Though his name had been attached to the Oriole and Acorn stoves,

Airflow advertisement in the Saturday Evening Post, December 16, 1933.
(Courtesy: Harry Ransom Center, The University of Texas at Austin)

it was unusual for Bel Geddes to lend his face as well. He appeared—in overcoat, brimmed hat, and leather driving gloves, a pipe stem in hand— in *The Saturday Evening Post*, framed in the Airflow's broad doorway.[273][

That spring, the Airflow ("not hog-tied to engineering traditions or artistic fetishes"[274]) was showcased performing stunts at the Chicago World's Fair, open for its second season. Chrysler's seven-acre pavilion included a quarter-mile exhibition track where a crew of "Hell Drivers" demonstrated what it and other Chrysler vehicles were capable of: taking banked turns, negotiating a forty-five-degree incline, making skid-free stops on a slicked down track. A front tire was shot with a rifle while the car was in motion. Other tests included a "Belgian Roll" (a "shimmy" machine that shook a car, running at full speed, "like a terrier might shake a rat"[275]) and a sandpit where cars were deliberately flipped at high speed. After a double somersault, the Airflow landed "on its feet, as nimble as a cat."

•

THE STANDARD EXPLANATION for the Airflow's demise is that the public took an immediate dislike to its unorthodox look. *Harper's* editor Frederick Lewis Allen described it as "bulbous" and "so obscenely curved" that it defied "the natural preference of the eye for horizontal lines."[276] (One wonders if Allen's taste in women ran more to angular Flappers or the Lillian Russell model.)

To the contemporary eye, the external differences between the 1934 Airflow (think New Coke) and Chrysler's conventional 1934 sedan (classic Coke) seem relatively subtle: the former is more rounded and tapered, its windshield slanted, its headlights more discreet. Critics—none of whom admitted to having taken a test-drive—compared the Airflow to a bathtub, its hood to the face of a basset hound, a rhinoceros, a burglar in a stocking mask.

"I now want to eat crow," Paul Merchant wrote Bel Geddes. An argument, months before, over the relative merits of GM and Chrysler, during which Norman "defended the Airflow lustily," had led Merchant, editor at a major New York publishing house, to explode with a string of expletives. Now, having finally driven the object of his scorn for 9,000 miles, he admitted it was "the most magnificent piece of automobile machinery it has ever been my good fortune to handle."

"It may interest you to know, Norman," the newly minted convert continued, "that the antagonism we have heard expressed . . . is confined entirely to hyper-conservative sonsofbitches who are against all change, and it may also interest you to know that . . . very frequently when [my wife and I] drive through towns, [people] dance up and down, shouting 'There goes *my* car!'"[277]

•

DESPITE THE MEDIA'S concerted attempt to denigrate it (whether out of conservatism or outside "influence"), the Airflow caused a sensation. According to engineer Breer, more orders were placed for Airflows at the January Auto Show than for any new car ever exhibited there.[278] New York Mayor Fiorello La Guardia could be seen tooling around Manhattan in one.

But the inventory wasn't there.

The Airflow's innovations required unprecedented expertise that many of Chrysler's workers lacked, at least that first year. The bridge-type steel frame had to be welded upside down. The wide, chrome-trimmed seats required a special assembly line. The 1934 Airflow was offered in four eight-cylinder-engine models, the largest of which sported the industry's first one-piece curved windshield, so difficult to install that four out of five broke. A tool-and-die-makers strike in the fall of 1933 exacerbated matters. Detroit's Big Three were as competitive as starlets. Taking advantage of the Airflow's predicament, General Motors mounted an expensive misinformation campaign.

The "smear" (a tactic GM would return to in the 1960s in an effort to discredit Ralph Nader) had been employed to dramatic effect by no less a personage than Bel Geddes's hero Thomas Edison. In 1903, the inventor publicly electrocuted—"Westinghoused" he called it[279]—a three-ton elephant in an effort of discredit Nikola Tesla's alternating current campaign, which threatened his patent royalties. When Topsy, a Coney Island elephant who'd killed a sadistic trainer, was put on trial and sentenced to hang, a humanitarian backlash ensued. Coincidentally, New York had just replaced the gallows with the electric chair. Edison, who'd previously arranged for several dogs, cats, and the occasional horse to "ride the lightning" in the course of his campaign, recognized an unprecedented opportunity and offered his services. As a precaution against failure—there were 1,500 spectators and the press was on hand with motion picture cameras[280]—Topsy was fed cyanide-laced carrots at the last minute.

Edison himself had been "smeared" by the gas companies when he first introduced his DC electrical system.

The Saturday Evening Post was the country's most widely circulated weekly, reaching three million.[281] Bel Geddes had touted the Airflow there; the competition would do him one better. In a barrage of double-page spreads, GM presented itself as a company that didn't make a move without "the priceless verification of the public itself," a company predicated on protecting the unwary "against ill-timed or dubious experiments." The reference to Airflows wasn't lost on John Q. Public.

GM had efficient, state-of-the-art vehicles, too (goddamn it), but theirs had "a mature refinement" and "beauty as well as speed."[282] Theirs "did not leap full-born into being," the hasty products of "abrupt inspiration," but were the result "of deliberate growth."[283] "An eye to the future," they promised, "an ear to the ground."

Going on about how ugly the Airflow was (leave that to the journalists) would have been unseemly; competing against its obvious advancements was dangerous ground. Just as Edison had presented AC current as a danger to be eliminated, the word went out that "the safest motorcar the world has seen" was patently *un*safe.

At the bottom of each double-paged magazine spread was GM's Silver Anniversary medallion, one side of which depicted a speeding, futuristic automobile backed by an immense vertical wing, the other an artistic rendition of an engine's combustion chamber. The irony was that Bel Geddes had designed it.[284]

Chrysler rallied with a spectacular publicity stunt. An Airflow was pushed off a 110-foot cliff into a Pennsylvania rock quarry. The car flipped over, landed on its wheels, and was driven away under its own power, battered but intact, all the doors and windows in working order. Then professional racer Harry Hartz was hired to run an Imperial Airflow coupe on Utah's Bonneville Salt Flats for twenty-four hours straight, covering more than 2,000 miles, where it set a series of new records,[285] after which Hartz drove the same car from Los Angeles to New York City to prove it had economy as well as speed. It averaged 18.1 miles per gallon, another record. A sister model, the six-cylinder DeSoto Airflow, averaged 21.4 miles per gallon during the same cross-country run. Filmstrips documenting these feats were distributed free of charge to movie houses, and miniature Airflows showed up as prizes in Cracker Jack boxes. The Airflow would go on to win the Grand Prix and Premier Prix at the Concours d'Elegance in Monaco.

The future is like heaven, James Baldwin would later write. *Everyone exalts it, but no one wants to go there now.* The Airflow was the car that subsequent cars would be based on. But production delays, combined with GM's pedantic spin ("the common sense of the common people!"; "he travels farthest in the right direction who is willing to listen as well as to lead") and smug observances in the press, won out over Chrysler's gamble on "a refreshing new kind of beauty"[286] that had evolved from the inside out.[287]

Only 11,292 Chrysler Airflows and 13,940 DeSoto Airflows sold in 1934, the numbers ratcheting down for the next three years, after which they were discontinued. (*Action Comics* fared significantly better, selling over 200,000 copies of its premiere issue, despite the publisher's doubts that anyone would believe that a guy in a cape and tights could lift a car over his head.) Chrysler survived partly because its Plymouth model sold like hotcakes and partly because it had hedged its Airflow bet by introducing the Airstream, a big, boxy, conventional car, trimmed to evoke a streamlined feel.

Though a number of Bel Geddes's ideas for the Airflow never got past the model stage (a tail fin gasoline container, a "crumple zone," radically indented grooves running from front bumper to back), and though he was only a contributing designer, detractors doted on the car's fall from grace—it would be held up, for decades, as a warning against adventurous pioneering and uncompromising standards—as yet another example of Bel Geddes's expensive impracticality.[288]

•

TEN YEARS LATER, Raymond Loewy would ask why Bel Geddes's Airflow, which embodied "the latest wrinkles" in aerodynamics, had found "few takers."[289] Automobiles were "ugly to the point of being repellent," he would write. "How long would the public put up with it?"[290]

Norman's perennial rival, Loewy had taken on a Hupmobile redesign, the "Aerodynamic" series, the first one of which was released the same year as *Horizons*—"a true air-line car, therefore beautiful."[291] The 1934 and 1935 "Aerodynamic" models (flared headlights, rounded corners, and a three-piece windshield borrowed from Paul Jaray's Tatra) were financed, in part, out of Loewy's own pocket, according to one source to the tune of $18,000.[292] But with fewer than 5,000 paying customers a year, the Hupmobile fared more poorly than its "repellant" foil.[293]

And then there was Harley Earl, whose Buick Y-Job for GM is often credited as the first "concept car." Its main claims to innovation, notes Christopher Innes—streamlined teardrop rear end, wraparound bumpers, horizontal radiator grille, fenders extending back into the doors, grooves along the sides—"all came from Norman Bel Geddes, who had introduced precisely these features to Chrysler more than five years earlier."[294]

The Airflow, an embodiment of a future to which the public said "no thanks," would prove profoundly influential, despite its brief life span. Half a century later, Airflows and other experimental designs (the Tucker Torpedo, the Edsel) would be highly sought-after by collectors.[295] "The influence of the Airflow on other automobiles was unmistakable," wrote Arthur Pulos in *American Design Ethic*. "The V front and the slant back became standard in the industry."[296] As did its lowered silhouette, all-steel frame, and unified exterior shell.

•

IN DAVID MAMET'S 1977 play *The Water Engine*, a struggling young Depression-era engineer, Charles Lang, creates an engine that runs on the energy released when the hydrogen and oxygen molecules of H_2o are separated. Cheap. Efficient. "Green." Revolutionary. It's his ticket, Lang thinks, to a better life. At first, the powers-that-be take him for a madman, a crazy dreamer. But when his engine proves itself, they quickly try to buy him off and bury it. When Lang refuses to relinquish the rights (the bad guys include one or more unnamed corporations), both he and his sister meet a gruesome end.

It's difficult to ignore the shadows of the Airflow (and the suppression of "green" technologies to come) in this cautionary tale. Set against the background of Chicago's 1933 Century of Progress International Exposition, where the Airflow was showcased, it's a haunting indictment of the American Dream, an evisceration of the Horatio Alger and "level playing field" myths that so many in the twentieth century were raised to believe in.

CHAPTER 14

Fickle Mistress

1929–1937

Mr. Geddes is not alarmed by pessimistic predictions that theater is on its way to the limbo of the dodo bird and the dinosaur.
—*DAILY MIRROR*, MARCH 8, 1937[297]

Broadway is a "harum-scarum business."
—*NEW YORK TIMES*, MAY 9, 1937

I n 1926, post-*Miracle*, post–Cecil B. DeMille, and post–*Lady Be Good*, Bel Geddes and Leopold Stokowski were part of an advisory arm to the League of Composers, founded by former pianist Claire Raphael Reis to facilitate productions of new musical works.[298]

The group, which included Leonide Massine, Aaron Copland, Agnes de Mille, and Martha Graham, usually met at Reis's home. One blustery February afternoon, assembled to hear "Stokey's" ideas for a full-scale production of Stravinsky's *Les Noces* at the Met, the maid rushed in screaming that the pantry was on fire. On inspection, a flame was crossing the ceiling; smoke was soon filling the drawing room. Norman bolted for the basement to retrieve an enormous fire extinguisher. Either someone inadvertently tipped it the wrong way or he and Stokey tried to lift it together toward the ceiling. In either case, the spray dripped down, soaking them through. One of the firemen, now on the scene, ordered them to wash down *immediately*. Together the two men—one tall and lean, the other considerably shorter and thickset—jumped into the shower. By the time they emerged, their clothes had begun to crumble. Once the fire was safely contained, the meeting continued, with the conductor and set designer both in ill-fitting golf clothes, courtesy of Mr. Reis's closet.

Two years later, Stokey asked his shower mate to design a trio of productions, ranging from ancient to modern, for the new Philadelphia Theater Association. He offered a budget of $100,000, plus eleven men from his own Philadelphia Symphony Orchestra to play, and he'd person-ally conduct on opening night. Norman chose *Lysistrata*, the 2,500-year-old antiwar satire, to start.

The play had never been mounted in the States. Norman coauthored an adaptation, modernizing the dialogue, adding laugh lines, and graft-ing in portions from other Aristophanes comedies. For help with the racy, diaphanous costumes, he brought in Lucinda Ballard, at the start of her career, to collaborate with Frances, who now, among other respon-sibilities, was the go-to "color" person on all Bel Geddes & Co. projects. For the female lead, he hired rookie Miriam Hopkins; for the reprobate senator, rotund Sydney Greenstreet. It was one of the British actor's first jobs in America.

The set, more *Oedipus Rex* than *Lysistrata*, made the comedy funnier. The lighting would be singled out for its beauty. The direction, also by Norman, fell somewhere between earthy and gleefully vulgar; notewor-thy was a shock-inspiring costume mishap involving Greenstreet's toga that Norman saw fit to leave in. Censors insisted on the removal of seven words from the script, and police were assigned to listening posts in the auditorium.

An unexpected smash, it ran for seven months, with seats going for as much as five dollars and fifty cents (about seventy-six dollars today). Cus-tomarily facetious Robert Benchley called it "the most important New York theatrical event" of the season, even if it *wasn't* in New York.[299] A road tour followed. In 1930, two years after its premiere, *Lysistrata* opened on Broadway to equally glowing reviews.

Bolstered by its success, Norman set his sights on another update, a modern-dress *Hamlet* (six years before Orson Welles's modern-dress *Julius Caesar*), shifting from bawdy ancient Greece to dour medieval Den-mark. He was particularly keen on having Noël Coward in the lead. They met half a dozen times and talked for hours, but in the end, Coward claimed never to have acted a role he hadn't written. He had, however, found a unknown Canadian working abroad. Had Norman ever heard of Raymond Massey?[300]

Opening night at Broadway's Broadhurst attracted a diamond-studded crowd: Astors, Vanderbilts, Hearsts, Bernard Baruch and Otto Kahn, Judith Anderson and Gene Tunney, Eddie Cantor and Somerset Maugham, Richard E. Byrd (back from the Antarctic), Clare Luce and exiled Russian aristocrat turned perfumer "Prince" Matchabelli. There were lots of curtain calls and some excellent reviews. But Shakespeare wasn't Aristophanes, and *Hamlet* wasn't a bawdy romp played for laughs. The all-powerful *New York Times* critic Brooks Atkinson, who'd called *Lysistrata* "a triumph," was brutal.

•

LIVE THEATER WAS, and would always be, what Norman called his "fickle mistress." But with *Hamlet*'s failure, he removed himself from theater entirely. Plenty of industrial projects to keep him busy. Until the day Sidney Kingsley approached him.

The playwright was fresh from his first Broadway play, *Men in White*, a head-on confrontation with abortion and medical ethics that would earn him a 1934 Pulitzer. Kingsley's latest offering was about the crushing destiny of the urban poor, about boys with futures preordained by circumstance. His title said it all: *Dead End*.

Prohibition had spawned a distinctly American variety of criminal—the gangster. Some saw these colorful hoodlums—Capone, Dillinger, Pretty Boy Floyd, Baby Face Nelson, and others—as heroic figures who'd pulled themselves up by their own bootstraps. (After all, went the reasoning, outlaws rob with guns, but politicians and businessmen rob with checkbooks and fancy lawyers.) *Dead End* eschewed the romanticism. It focused on a group of mostly Irish youth, shrill, attractive urchins who inhabit the alleys and wharves near the Queensboro Bridge, swimming in the filthy river, playing cards on a stoop, roasting potatoes in a can, and generally trying to avoid their poverty and abusive fathers. Implacable circumstances lead to shenanigans that escalate into crime. "When I was in school, they used to teach us that evolution made men out of animals," says Gimpty, the unemployed young architect who grew up on the play's dismal streets. "What they forgot to tell us is that it can also make animals out of men."

The script was more journalism than fiction. Dead End was an actual place in Manhattan, a slum adjacent to the river perfumed with the wafting stink of nearby slaughterhouses. Everyone, even the owners of luxury apartments and private yachts who inhabit the play's periphery, was portrayed as a victim.

Kingsley intended it to shock. He saw theater as an opportunity to effect change; he'd go on to dramatize the horror of Stalin's purges. But his Pulitzer was no guarantee that people would pay good money to be reminded, in the midst of the Depression, of the desperate conditions around them. Most preferred the escapades of Mrs. Simpson and her Prince, or the Old Gold Cigarette Contest, with its astonishing $100,000 prize. For two bits, anyone could watch Nick and Nora Charles solve crimes over cocktails, see Clark Gable cavort with braless blondes, or escape into the candy-colored world of Disney's *Snow White*. Each week, 85 million Americans did just that.

At the same time he was working on Pan Am's Clippers and a diesel yacht for billionaire industrialist Axel Wenner-Gren, Bel Geddes gave nightly readings of *Dead End*'s script to potential backers, telephoning some as far afield as Europe. As producer, he was responsible for raising $25,000.

He also needed a venue. Morris Gest had inherited the Belasco Theatre from his father-in-law's estate. Twenty years before, the Bishop of Broadway, with his absurd priest's collar, had usurped the naive young Norman's lighting ideas. The temptation to outdo the famed arch-realist on that very same stage was irresistible.

Memories flooded in as Norman took the private elevator to the former impresario's office; with a two-year lease, it was his. Bereft of antlers, banners, Japanese armor, and Borgia trinkets, it was now a barren attic; farther up, off a narrow balcony, was the notorious bedroom where numerous amours had reputedly taken place.

•

ON SILVER SCREENS across the globe, New York was being depicted as a glamorous, idealized landscape—thrilling, mythic, more "New York" than the actual city could ever be. Norman set to work designing something in vivid contrast, an expressionistic urban slice that emphasized the disparity between haves and have-nots.

He spent hours with Kingsley studying the East Side, riding along in police cars at night, soaking up atmosphere and details. But his genius, in this case, was understanding that realism wasn't a matter of "duplicating reality." Replication was a recipe for dullness. He understood that in the same way that his abstract sets hadn't been strictly dictated by elimination or bareness, Kingsley's stark world required suggestion and illusion. The goal was to convince the audience that they were there. The solution was a heightened, hyperrealism based on exaggeration and distorted (even reversed) perspective—a solution, in this case, partly inspired by necessity. With a depth of less than twenty-three feet, the Belasco stage was impossibly shallow.

Hyperrealism, Henry Dreyfuss would later note, was something Norman used to laugh at.[301]

The original script simply called for waterfront *atmosphere*, but clearly the East River was as important a character as any of the actors. It just so happened that, thanks to a loan, Norman and Frances were now living at River House, a massive new art deco structure on East Fifty-Second Street, a fashionable address that ten years before had been a slum. The view from their second-floor apartment gave the designer an idea. Kingsley had envisioned the East River *upstage*, at the back. Norman convinced him to move it downstage, *all* the way down, into the orchestra pit. That way, the actors could dive into "the river" toward the audience (who sat, theoretically, in the river itself).

Most of all, he wanted the audience to feel the river's *presence*—its wetness, its sound—without having to harness the set to a hydroelectric plant. He spent four days in a sound truck recording steamboats tooting their horns, the slap of the river against the pier, the splash of a boy diving, the sputtering of a speedboat, street traffic passing along First Avenue. He organized a party on a local man's yacht anchored off East Fifty-Second for the boat party scene. Edited down to forty-five minutes, Norman's tape became the first prerecorded soundtrack used as continuous background ambience in a stage play.

Speakers were installed on both sides of the stage, in the orchestra pit "river" and in the auditorium, creating a complete aural environment. The set's "sidewalks" were paved with asphalt and the "street" with slate, so that the sound of a pedestrian's heels could be heard in the theater's back row. Norman's tape, with multiple soundtracks, would play on a loop. Sounds would also be keyed to specific actions, like the crackling of a bonfire in a trash can. Police radios and car sirens would call the audience back to their seats after intermission.

Kingsley, who was directing, also took over casting. There were forty-five roles, including G-men, policemen, ambulance drivers, and a bevy of adolescents, "obscene little toughies" who would stun both critics and the public with their "shocking" jargon and shameless "half-naked" naturalism.[302] Among them was eleven-year-old Sidney Lumet, the future filmmaker.

The role of gangster Baby-Face Martin's mother went to Marjorie

Main, an actress who, unbeknownst to anyone, was planning to kill herself. Devastated by the death of her husband, her career in a shambles,
she arrived at the audition unwashed, unkempt, unfocused. The part
called for a woman completely beaten down by life, her dead face punctuated with "dull, horrible eyes." Kingsley saw the potential within her
obvious distress. The role would save her life and relaunch her career.[303]

Lodged into the shallow stage, the exaggerated, foreshortened set gave
the impression of several city blocks, exploding the viewer's sense of
space. It featured a trio of dismal, decaying wharf-side tenements strewn
with junk. Clotheslines stretched from one side to the other, draped with
wrinkled laundry, and a sidewalk was cluttered with newspapers, empty
milk bottles, and coal. There was also a sand hopper, a steam shovel,
and a scaffold blocking out the sky. Then flush against the tenement and
dingy wharf was a palatial high-rise with terraced gardens—Norman and
Frances's River House. A ramp ran diagonally back from the orchestra
pit, gradually rising and narrowing up to a corner of the rear wall, adding
to the distorted perspective. It all seemed impossible, given the tiny stage.

Impressionistic lighting created a documentary quality. For the river,
eight 1,000-watt spotlights were focused on mirrors set into four pans of
water on the orchestra pit's floor. The water was agitated by electric fans;
the reflected lights created the shimmering effect of a lapping river. A recessed tank allowed the young hoodlums to literally dive from the "dock";
nets were rigged to catch them. Appropriate sound completed the illusion.
As they climbed back up a ladder, the boys were sprayed with oil and water
and draped with bits of garbage to simulate having swum in the polluted
river. Set, lighting, and sound were all carefully sculpted to enhance action
and character. The evening scene was illuminated by a realistic lamppost
at one end of the street. It had all been done for less than $3,000.

•

FRANK LLOYD WRIGHT, now in his seventies, was busy reinventing his career with his unavoidably brilliant Fallingwater house and the Johnson
Wax building. Over the years, when his contentious young friend had
plays in the works, he'd sometimes motored from Taliesin to Manhattan
to stay for a week or more attending rehearsals, afterward sitting "far into
the night" with Norman, cast, and crew, talking shop.[304] *Dead End* was
one of those productions.

Opening night was October 28, 1935. Rumors of the young actors' sordid language (the f-word was invoked) inspired Mayor Fiorello La Guardia to post plainclothesmen throughout the theater, as he had with *Lysistrata*, with instructions to shut the production down if need be.

Despite what would today be deemed melodrama, *Dead End* proved box-office gold, a 19-month, 687-performance, standing room only sensation—the longest in the Belasco's history, the tenth-longest run in New York theater—followed by a road tour to Boston, Chicago, San Francisco, and Los Angeles.

Norman's set would go down in theater annals as one of Broadway's most famous designs. Some nights when the curtain rose, the audience burst out in spontaneous applause. Other nights, they audibly gasped. Along with a set inseparable from the play was his magnificent lighting. A scene in which the half-naked boys clustered around a card game was, wrote one reviewer, "suffused with a kind of amethyst light in which the gleaming of the flesh was as realistic as Murillo . . . as sophisticated and beautiful as one of El Greco's effects."[305]

Norman was convinced that Brooks Atkinson, in his role as dean of New York critics, had almost singlehandedly undone his *Hamlet*. Not this time. His review called *Dead End* "enormously stirring"; Bel Geddes's contributions were "solid down to the ring of shoes on the asphalt pavement . . . push[ing] the thought of the author's drama ruthlessly into the audience's face." Not prone to using exclamation points, Atkinson remarked that the East River illusion was "so real" that "sitting there in mid-river" he found himself "paddling to keep afloat!" Taken all together, it was "a practical masterpiece."[306]

Even more gratifying was Norman's triumph over the imperious David Belasco. The designer's "surprising, enthusiastic and reckless duplication of actuality" had, wrote Stark Young, taken the master's realism legacy to new heights.[307] And in the process, it saved the theater, which Morris Gest had been in jeopardy of losing.

After attending a performance, Sam Goldwyn bought the screen rights for a record $165,000, half of which reputedly went to Norman. Five of the original Broadway "kids" were put under contract and ended up working with Jimmy Cagney and John Garfield. Marjorie Main was hired to reprise her role, opposite "son" Humphrey Bogart (who'd played a gangster the year before in *The Petrified Forest*). The result, based on a sanitized, romanticized script by Lillian Hellman, was nominated for four Oscars, including Best Picture.

In direct response to the play, the New York Police Department opened a Community Youth Center, sixty local clergymen requested copies of the script for use in their sermons, and Eleanor Roosevelt hosted the young actors at the White House, showcasing them in her radio and newspaper appeals for her charities, which led to the passing of the Wagner Housing Act, requiring safe, clean, low-income housing.[308] The play inspired citizens across the country to contribute to Boys' Clubs, introduced "dead-end kid" (rowdy innocents growing up in squalor) into the lexicon, and inspired a comic strip and, eventually, an entire genre.

•

IN FEBRUARY 1935, British author Rebecca West introduced Anaïs Nin to Norman and Frances. West, who was fond of Nin's banker husband, Hugo, would soon regret it. Joined by Raymond Massey, the five attended an ice hockey match at Madison Square Garden, followed by whiskey and sodas at the members' roundtable bar. Nin sized up the designer as "good-natured," "weak," and, like both Massey and director John Huston, who'd joined their table, "aflame with interest." When Norman told her that she looked like the women in Persian miniatures, Mrs. Bel Geddes's "weasel" nose had seemed suddenly longer, her "viperous" tongue "more venomous," because, of course, she was jealous.

"So many men there in love with me that I did not taste," Nin later wrote in her famous diary. Along with Massey, Huston, and Bel Geddes, there was a publisher, "an extremely handsome Russian Jew," a Cuban vice-consul, "an attractive blond giant," and a friend's son-in-law. Those she *was* tasting included her therapist, Otto Rank; Peruvian artist Gonzalo More; and her ongoing obsession, writer Henry Miller.

A year later, back in New York, Nin resumed her liaisons with Rank and Miller and attended a couple of "suppers" with Norman and Frances. "Bel Geddes was disappointed to hear I was not alone here and wanted to take me to Harlem."[309] The Harlem Renaissance was winding down but uptown was still *the* place for music, dancing, drinking, and brilliant floor shows. Maurice Ravel had asked Gershwin to take him there; Norman had taken Poiret and Eisenstein.

Americans, Nin was convinced, drank because they were terrified of intimacy; in her so-called Elysian world, stimulants were unnecessary. *Her* terror was her inexorable attraction to short, bald, myopic Henry. Deceiving Henry, who had his own infidelities, made her feel "less sentimental," even happy. Sexual "gymnastics" was her recipe for numbness, liberation from "the pain of love."[310]

There was no lack of options. Her current tasting list included a literary agent, a novelist, and a ménage à trois with a publisher and another woman. As for Bel Geddes, she didn't want him "as a man." Still, he was an artist, and a highly successful one—in the process of moving to a spacious nine-room flat on Park Avenue!—a designer of nightclubs, airplanes, and yachts, author of a bestseller. The papers and magazines were full of him. There was something else, too: Henry's great wish "to be on terms of fraternity," as Nin put it, with Broadway and Hollywood,[311] something Bel Geddes had in spades. People still talked about *The Miracle, Dead End* was breaking box office records, and Goldwyn's recent film rights deal was unprecedented (ergo the Park Avenue upgrade). Seducing Norman would make her "betrayal" of Henry that much more profound.

According to her diary, on February 15 she was in Miller's apartment, lying on her stomach on the bed, on the phone with Norman, "refusing for the eleventh time to go to Harlem . . . because I don't enjoy the debate over the bedtime plans," when Henry came in, slipped his hand between her legs, and took her, "with swift excitement," from behind.

By March, the novelist was letting his jealousy show. "Don't go to Harlem," Henry decreed. *Now he knew how it felt.*[312]

That Bel Geddes was about to embark on a ten-week trip through Mexico with Frances (for their third wedding anniversary) contributed to Nin's strategy. The longer she put him off, the more she could control how it would all play out.

Shortly before he was scheduled to leave, she finally agreed to Harlem. After drinks at the Ritz Bar, they headed uptown to the Kit Kat Club to see the floor show, and to dance. Everywhere they went, everyone seemed to know him. The entertainers sat at their table.

"*Down into adventure . . . to light and lascivious Negresses, to champagne and promiscuity.*" She leaves her phrases "*unfinished,*" spinning her web. Norman is tired of "*baby faces and uninteresting Broadway beauties.*" She is amused by his conversation, shot through with names, "*like on a marque.*"

At one in the morning, they drive to his office. *"The first collision is fiery,"* she will report. *"I am all body, all flesh . . . aroused by his vigor, by his sensuality . . ."* He is angry that she'd resisted him for so long. Time wasted. Places he wanted to take her. Sometime before dawn, dropping her at her hotel, he murmurs, *"You're marvelous."*[313]

Three days later, with Norman's departure imminent, Nin found an excuse to put him off. She wasn't in the mood. "Getting to know" the men one slept with only led to disagreements.

To hell with knowingness. Fuck. Fuck. Fuck.[314]

It was an elaborate itinerary for someone who rarely had use for a passport—Mérida, Chichen Itza, Uxmal, Oaxaca, Orizaba, Puebla, Tehuacán, Tehuantepec, Veracruz, Acapulco, Taxco, Mazatlán, Cuernavaca. Starting out in New Orleans, the couple proceeded to Mexico City to visit Diego Rivera. (Rivera's controversial mural for Rockefeller Center had since been destroyed.) They found him in low spirits, his head bandaged, battling a mysterious illness that was wreaking havoc with his eyes and kidneys.

Yucatán was a revelation, the most gloriously unexplored place in the world. A terraced pyramid, recently unearthed at Monte Albán, extended for three New York City blocks. The ruins at Chichen Itza and Uxmal left Norman spellbound, "bowled over" by their similarity to Barnsdall's Hollyhock House. He was pretty sure that Wright had never been here, yet his signature style was everywhere. Several picture postcards were promptly mailed to Taliesin: *"When did you do this?!"*

By mid-June, he was in Adrian, Michigan, where he'd lived until the age of six, to receive an honorary LL.D. from the local college. "As you probably know," he began, resplendent in cap and gown before the assembled, "I am, scholastically speaking, an uneducated person."[315]

•

ACCORDING TO REBECCA WEST, Nin followed up her Harlem adventure with a series of "quite obscene" letters.[316] Those that survive are merely flirtatious.

> *I heard from Pierre Matisse that you were back from Mexico very glowing. Aren't you coming to Paris now that you're inoculated?*
> *. . . [I'm] taking an apartment on the Seine—Always loved the river, it makes you feel you can skip down and out to other countries so easily. So I look at the river and . . . ask, what is Norman doing at this*

hour—He is a very serious man who knows how to laugh—I like that in Norman. We didn't talk enough or maybe we had other things to do. Would like to have that letter you tore up, or which perhaps you never wrote . . .

When will you take me out . . . dancing . . . again?[317]

•

BACK FROM ADRIAN, diploma in hand, he lost no time. His latest venture was scheduled to open at the Longacre in mid-October. But he managed to squeeze in an interview with *The Chicago Daily News*, during which he explained that his middle name was Belmont. As a young man he'd shortened it to Bel, only to find out later that he couldn't drop it without entangling his business affairs.[318]

His "steel as a protagonist" idea now manifested as *Iron Men*, "a pulp yarn," one reviewer would chide, about skyscraper construction workers and the "golden hearts that beat beneath their dirty jumpers."[319]

Doing triple duty as producer, director, and scenic designer, Norman created blueprints in lieu of set sketches. The actors would construct a "practical" derrick onstage ("not to exceed 2,000 pounds," his notes read), swinging steel girders, I-beams, and columns into place, and walk, in convincing peril, along a four-inch edge.

Iron Men had the advantage of opening a little over two weeks after *The New York Herald Tribune* ran a soon-to-be-iconic publicity stunt photo by Charles Ebbets, "Lunch Atop of Skyscraper," eleven ironworkers non-chalantly posed—no harnesses, netting, or protective gear, two of them shirtless—850 feet above Forty-First Street, feet dangling from a beam of the future RCA building. But what caught the press's attention was Norman's unorthodox casting. The lead role of Andy, a smart-aleck-but-idealistic blowhard, went to William Haade, six-foot-three, 220 pounds, an actual construction worker with no acting experience but movie-star looks ("astonishing . . . undoubted catnip") who handled the part "with all the confidence and force of an old Equity boy."[320] If Hollywood has any sense, noted one reviewer, Haade had driven his last rivet.

Rudolph Kommer had warned Norman, back in 1924, to avoid theatrical producing.[321] The set shared Haade's accolades, but not so the play, which lurched along for a grand total of sixteen performances, at a loss of "some $32,000" ($547,000 in 2017).[322]

•

MAX REINHARDT HAD headed Berlin's Deutsches Theater for three decades when Adolph Hitler offered him the title of Honorary Reich Artist, inviting him to take charge of all Reich productions—a magnanimous offer given that Max was a Jew, if not a practicing one. Upon refusing, he was banned from Germany. He left light-handed, his finances having been depleted by, among other things, an open-air theater (sketched by Norman during his 1924 visit) he was planning on Leopoldskron's grounds to the tune of 50,000 Reichsmark. (Max's bookkeeper compared him to Marie Antoinette.)[323]

Similarly, Fritz Lang left Berlin the day Goebbels told him he and Der Fuhrer had seen *Metropolis* and decided Lang was just the man to create the quintessential Nazi film. Max and Fritz, both of whom headed for the States, were among the lucky ones. Between 1933 and 1945, the quota of Eastern Europeans allowed into the United States was blocked.

In April 1933, a group of American producers, actors, drama critics, and teachers, Bel Geddes among them, sent Hitler a cable "expressing indignation" at Reinhardt's ousting. Having read the news, Meyer Weisgal sent a cable to Paris, where the exiled maestro had stopped to direct *Die Fledermaus*. "If Hitler doesn't want you, I take you."

Weisgal, an American Zionist, had recently produced The Romance of a People on Jewish Day at the Chicago World's Fair. Increasingly concerned about the rise of National Socialism, he had his eye on a play he hoped would arouse public interest in Jewish traditions before they were crushed entirely—Franz Werfel's *The Eternal Road*, a five-hour retelling, with dance, music, and pageantry, of Old and New Testament stories. Set in a synagogue during a pogrom, it began with a congregation awaiting news of their possible exile and ended with their exodus, the start of a new chapter in an ongoing diaspora. The parallels to what was happening in Germany were obvious.

Max wasn't interested. "An opportunity for DeMille which, alas, I could not seize for all the De Millions in the world," he confided in a letter to his son. He wasn't averse to the theme but didn't see how the current Jewish plight could be helped by "rose-colored trash."[324] From Paris, Max made his way to Los Angeles, joining an extraordinary brain trust of European emigres (Jascha Heifetz, Aldous Huxley, Billy Wilder, Thomas Mann, and others), who were waiting out the war amid stately palms, ocean views, and night-blooming jasmine.

Despite been assured by friends that Reinhardt lived on champagne and caviar, smoked three-dollar cigars, and traveled with a retinue "rivaling a Turkish pasha,"[325] Weisgal was determined to have him.

Max wasn't the only one with doubts. The playwright's wife, the infamous (and militantly Christian) Alma, "the Midwife of Genius"—her lovers and husbands had included Rainer Maria Rilke, Gustav Klimt, Oskar Kokoschka, Gustav Mahler, and Walter Gropius—also opposed the venture.

But promised a free hand and generous terms, Reinhardt eventually capitulated (money was tight, his days of champagne and caviar behind him) on the condition that Kurt Weill, the raffish creator of *Three Penny Opera*, be imported from Paris (he, too, had fled Germany, within hours of being tipped off by friends of his imminent arrest) to compose the score, and that Bel Geddes come aboard as set, costume, and lighting designer. Alma's opposition was placated with Benedictine brandy (her favorite), an ample supply of which was made available, despite war restrictions, thanks to Weisgal's connections.

So it was, observed Gottfried Reinhardt, that his father, a "skeptical, left-wing materialist," found himself in league with a Roman Catholic "zealot" (Werfel), a "fiercely nationalistic" Zionist (Weisgal), a "Jewish atheist with Marxist leanings" (Weill), and Norman, "a waspish American"[326] to create an Old Testament drama. From the ill-matched team's first meeting, the writing was on the wall. Weill, who envisioned an operatic oratorio, was offended by Werfel's pretensions, his huge bulk, and his idea of music as "background." Weisgal, meanwhile, saw the venture as a fund-raiser for victims of fascism, putting the whole issue of salaries into question.

Max and Norman considered staging it in a tent in Central Park. Spectacles were traditionally staged in tents. But city authorities refused to issue a license, and besides, how would they heat it? Then there was Radio City Music Hall, with its revolving stage and elevator-operated orchestra pit. According to Weisgal, Max would sit in the front row "all but shedding tears" at the thought that it was being desecrated for "piffling amusements." The Hippodrome was a more likely option, but no sooner had Weisgal put down a $5,000 option than Billy Rose nabbed it out from under him for *Jumbo*, a musical about a financially strapped circus. The Manhattan Opera House on West Thirty-Fourth Street, which Max dubbed a "living corpse," was the only available venue.

Kurt Weill arrived with his wife, Lotte Lenya, who would play Miriam. Sidney Lumet, now thirteen, was cast as the young Isaac, who cries out "Father!" when Abraham lifts the sacrificial knife. Sam Jaffe would play the Adversary; future playwright Horton Foote was in the ensemble. Also on board was Rosamond Pinchot, cast as Bathsheba, King David's queen, happy to be working again with Norman and mentor Max.

After two years playing *The Miracle*'s Nun in New York and at the start of the national tour, making headlines as she went, Rosie had performed as the Amazon Queen in *A Midsummer Night's Dream* at the Salzburg Festival, a role for which she'd mastered German, and as the paramour in Von Hofmannsthal's *Everyman*, staged in an Edinburgh abbey. Once back in the States, determined to succeed on her own merits, without the Pinchot name, she'd traveled incognito to California, even worked in a cannery for a while. In the end, feeling rudderless, she'd returned home, married, and given birth to two sons. But she was restless and deeply unhappy. Her husband, Big Bill Gaston, so perfect on paper, had proven himself a toxic bully.

•

THE ETERNAL ROAD'S opening night was originally set for December 23, 1935.

Werfel's script called for ninety-seven different scenes; he'd done his best to cram the entire Old and New Testaments in. Norman designed an enormous series of platforms and ramps (the eponymous "road") that began twelve feet below what had been the orchestra pit (now the synagogue), then rose up another five stories to the portals of Heaven, where a choir of 200, in angelic robes, would sit in full view.

Elevators were built in a semicircle at the rear of the stage to transport scenery. Some twenty-six miles of electric cables and wiring were installed, and thousands of custom lights, $60,000 worth, enough current, the *Times'* Bosley Crowther would note, to illuminate the *Queen Mary* on a dark night. Speakers were installed around the stage, in a vertical line up and over the audience and under the central seats, in anticipation of the terrifying voice of God, the play's penultimate moment. A prerecorded, high-fidelity score was ultimately abandoned, replaced (at the insistence of the musicians' union) by a live orchestra, at an alleged cost of $68,000.[327]

Everything had been arranged based on a synopsis. The first pages of the final script arrived in German. Conferring with the translator was problematic, as he was living in Vermont with his mistress, in retreat from a wife waiting in New York with a summons.[328] Then Reinhardt announced that the synagogue was too small and shallow ("the contrast between the world and the underworld must be emphasized"), a decision that required the removal of three or four orchestra row seats—Weisgal's "best" seats, his profit margin. In the course of this expansion (which allegedly consumed $75,000), Norman's workers hit a water main, a scene the beleaguered Weisgal compared to Moses splitting the Red Sea. A twenty-one-foot geyser flooded the pit, the auditorium, the costumes and the set before it could be capped and pumped out.[329]

"Even genius has its limitations," Weisgal protested. Norman had "little feeling" for the play's spiritual content or relevance, though admittedly he "threw himself in with ardor," his "techno-megalomania" being "akin to God's alteration of the universe during the six days of creation, except that God's was cheaper and faster." At his wit's end over delays and expenses, Weisgal fired Bel Geddes "before old age overtook us both."[330] Not long after, rehearsals were shut down and the cast dispersed.

It took a year for Weisgal to rally with additional sponsors. Albert Einstein was recruited to speak at a Waldorf fund-raiser. In the interim, Reinhardt returned to Los Angeles to direct a lavish film version (cast of 1,000, with a young Mickey Rooney as Max's "best, pagan-wildest" Puck,[331] and James Cagney as Bottom) of his recent *Midsummer Night's* triumph at the Hollywood Bowl, which had played to nightly audiences of 12,000. Back in New York, Max telephoned Norman. He was skirting Weisgal's veto. They needed him. It was just after Christmas 1936, a matter of weeks since *Iron Men* closed, though *Dead End* was enjoying a second bonanza year.

Eternal Road, which had by now been postponed ten times, was slated to open in five days. Eight, Bel Geddes countered, if we're lucky. Weisgal had managed to raise $150,000. Norman added another $5,000, an advance on the $11,000 the show's original backers owed him, and he relinquished any additional salary.

Dress rehearsals were long and arduous; one lasted until four in the morning. There were 245 actors, 40 dancers, the celestial choir, and a total of 1,772 costumes (designed by Frances) to contend with.[332] Sidney Lumet, whose voice had changed, was relegated to crowd scenes, replaced by eight-year-old Dick Van Patten. The morning of the premiere, the theater still lacked carpets, a box office, and, worse, public bathrooms, and the elevator installed to raise the rafter-high columns of Solomon's Temple became irretrievably stuck halfway. Max had the wisdom to eliminate the scene entirely. At 7:30, the final dress rehearsal still in progress, a skirmish broke out between the dancers (mistakenly rehearsing a number cut the week before) and the costumed warriors, the confusion made worse by the flickering chiaroscuro—Norman was still in the process of determining the complex spots, gels, mixes, and fades of Act Three.

Then, with half an hour to go, the fire department turned up, claiming inadequately treated wood (the staging involved considerable candle-light) and questionable wiring. At this point, Weisgal was next door at the New Yorker Hotel (he'd rented a room for opening night), preoccu-pied with the fact that his wife had neglected to bring his patent leather shoes—perhaps there was hope, if he could line up all the bellboys and try on their footwear—when news arrived of the imminent shutdown. Weisgal's attorney, the estimable Louis Nizer, managed to track down Mayor La Guardia just in time to have him green-light the evening. However, forty firemen were ordered to be on hand, extinguishers at the ready, at every performance until the details could be settled, another not-insignificant expense.

Despite the fact that everyone was exhausted before the lights even went up, all went brilliantly, breathtakingly, for the first three acts, by which time it was midnight, and intermission. Act Four ran until 2:00 a.m. Fortunately, the reviewers had left after Act Three in order to get their ecstatic reviews in on time.

Atkinson deemed the production "a deeply moving experience." Bel Geddes, he wrote, was "a superman in his own right," and though the resulting "ballyhoo" focused more on the massiveness of the enterprise than the music and text, the "grandeur of line and imagination" charac-terized his "mightiest" work.[333]

Against the odds, Weisgal had a hit on his hands.

But *Eternal Road*'s eternity proved short-lived. The $28,000 weekly "take"—every performance a full house, with standing room only—was exceeded by weekly expenses. Weisgal worked valiantly to cover the defi-cit, even persuading the archbishop of New York's diocese to lift the ban on Catholics attending the theater during Lent. Privately, he admitted that the costs, more than half a million, hadn't entirely been Bel Geddes's doing.

About three-quarters of Broadway shows fail financially, *The New York Times* would note seventy years later.[334] And *Eternal Road* wasn't *Fiddler on the Roof.* If Weisgal hadn't known from the get-go, Norman and Max certainly had: Broadway is a devil's bargain. Neither the press nor the public harbor sympathy for those who take the risk.

The play opened January 7. Weisgal's dream came to an end on May 15. The final performance, number 153, was a benefit for him; everyone worked without pay. His resources depleted, Weisgal had been forced to give up his home, moving his family into a fleabag hotel, with cockroaches to keep the fleas company. He'd even sold off his brother-in-law's beloved stamp collection, a decades-long endeavor. He was being sued for unpaid bills and unpaid salaries. That his children would have college educations was, at this point, questionable.

Billy Rose's *Jumbo* at the usurped Hippodrome—the 5,000-seat venue converted into a vast circus tent, complete with acrobats and wild animals—had closed after 233 performances, with a loss comparable to Meyer Weisgal's entire much-inflated outlay. "My ultimate consolation," Weisgal recalled years later, "was that critics said *Eternal Road* made *Jumbo* look like a pygmy."[335]

CHAPTER 15

Birth of a Classic, Death of a Beauty
January 1938

In our town, we like to know the facts.
—NARRATOR IN *OUR TOWN*

omewhere between his ill-fated *Hamlet* and the fortuitous meeting
with Sidney Kingsley, Bel Geddes had approached Thornton Wilder
about acquiring something "modern." The playwright, who'd seen
a number of Norman's productions (including *Hamlet*, which he
liked), agreed that his next project would be something specifically for
him, perhaps something in radical contrast to Norman's reputation as
a scenic wunderkind—*something that required no scenery at all!* Not that
Norman hadn't already done that, using only light and props. But he took
it as a promise and a challenge.

A year later, Norman's "obstinate insistence" on a modern piece still
had Wilder stumped, though he remained intrigued.

"You are probably the greatest designer since the Baroque Age," he
wrote from New Haven in the fall of 1932. "But you start being creative
too promptly after reading your text. You should count ten."

That same year, Wilder wrote a British friend:

*I have a sort of urchin's hero-worship—urchin watching brilliant lion
tamer in spangles—for Norman Bel Geddes.*[336]

Four years later, in the wake of *Dead End*'s huge success, Norman
sent off another reminder. "I wish I had a play to send you, but if I had,"
Wilder wrote, skirting around the notion of a "modern" piece, "it should
be full of costumes designed by yourself, so handsome that they would
never be forgotten. You should keep your hand in for such investitures,

because Christian Berard is getting to be the best in the world, merely by your default."

"Now that you're so rich," he added, somewhat contradicting his "count ten" advice, "you should let loose the thunders and lightnings of your imagination."[337]

The following September, Wilder wrote again, this time from Switzerland. After visiting with Gertrude Stein and Alice B. Toklas in Paris, he'd headed for the Salzburg Festival and enjoyed late-night suppers with Max Reinhardt at the Schloss, where Norman's name had been spoken of with enthusiasm.

Now all I'm doing is work. Haven't spoken to a soul except waitresses (Swiss gals are so plain, they don't constitute no distraction. Very restful.) . . . Had a great time at G. Stein's having foolish ideas driven out of my head. A juggernaut of commonsense.

He was, he added, returning home with three plays.

Until then, I have a superstitious feeling that it's bad policy to even discuss 'em. Maybe they're all great big fat lapses of judgement. All I'll say for now is that when I build cloud castles, I know which one I want you to read.[338]

What he didn't say was that before leaving Paris he'd met with enfant terrible Jed Harris, whom he'd known since their student days at Yale, and signed a deal for a still unfinished three-act. One whose opening line would read "No curtains, no scenery."

•

NINE YEARS BEFORE, Jed Harris had leapt, overnight, from anonymity to theater legend and millionaire with an unprecedented coup—four Broadway productions running simultaneously, and every one a smash hit, a feat never to be matched. A risk taker, he burst on the scene not unlike young Orson Welles a decade later. (Norman, who'd designed sets for Harris's production of *Spread Eagle* in 1927, thought him an "extraordinary" interpreter, with talent to burn, and had considered going into partnership with him.) That September, Harris made the cover of *Time*—

slouch-shouldered, hands in his pockets, hair slicked back, an enormous sensual slash of a mouth. He was twenty-eight years old.

When Wilder arrived back in the States, Jed met him at the dock. Then he "imprisoned me," the playwright wrote Stein, in a Long Island cottage he was renting, "and commanded me to write . . . the play's getting better every day."[339] The cottage in question was in Old Brookville, an easy drive from Manhattan. It was also, not coincidentally, a few doors down from the low, rambling farmhouse on Valentine's Lane that Rosamond Pinchot, estranged from Big Bill, was renting for herself and her two young sons, now nine and six. On Sundays, she'd taken to hosting luncheons there, attended by scores of theater, literary, society, and political friends.

Once Wilder realized the subterfuge—that he was there to cover the fact that Jed was seeing Rosamond—he bolted.

•

THE SIGNS HAD been there early on, had anyone had the prescience to decipher them.

In 1926, as a favor to Rosamond, Norman had designed the sets and lighting for her fiancé's first and only play. Big Bill Gaston's *Damn the Tears* was a semi-autobiographical hodgepodge about a lawyer from an old Boston family, a former baseball player who becomes homeless and loses his mind. The reviews were scathing, the play's only redeeming factor being Norman's sets, "so uniformly fine," wrote one reviewer, "that I left with a wild hope that someone would write an actable play in which they might be used again."[340]

Once married, Big Bill revealed his true stripes—selfish, angry, and cruel, a philanderer with little interest in his profession or his marriage. Devoid of creative talent, he resented his wife's successes. Short on friends, he resented hers—Eleanor Roosevelt and Elizabeth Arden, George Gershwin and Cole Porter, Gary Cooper and Myrna Loy. It was Rosamond, not Big Bill, who got invited to swim and play croquet with the Algonquin crew on Woollcott's private island. He couldn't even claim superior sports prowess with a wife so effortlessly athletic. In the summer of 1932, following an abortion, she went to Reno seeking a divorce, an exceedingly bold move for a young society woman and almost impossible to keep discreet. *Time* noted it in its Milestones column, just below the birth of the Charles Lindbergh's second son.[341]

But William Gaston, of the venerable Boston Gastons, wasn't interested in divorce. As the estranged husband, he had everything he needed—freedom, social standing, access to Pinchot money. He had a lifestyle to maintain.

He was adept at keeping "the sugar-coated bitch" off-balance—giving out his home phone number to women, leaving lipstick-stained handkerchiefs where they'd be easily found, orchestrating reconciliations, then berating his wife for getting fat. In an effort to placate him, she subjected herself to a brutal lettuce, buttermilk, and date diet, augmented with "glandu-

Rosamond and a friend playing leapfrog in Bermuda. Town & Country Magazine, 1932. *(Courtesy: Bettmann/Getty Images, photographer unknown.)*

lar" treatments—injections of monkey or fetal sheep hormones—the cure de jour for high-society moms.

Rosie and Norman had kept in touch since *The Miracle* days. He was on her party list, and she attended his events. Among the latter was a lavish supper dance celebrating Bel Geddes's completion of the Hotel Pennsylvania's Manhattan Room, where Benny Goodman would later play. "It has all the virtues," cooed the *New York World Telegram*, "modernistic" without being "daffy."[342] (Synthetic daylight streamed through Venetian blinds.)Rosie, already separated from Big Bill, mingled with an elegantly dressed constellation of guests that included Mr. and Mrs. Irving Berlin and Mr. and Mrs. William Paley, Edna St. Vincent Millay and Margaret Bourke-White, Rudolph Kommer and Cole Porter, Walter P. Chrysler, Jr., and Conde Nast, Harold Ross and Bennett Cerf, Ernest Lubitsch and Otto Kahn.

After nearly a quarter century, Kahn had retired from the Metropolitan Opera, tired of fighting the good fight.[343] It was the last time Norman saw him; Kahn died the following March, collapsed at his desk.

•

SHORTLY AFTER THE Manhattan Room soiree, Rosie had headed west again, leaving her sons with their grandmother. Though too tall by movie industry standards, she was cast as Queen Anne in RKO's remake of *The Three Musketeers* (young, vibrant Lucille Ball played her lady-in-waiting). Which was when she met Jed Harris, at a Hollywood party.

Harris, thirty-five to her thirty, was awaiting his next big break. Brilliant and successful as he was, he'd recently staged the worst fiasco of his career, a play featuring Kate Hepburn as a spoiled young woman whose husband, whom she doesn't love, drowns on their honeymoon—the role that elicited Dorothy Parker's famous quip, "Miss Hepburn's performance ran the gamut from A to B."

The former Jacob Hirsch Horowitz was known as much for his tyrannical methods and demonic rages as for fashioning theater gold out of "dead" scripts. Noel Coward, who early on dubbed Harris "destiny's tot," compared him to a ruthless praying mantis. Rumor had it that Harris had been the inspiration (both his temperament and looks) for Walt Disney's Big Bad Wolf.

But Jed was also witty and charming, a brilliant mimic and infectious storyteller who could speak authoritatively on everything from baseball and French cuisine to Ibsen and Chekhov. He played violin and piano, could dazzle with card tricks, could even be a mentor when it suited him.

Rosamond, who never lacked for suitors (George Cukor, "Dave" O. Selznick, Bennett Cerf, and Sinclair Lewis among them), wasn't interested in all the horror stories.

It wasn't long before Big Bill got wind that his wife was "seriously" involved with Harris, who was, among other infuriating things, Jewish.

That was when Rosamond's diaries went missing.

Returning home from California, she learned that in her absence, someone had broken into her mother's townhouse on East Eighty-First Street and stolen all her leather-bound notebooks, a decade's worth of private thoughts. Few people knew she even kept them, fewer where they might be.

Bill had thwarted his wife's previous divorce efforts with a cocktail of mind games, character smears, and sex. More than jealous of her independence and success, he was afraid of what she might say about him. Divorce would necessitate the public airing of his considerable dirty laundry. There was nothing illegal about an estranged husband sowing his oats, but few knew he preferred his women black, and from Harlem.[344]

Now, with a potential groom in the picture, he reasoned that the diaries would provide evidence of *her* infidelities. If worse came to worst, he could release them (carefully excised to his advantage) to the press. She would, he knew, do anything to protect the Pinchot family name. This time, he had her.

•

OUR TOWN, THE play with no curtains and no scenery, was scheduled to premiere on Saturday, January 22, 1938, at the McCarter Theatre in Princeton, New Jersey, followed by a short Boston run, before opening in New York. Rosamond had handled props for Jed's recent production of Ibsen's *A Doll's House*, but now there were no props to speak of, and no parts for a tall, patrician, thirty-three-year-old female, so she took over the sound effects—train whistles, thunder, cricket chirps, owl hoots and rooster crows, the chiming of the town clock, the clomp of horses' hooves, the jiggling of milk bottles.

Rehearsals began just after Christmas. As did the skirmishes between writer and producer. Harris was convinced the play was "terribly overwritten" and set about mercilessly pruning. Wilder was convinced Harris was "shipwrecking" his intentions, robbing the script of its "cosmic overtones."[345]

Jed, like Norman, was a perfectionist known for running on adrenaline, in lieu of food or sleep, during the final days of prepping a show. Adding to that, he'd invested over $400,000 of his own money in the production.[346] Miserable weather, a scheduling mistake, and technical problems caused the Princeton dress rehearsal to be repeatedly postponed. It finally began at midnight, running until eight in the morning of opening day. The stagehands union, meanwhile, had gotten wind that the actors were carrying planks and chairs, and threatened to shut the production down. Jed's explanations of "experimental" staging fell on deaf ears.

Everyone was exhausted, but Rosamond appeared to be in bright spirits. She and Jed had been up all night on Benzedrine.

It was a sold-out house. The stagehands the union forced Jed to hire were sequestered in a room with a pinochle deck, ordered to stay clear of the stage. But moments before the lights were scheduled to go up on

Act One, two burly men stood on either side of the stage, each holding a chair. Jed approached one and told him to put it down. Couldn't, he said. Union rules.

There was little time for amenities. Harris punched him, hard, in the jaw, raced downstairs, crossed under the stage, darted up the stairs to the opposite side, yanked the other stagehand back by his shirt, then crooked an arm around his neck and squeezed.[347]

All things considered, the premiere went remarkably well. *Variety* called it "the season's most extravagant waste of fine talent" and wondered what the "wonder boy of Broadway" saw in it, but the audience laughed and cried. Seated in the audience was actor José Ferrer, who would re-member the performance as "a miracle of understatement" with Jed's unique stamp all over it.[348] (It was Harris who invented one of the play's most memorable images, mourners standing together in dark clothing, holding black umbrellas, their backs to the audience.)

Shortly after the final blackout, Harris asked "Thornie" what he thought. To his surprise, the playwright said he was pleased overall . . . but there was still something about the final scene that Jed just didn't understand. Harris flew into a rage. He didn't understand the cemetery scene? He'd "spoiled" Wilder's "shining prose"? Just then, Ferrer came backstage to find Harris looking like a train wreck, beyond exhaustion, ashen-faced, his eyes bloodshot, his hands around a cable as if holding himself up by sheer will. Rosamond's congratulatory kiss was rebuffed, as was her expectation that they'd head out for the cast party.

It had been "one long fight" to preserve his play from Harris's "in-terpolations," Wilder would complain to Gertrude Stein a few days later. That's why it had failed. He washed his hands of the whole business. "All I want out of it is money."[349] Rosamond drove home, alone, to the Long Island farmhouse.

•

SOMEWHERE BETWEEN DESIGNING his Skyscraper Cocktail Set and other as-sorted pieces for Revere, organizing an ambitious "City of Tomorrow" campaign for Shell Oil Co., and working on a venue to replace the Al-gonquinites former meeting place for Alex Woollcott, Norman designed and produced an Irwin Shaw three-act, *Siege*, about the ongoing Spanish Civil War.

"He made the goddamndest set you've ever seen," noted George Woods, who'd helped obtain backers. A mountain and a cave on a turntable—by the time the actors descended from the first set, the scene had rotated and they appeared in the second. Woods claimed that two nights before the opening Norman had kept the crew up until all hours boring holes and fixing "rocks" with wires so that they'd fly off, for added realism.[350]

Siege opened December 7, 1937. The following night, Norman attended a costume party dressed as an undertaker, the play's reviews pasted on his clothes.

The chances are good that, fresh from that latest disaster—as producer, he lost a lot of money—Norman attended the Princeton opening to see the "no scenery" play he felt he'd been promised, the so-called cloud castle he and Wilder had been talking about for more than five years, only to be passed over, and to see what wunderkind Jed Harris had done with it.

If Norman *wasn't* at the McCarter Theater that evening, then Rosamond phoned him at some point. Or possibly he phoned her. Because he knew she was upset at having been "stood up" by Jed and that she saw no point in going on to Boston, where the cast and crew were scheduled to rendezvous the following day. The fight to get the Pinchots to accept both the prospect of divorce (from a "respectable" diary-stealing bully) and remarriage (to a theatrical Jew) had been protracted and emotionally draining. Now here she was, alone, on opening night, on Valentine's Lane.

Norman arranged to take her out dancing the following evening, somewhere elegant, to make up for Jed's slight.

Rosamond's cook and the governess would later verify that she spent Sunday, the day following the opening, at home. After dinner, she put William and James to bed, changed into a white evening gown, silver slippers, and an ermine wrap, and drove into Manhattan.

Where would he have taken her on a Sunday night? Billy Rose, fresh from his *Jumbo* disaster, was still putting the finishing touches on his lavish Diamond Horseshoe Club. Café Society, where Billie Holiday would headline, was also getting ready to open.

Perhaps downtown to the Stork Club, with its famous golden rope, or to El Morocco, with its zebra-striped walls, places he, Frances, and Rosamond knew well. Or uptown to Harlem, where they were less likely, given Rosie's state of mind, to run into familiar faces. To the Cotton Club, with its "50 Sepia Stars" floor show, to Connie's Inn or the Kit Kat Club, scene of the ill-fated tryst with Anaïs Nin. Years later, in his private notes,

Norman would specifically recall getting his eye scratched that evening by the rim of someone's hat.[351]

Sometime before dawn, Rosamond's Buick pulled into the garage of the Long Island farmhouse. Whether she wrote the two notes then or had already composed them during her leisurely afternoon, whether she'd said anything to Norman that evening that might have remotely presaged her plan, will never be known.

She placed the notes on her bed and may have looked in on her sleeping boys. Or not. Then she returned to the garage, her silver shoes crunching against the frozen gravel, the ermine wrapped tightly around her—statuesque, fair-haired, a figure in white moving through the cold, dark quiet under a three-quarter moon.

She closed the garage door, connected the garage hose to the Buick's exhaust pipe, then threaded it through the left rear window into the passenger compartment.

After stuffing the window chink with a piece of burlap (a Montreal paper would elevate the burlap to a "lap robe",[352]) she started the motor and lay down on the backseat.

The first press reports were quick to focus on her clothing, suggesting a connection between the white evening gown and the wedding dress in *Our Town*, the one Emily wears to return, for one day, from the dead. "Miss Pinchot's Suicide Follows New Play Theme" proclaimed *The New York Daily News*. "The play . . . makes death a beautiful thing."[353] The *Daily Mirror* spoke of "the self-slain" Rosamond's "self-made auto death chamber" and ran several pages of photos under the headline "In Ermine, Rosamond Pinchot Exits from Life."[354] Amos Pinchot managed to control his shock and grief long enough to "correct" the official account. Aside from the fact that he was in the midst of a political campaign and that his favorite child had just done something unthinkable, the clothing issue was entirely too provocative and raised unanswerable questions.

The nineteen-paragraph report that ran front-page-center in *The New York Times* noted that Rosamond had remained at home that last evening with her young sons, that she was found wearing "slacks, an old sweater, woolen gloves and bedroom slippers," and that she and her husband had "separated amicably" two (not five) years before. The article went on to mention her "sports clothes" a second time, as if repetition would allay any suspicions.[355]

The story quickly went national, with the actress's "final curtain" a favorite metaphor. The news reached the cast, in rehearsal in Boston, Monday afternoon. There were rumors that Gifford Pinchot, the governor, was arranging "a mob hit" on Jed Harris, who, for his part, was maintaining a very low profile.

Despite Amos's best efforts, Rosamond's attire took on a power and life of its own. Conflating Grey Towers, the Pinchot's Pennsylvania country retreat, with the Long Island farmhouse, Max's son Gottfried would later insist she'd been found in her riding clothes, suggesting a last, mad, icy gallop, like Artemis beneath the moon.

The funeral took place on the Gastons' tenth wedding anniversary. Big Bill flew in from Colorado, sporting an expensive raccoon coat on one arm, a blond "friend" on the other. He reportedly wept through the service.

The bronze casket was covered with a carpet of snapdragons, white orchids, lilies, and branches of apple blossoms, the latter imported from God knows where in the dead of January. Encouraged by pressure from Amos and Gifford, the police file had been quickly closed: suicide caused by depression and fatigue. There was no autopsy, and no eulogy.

Max Reinhardt who, while devoted to his second wife, had been in love with his "discovery" for years, was unavailable for comment. Less than a year before, Rosie had been "walking the boards" as the exquisite Bathsheba—her third "queen" after Hippolyta and Anne. Had Max done her a favor, launching her out into a world so different from the one she'd been born to? Rudolph Kommer, Max's unflappable longtime lieutenant—he'd seen Rosamond just five days before, "in the best of health and in her usual high spirits"—was in shock.[356] Morris Gest exploded in front of the reporters gathered outside the church. "I can't understand it. My God, she was dynamite! She could run faster than someone else could skate. She was so full of life!" Bel Geddes was not among the mourners.

Having read the obituaries, a witness contacted the family. Driving back from New York that last time, at about 3:00 a.m., Rosie had picked up a hitchhiker, a retired World War I flying ace with a lame leg. The air was brittle, freezing. Colonel Harold Hartney had missed his ride and was walking in the snow. She'd asked for identification, then insisted he sit in the backseat. They got to talking and discovered a couple of mutual friends. When she stopped to light a cigarette, he noticed her hands were trembling. When she dropped him off at the train station, he'd told her she'd probably saved his life.[357]

Still, no one knew why, shortly before her death, Rosamond Pinchot was driving around in the middle of the night on dangerously icy roads, dressed in an evening gown and silver shoes. Norman, grappling with his own shock and grief, had no good reason to step forward.[358]

•

THANKS TO TEPID reviews (and news of the Pinchot suicide), *Our Town*'s Boston run was cut short. On February 4, 1938, three days after Wilder wrote to Gertrude Stein bemoaning his fate, the play opened at Manhattan's Henry Miller Theatre.

The final blackout was met with an eerie silence. Then the audience erupted in cheers, whistles, and a standing ovation. Sam Goldwyn and actress Bea Lillie, both in the audience, were spotted weeping.[359] The play would run for 337 performances, during which no stagehand touched a chair. Thanks, in part, to Jed Harris's interventions, the play won Thornton Wilder his second Pulitzer. It would go on to become an American standard, one of the century's definitive plays, earning Wilder more attention and money than anything else he would write.

CHAPTER 16

Too Goddamned Waldorf Astoria: The Exhibit That Nearly Wasn't

1936–1940

Why do it if it can be done?

—GERTRUDE STEIN

uturama, the most iconic World's Fair exhibit of all time, had its start in Bel Geddes's office on East Thirty-Seventh Street, "an awful dump . . . a firetrap," his secretary would recall, with lousy lighting and "plaster falling out of the ceiling."[360]

Instead of downsizing during lulls between jobs, Norman had made a habit of giving his staff "development work," hypothetical problems like the ones that had evolved into his Airliner #4. *Get a thousand luxury lovers from New York to Paris fast. Forget the limitations.* Now he set them the daunting task of redesigning Manhattan's traffic.

How could cars switch highway lanes without crashing? What was the best way to turn left at a cloverleaf? What could be done about road hogs, poor night vision, pileups?

In 1936, the J. Walter Thompson agency began looking for an idea for Shell Oil Company's next national advertising campaign. The goal was to spur a five to ten million annual increase in auto sales, with every vehicle thirsty for Shell's new "motor-digestible" gasoline. The sketches and blueprints Norman's draftsmen had created—high-speed lanes, viaducts, control towers, and a variety of bridges—were an ideal "fit," if done on a large scale.

It was decided that a section of midtown Manhattan would be recreated as a six-foot-long model. Norman and his staff pored over a decade's worth of data on auto registration, population trends, city planning, and street traffic. The project quickly grew to include issues of housing and lifestyle.

A longtime advocate of tall buildings, Bel Geddes imagined structures like the Empire State Building expanded to occupy a full block, containing the same floor space as numerous scattered ten-story buildings, an idea that would free up space for parks and open country; landowners could pool their interest and receive dividends. Everything a small city needed to function—fire, hospital, and police departments, restaurants and picture shows—could be housed under one roof. Elevated walkways could separate pedestrians from the flow of traffic, increasing sidewalk capacity by more than half, even facilitating window-shopping, all told an elegant solution to civilian traffic casualties.

The City of Tomorrow gradually coalesced. Enormous, widely spaced towers replaced low-level housing. Conveniently located underpasses and express streets made traffic lights obsolete. Interchanges, overpasses, and gently curving access ramps replaced traffic circles and cloverleafs.

The completed model was photographed in a warehouse spacious enough for aerial camera maneuvers. In lieu of chain-smoking assistants, smudge pots provided the requisite urban haze.

•

WITH THE COUNTRY mired in the worst economic crisis since the Civil War, a group of New York businessmen decided that a World's Fair would be just the thing to lift "psychological doldrums." The 150th anniversary of George Washington's inauguration was chosen as an auspicious target date. A malodorous marsh in Flushing, Queens, that eleven years before had entered the literary landscape in *The Great Gatsby*, would provide the requisite 1,200 acres.[361]

The classical motifs of previous fairs would be jettisoned for modernism. Not the ivory tower modernism of the Bauhaus, with its left-leaning, "intellectual" politics, but a "plain American" modernism: an amalgam of Flash Gordon, Buck Rogers, H. G. Wells, and Jules Verne, with a touch of Fritz Lang's *Metropolis* (minus the class divisions) thrown in. A shiny, sanitized version that would lend itself to consumerism and the primacy of industry.[362]

As with all major fairs, the latest technological breakthroughs would be showcased, along with the materials that made such progress possible. "Synthetic" had evolved from Twenties slang word to Thirties buzzword—Science improving on Nature. Cellophane, the long-sought-after "flexible glass." Bakelite. "Artificial silk" nylon stockings, spun from coal.

The goal was visionary commerce. But as Bel Geddes knew only too well, being "visionary" was a mixed blessing.[363]

Among the Fair's highlights would be Philo Farnsworth's "picture radio" (also known as television); more than a decade would pass before it would become commercially available. It would take seven years for Atanasoff and Berry's Fair prototype to be introduced as ENIAC, the world's first general-purpose computer. The designer of the Fair's "dry copier" (an arthritic office worker who'd spent years concocting foul-smelling mixtures in his kitchen) would be turned down by IBM, GE, and RCA, none of whom saw a market for it. Xerox's first copier would appear in 1959.

The microwave would also debut at the Fair. Developed for detecting nighttime air attacks, it would take twenty-seven years to surface as a domestic cooking device. The jet engine, which could have changed the course of the brewing European conflict, was also ignored.[364]

Charles Goodyear had been one of the lucky ones. While struggling to convert tree gum into a stable elastic, he endured years of crushing debt. By 1938, his eponymous tire company was a household name.

Still, he would be crazy to not participate. The Fair could provide an audience vastly larger than any theatrical production, even a Reinhardt spectacular. Alexander Calder, Alvar Aalto, ex–business partner George Howe, and former student Russel Wright would be contributing, along with the rest of the Big Four. Dreyfuss, now a youthful thirty-four, would have the Perisphere—its massive hollow interior more than twice the size of Radio City Music Hall——as his venue. Teague, fifty-five, would design an immense, 100-ton turntable (on which eighty-seven Teague-designed puppets turned raw materials into car parts) for Ford, a 105-foot tower for Dupont, a 40-foot cash register that rang up fair attendance, and a gilt-trimmed Steinway piano. Loewy had commissions to design Chrysler's Frozen Forest (steel palm trees with refrigerant inside), vitrines for the House of Jewels, and the Rocketport, a simulated trip to outer space.

The Shell model struck Norman as ideal grist for an exhibit. Suburbs were growing three times faster than urban centers, drive-ins were sprouting up from Camden to Corpus Christi, and gas was reasonable at ten cents a gallon (a penny more than a loaf of bread). Who could argue that freedom (as in mobility) wasn't a core American value? To emphasize the imminent hegemony of the automobile, the Chrysler, Ford, and GM pavilions were situated near the Trylon and Perisphere, the Fair's focal point.

Shell, the obvious first choice for a sponsor, wasn't interested; as far as they were concerned, the model had fulfilled its purpose. The Fair's general manager had no interest in a non-partisan exhibit covering the entire automotive industry. Norman contacted Richard H. Grant, General Motor's Harvard-educated vice president in charge of sales, as well as William Signius Knudsen, GM's president. Both knew Bel Geddes from five years before, when he'd designed the corporation's silver anniversary medal. Both turned him down. Chrysler declined. Trying Grant a second time, Norman received an emphatic, unequivocal "No."

Success came when Goodyear asked him to develop some promotional ideas. (Rival Goodrich had just come out with clear, synthetic raincoats.) He submitted ten. Goodyear responded with an offer of $5,000 to refine and repackage his City of Tomorrow, along with a Fair-worthy building to house it in.

Making a brilliant leap from a static model to a kinetic aerial "ride," Norman added a long ribbon of armchairs that would travel high above the cityscape, its passengers looking down as if from a low-flying plane. In theater, the audience was stationary and the play "moved." He would reverse the equation.

"The airplane is rapidly removing the letter *T* from the word 'there,'" *Popular Mechanics* had announced in 1931. Photos taken from dizzying heights filled magazines and newspapers, offering the thrill of a new perspective. The airplane was elevating things, literally and figuratively, into what Le Corbusier was calling the "bird's-eye view." *Bail out, zoom, fly blind, air pocket, blow up, nosedive,* and *tailspin* were already embedded in the vernacular. But the populace at large had yet to travel "by air."

A master of "immersion theater," Bel Geddes set out to devise an exhibit in which visitors were participants, not merely observers.

Rather than just circling around, the plush, wing-backed chairs of his "ride" would "change altitudes." The model, in turn, would be built to differing scales, then painted and lit accordingly. With the audience in motion, moving through it, the experience would be *cinematic.* The chairs would rotate at predetermined points, shifting their occupants' line of sight, their "wings" limiting peripheral vision. Individual speakers—on individual soundtracks that coordinated with a chair's position in the loop—would narrate what was being seen, keeping everything "on message." The moving chairs necessitated that the exhibit be viewed in a particular sequence and with predetermined pacing, a brilliant solution to crowd control—always a huge problem at world's fairs. There would be no trailing after errant children, no wandering in and out. In short, a comfortably captive audience.

•

SEVERAL MONTHS LATER, with all the bids in and the designs approved, Goodyear's president and Fair president Grover Whalen (known for his gardenia boutonnieres) had a falling out. Within half an hour of receiving the news, Norman was on the phone to Richard Grant for a third try.

It had been over a year since "Dynamic Dick" last turned him down. The Fair was a trade show on steroids. Entertainment and education were fine only insofar as they served to "move" merchandise and promote the corporate image. But what if Norman could expand GM's presentation beyond the predictable "money grab"? *What if the goal was to have the public wedded to GM's "vision," and to make that vision so attractive and accessible that the average Jack and Jill would have a hard time imagining a future apart from it?*

It was all theater at heart, wasn't it? "Multiformed" and "many-headed," as Reinhardt liked to say. Extravagant Fair pavilions, constructed of wood, wire, plaster, and paint, were like scenery for a play. Norman would incorporate every possible technique, from lighting, camouflage, sound, music, and color to changing perspectives and scale, to induce an emotional response. He'd transported audiences with his vivid depiction of the present in *Dead End* and of the distant past in *The Miracle*. Why not the not-so-distant future?

With a reluctant Richard Grant on the line, he explained Goodyear's interest and subsequent withdrawal, insisting that his updated concept was *guaranteed* to capture the public's imagination beyond mere merchandising.

Grant coolly informed him that GM's exhibit plans had just been approved by the Fair's Design Board (which Bel Geddes had declined to be part of, considering it a waste of time), and would he be so kind as to *not* query him again?

He'd never broached his proposal to Alfred Sloan, GM's president, chairman of the board, and CEO. Surely the man who'd stolen the market away from Henry Ford would at least *consider* Norman's plan. "Dynamic Dick" assured him that Mr. Sloan would simply refer the matter back to him without bothering to look.

Then perhaps Mr. Grant would at least allow a small favor?

Would he agree to *not* say anything, so that Mr. Sloan might make up his own mind?

Why on earth would he mention it?

That left the small matter of gaining access.

Unlike Grant, Paul Garrett was a friend. Norman knew GM's vice president in charge of public relations well enough to know he was a frustrated man. Try as Paul might to pursue the "public relations" aspect of his title (by soliciting stockholder feedback, especially from those who *sold* their stock), his bosses saw his role as one of cranking out publicity. That frustration might be used to advantage.

Norman phoned to say he'd be in Garrett's Manhattan office in ten minutes and that whatever luncheon plans Paul might otherwise have, regardless of how important, he should cancel. Norman promised to trump them; what he had to say couldn't wait. Lunch was on him, at 21. Then he put down the receiver and headed for the street. Paul would be either infuriated by his impudence or amused. No matter. As long as he was curious.

Norman's City of Tomorrow, with its "streamlined" traffic (100,000 cars traveling at up to 100 miles per hour, with four times current efficiency), was designed to encourage public support of road-building through taxes, and without being preachy. Also in his favor was GM's aversion to mass transportation. Bad for business. The corporation had a history of buying up urban streetcar lines in order to scrap them. There were no streetcars or trains in Bel Geddes's metropolis.

The sobering number of annual fatalities, many of them pedestrians and children playing unsupervised street games, also worked in Norman's favor. During the first four years following World War I, more Americans died in car accidents than had fallen in battle in France. Driving tests didn't exist. By 1934, there were 25 million registered cars in the United States.[365] It wasn't uncommon for reckless motorists to be attacked by angry crowds. GM had gone so far as to produce a cartoon driver safety film screened in 7,000 theaters nationwide. [366]

Garrett managed to arrange an informal meeting with his boss.

The exhibit would be the largest, most ambitious, most realistic diorama ever built, Norman told Alfred Sloan, a scale model presenting a future predicated on automobiles, *and the freedom, autonomy, and mobility they embody.* A visual dramatization of the complex tangle of American roadways—and its solutions!

Rather than asking the public to buy GM's products—the aim of every corporate exhibit—it would convince them that safety, comfort, and speed were compatible, that progress in transportation was inseparable from progress in civilization.

Less than three generations ago, most of this country was still new, and its roads were still in the Dark Ages. Today's great cities were largely planned and built before the motor car was even a dream. Estimates show a future demand for 100 billion more car travel miles than are possible today. This exhibit is about starting fresh instead of patching up an old system that doesn't work.

After a loud twenty-minute synopsis (the chairman of the board was losing his hearing), Bel Geddes was offered a paid commission to develop and expand his Shell-Goodyear model into one with a national focus—*if* he could do it in four weeks. He promised to be ready in three.

•

THE PITCH, ROUND two.

Assembled in the Director's Room of GM's Manhattan headquarters were Alfred Sloan, Mr. and Mrs. Grant, William Knudsen, and several other company employees. Norman brought along Worthen Paxton (Pax), his business partner and right-hand man, as backup. Privately, Norman had told Paul Garrett that it would cost $5,000 out of pocket "just for you people to understand" his concept.[367]

Given Sloan's deteriorating hearing, and his vanity regarding hearing aids, Paul seated the three principals behind a large table with Norman's visuals aids, sixteen three-foot-by-two-foot charcoal drawings, laid out in front of them. As per his mandate, the City of Tomorrow had been refined and expanded to include the country as a whole. Bel Geddes would give his presentation standing *behind* the semi-deaf board chairman, increasing his volume or slowing his pace based on prearranged signals from Garrett, stationed across the room to gauge Sloan's reactions.

According to GM employee Frank Harting, also at the meeting, Bel Geddes "never expected to sell it."[368]

THIS EXHIBIT DEMONSTRATES that the world, far from being finished, has *hardly begun, and that the future we create today will create richer lives and greater opportunities tomorrow.*

Visitors descend a switchback ramp into a twilight-lit "map lobby" beneath an eighty-foot rotunda. On the wall is an immense map of to-day's roads—designed for yesterday, based on buffalo paths and Conestoga wagon trails. The haphazard planning and resulting congestion of 1939 traffic systems are picked out in lights; varying widths illustrate the volume of traffic flow.

Now the map lights change, showing how traffic is destined to increase—as many as 38 million cars by 1960.[369] The lights change a third time, showing how that congestion can be relieved by a great arterial web of fourteen-lane, high-speed superhighways, "motorways" bearing traffic cross-country in comfort and absolute safety.

Now the tour of the World of Tomorrow begins.

Visitors are seated, in pairs, in luxuriously upholstered, spring-padded armchairs—550 of them mounted on a continuous moving escalator that glides, silently, along the upper reaches of a vast, darkened theater.

Each chair has its own speaker, with narration synchronized to the chair's movement. Once seated, a confidential voice at the visitor's shoul-der explains what he's about to see—an airplane ride across America in 1960. Moving at ten miles per hour (twice the speed that a person can walk), the chairs began their seventeen-minute glide.

Everything is seen through a continuous glass panel curved around and above a beautiful three-dimensional landscape. Rolling farm country in springtime, fresh-leaved trees by streams, cattle grazing, goats nibbling. The spectator's chair climbs over a hill, then descends into a valley.

1940 is twenty years ago!

Ahead is a dairy, and an experimental nursery with fruit trees growing under domes. An aeration plant purifies lake water, then distributes it hundreds of miles throughout the countryside. The narrator's voice calls at-tention to a truck coming out a farm driveway. It swings into a secondary road that merges into a fourteen-lane superhighway. The driver advances to 50 miles per hour, then shifts to the 70-mile-an-hour lane. If the jour-ney is a long one, a gap will open in the 100-mile-per-hour lane. (At that rate, he can drive from the Atlantic to the Pacific in twenty-four hours!) The widened curves require no reduction in speed. "Switch lanes," the narrator explains, are far safer, faster, and more cost-effective than today's bottlenecked cloverleaf intersections. Thanks to these super highways, the landscape has opened up. Suburbs now take the place of congested areas. Our truck merges with thousands of teardrop-shaped, rear-engine motor cars (no motor noises, no gas odors) with wraparound windows. Control towers monitor traffic with radio signals. It's impossible for a drunk or a sleeping motorist to leave the road or swerve into another vehicle.[370]

NIGHT FALLS.

AS WIVES SERVE *supper to hungry families, the motorway lights up automatically. Gas-filled tubes, controlled by photoelectric cells, are built into the traffic separators every 400 feet—well below the driver's eye, and in a color that minimizes eye strain. Alerted by radio waves a quarter mile ahead of an oncoming car, the lights switch on, then off a quarter mile behind, at one-sixth the cost of incandescent lighting.*

What's that up ahead? An amusement park in full swing, boys and girls shrieking with glee on a pretzel-like sky ride. There's a thriving steel town, its glowing furnaces brightening the sky, and a "floating" airport.[371]

DAWN BREAKS.

THE CHAIRS BEGIN *an uphill climb, 500 feet above earth, now 10,000, to a thrilling bird's-eye view of snowcapped mountains and fertile valleys. We're on top of the world—the world of 1960! A thick cloud momentarily blocks the view. The express motorway crosses a lake over an elegantly cantilevered double-deck suspension bridge supported by a single cable. High-speed traffic keeps to the upper deck while slower traffic moves below. Beneath us, in the valley, is a resort town. And ahead, a giant lake dam, a miracle of engineering that provides hydro-electric power for hundreds of miles around.*

Another city, a great metropolis, looms. From the safety of our upholstered chairs, we bank high over glass-sheathed skyscrapers with landing platforms for "autogyros" and other airborne commuter craft. No more overcrowded streets. Modern, efficient. More parks, playgrounds, and abundant sunshine for everyone, rich and poor.

We fly lower, approaching downtown. Soon we're floating above an intersection surrounding an open plaza. There's a Department Store resting on enormous glass pillars, a Science Auditorium, a four-story Apartment House, and a handsome Automobile Salon.

As our chairs swoop in for a closer look, the intersection looms larger and larger until the four buildings are rising several feet into the air. People are walking, in safety, fifteen feet above the ground on the elevated sidewalks that link the buildings. They're looking into bright shop windows and lounging in rooftop gardens. Below, cars, buses, and taxis (all by GM) are moving six abreast.

But before the spectator can take in all the details, their chairs suddenly swing around and burst out from the building into the open air *as the narrator announces "All eyes to the future!"*

Leaving their chairs, the visitors find themselves standing on a full-sized elevated sidewalk, transported into a life-sized version of the exact scene they've just "flown over"—*the same busy intersection, extending for nearly a full city block, the same buildings and signage.*

It's the World of Tomorrow *in all its wonders!*

No longer spectators, they merge with the crowd, enter any building, walk the elevated walkways, sit in the roof gardens, witness the thrilling stage shows in the Science Auditorium, see GM's latest offerings displayed in the elegant automobile showroom, even ride in them on the streets.

The model has come to life.

•

ALFRED SLOAN TURNED to his colleagues, unaware that they'd already rejected the project three times. *Isn't this tremendous?*

"It won't sell a single automobile." Richard Grant sniffed.

"It sells the future," Norman replied. "With the promise that every citizen can own a *piece* of that future for the price of a General Motors' automobile."

"One-hundred-mile-per-hour lanes? Won't we look irresponsible?"

"The Germans have already done it," Norman countered. "You've heard of the Autobahn?"

"There was no outcry when the new streamlined trains reached 120 miles an hour," he added. "But because cars evolved from horse carriages, people see them differently. The problem isn't speed, gentlemen, it's the haphazard, unplanned growth of roads, combined with human factors— fatigue, indecision, poor judgment."

"It's an amusement ride," Grant muttered, "*Alice in Wonderland* meets *Gulliver's Travels.*"

In truth, it was closer to Bel Geddes's War Game, with considerably more landscaping and architecture.

"You've seen the Rockettes at Radio City, weaving in and out in perfect precision?" Norman continued. "Cadets performing an infiltration pattern? In Grand Central Station, a dispatcher feeds trains from thirty-eight tracks into two without a bump. How? By automatically controlling speed and spacing. With everyone at a constant speed, collisions are impossible. The same thing can be achieved on intersections, without traffic lights or cops.

"Everything I've proposed," he added, "has been endorsed by leading highway engineers."

Even if he liked it, William Knudsen objected, it was impractical. Taking 250 people at a time on a "ride" presented a logistical problem, and Geddes was talking about more than *twice* that number. Beyond that, there simply wasn't enough time. The Fair was scheduled to open in less than eleven months and the Design Board had approved Albert Kahn's exhibit plans three weeks ago, as Geddes knew full well. Under the circumstances, was it advisable to set those decisions aside?

"Is it true," Norman asked, "that the approved plans feature a Chevrolet assembly line, a repetition of your exhibit at the 1933 Chicago Fair?"

It was.

"In that case, do you think it's *advisable* for General Motors to admit it hasn't had a new idea in five years?"[372]

A smile crossed Alfred Sloan's face.

The bigger truth was that Henry Ford had initiated the assembly line idea, to great success, at its 1915 Panama Pacific pavilion. (The plan had been to repeat it in Chicago in 1933, but GM beat him to the punch, to Henry's fury.) Which made the concept not only stolen but one that dated back a generation.

Knudsen frowned, clearly irritated. The six-foot-three Dane had presided over the Chevrolet division for thirteen years. The assembly line was a point of pride; he'd been personally responsible for building it to its present level of efficiency. As for GM's Chicago Fair exhibit, the public had *loved* it. Groups of 1,000 visitors watched raw materials enter through a door. By the time they exited another door, they'd been turned into finished cars. Not only that but a visitor could select the materials for "his" vehicle, watch its progress along the assembly line, then get in and drive it off. What could be a more powerful sales tool?

"Geddes's exhibit is complicated and experimental," Knudsen complained. "It calls for an enormous display with tens of millions of parts and a life-sized street intersection with elevated sidewalks. Plus landscaping, untried exterior illumination . . ."

"The only difference between building one section and two hundred," Norman countered, "is multiplication, orchestrating the work into units." There were fifty men in his office, himself included, who could perform all the necessary tasks, or teach others to. He'd already crunched the numbers. Five crews could complete the floor model in four months.

"And that conveyor belt of moving chairs?" the Dane continued. "Conveyor belts *do not* turn corners. They operate in straight lines, changing direction by dumping from one straight belt to another. Are you planning to 'dump' our visitors?

"And what about the noise? That conveyor belt will have to run fourteen hours a day for six months at a stretch!"

Norman produced a letter from the Westinghouse Elevator Company stating that they'd examined his designs and saw no obstacle to constructing them. The belt would turn corners, rise up ten stories and lower back down, and all without clatter, swaying, or jerking.

"What about that voice of yours," Dick Grant wanted to know, "the one coming through the back of each chair? Five-hundred-plus pairs of microphones, each playing on a different loop? Everything perfectly synchronized to what each person is seeing at any given moment?"

"The Polyrhetor," Norman explained. "Latin for 'many orators.' It will be intimate, personal, instead of a blaring loudspeaker. Research suggests that the public might object to earphones or hoods."

Knudsen's patience was flagging. "Another mechanism that doesn't even exist. It's ludicrous."

Norman assured the executive trio that he'd take full responsibility.

Alfred Sloan put an end to the jousting. Unlike his rival, the autocratic Henry Ford, he believed in collective decision-making. "This is something for Kett," he said and left the room.

Anticipating the pleasure of Bel Geddes's comeuppance, Knudsen offered to personally arrange a meeting with the corporation's VP in charge of research at 9:00 a.m. the following morning. Which meant leaving for Detroit that night. Norman was in the midst of overseeing commissions for a line of Majestic radios, stainless steel auto bumpers for Rustless Iron, new typography and formats (a $100,000 commission) for the *New York Post, Collier's, Woman's Home Companion,* and *American Magazine,* plus last minute details on an upscale private club commissioned by Alex Woollcott.

"Mr. Knudsen," he said, "I have a full day ahead of me."

"You don't want this job?"

"I didn't realize you were giving it to me."

"Bring those drawings of yours along. If you sell Kettering, you've got it."

●

AT SIXTY-TWO, CHARLES FRANKLIN KETTERING, also known as Boss Kett, was a force to be reckoned with.

He personally held some 140 patents, from Freon for refrigerators, a copper-cooled engine, and the all-electric self-starting ignition (a replacement for the hand crank, it was "blamed" for putting women behind the wheel) to an eponymous infant incubator, the Kettering Bug. He was funding photosynthesis studies with the aim of converting sunlight into a usable power source. He could pilot an airplane and liked to study weather and air currents while aloft. Just the year before, he'd been made a Chevalier de Legion d'Honneur, France's highest civilian honor[373] for his research on artificially induced fevers as a treatment for a range of serious ailments. Seven years hence, he and his boss would establish the Sloan-Kettering cancer research center.

Norman took Pax and his staff engineer, Roger Knowland, to Detroit with him. The assumption was that they'd meet with Knudsen, Kettering, Grant, and perhaps a few others. Forty-four men awaited them in the conference room, fresh from an 8:00 a.m. briefing on the status of GM's Fair plans and the unschooled "artist" in their midst.

Half of the hostile assemblage were corporate officials and directors—a sea of dark suits, boring ties, and mostly thinning hairlines—the rest, the design staff that had created the already-approved assembly line exhibit. Albert Kahn was also there. The architect, sixty-nine, had set up shop when Bel Geddes was still in diapers. He'd designed both GM's 1933 Chicago Fair building and assembly line display, and he and Knudsen happened to be great friends.

An unacknowledged elephant in the room was the Chrysler Airflow disaster, a not-so-distant memory. Four years before, when Norman was helping refine the car (whose unjust fate GM largely orchestrated), he'd criticized "the conservatism of General Motors in general and Kett in particular" to a friend.[374] Now he was about to face off with GM's resident genius on a project that few other than Norman were invested in.

All but half a dozen faces were unfamiliar, and no one bothered to introduce himself.

GM's personnel weren't the only ones looking forward to the upstart's demise. Walter Dorwin Teague, at work on Henry Ford's pavilion, had seen Norman's early exhibit plans. Four days prior to the meeting with Kettering, in a letter to a Ford official, Teague predicted that Bel Geddes's grandiose scheme, with its "enormous difficulties," wasn't likely to materialize.[375]

Laying out his drawings, Norman reprised his presentation for what was now being called "the futurama."

Boss Kett—seventeen years Bel Geddes's senior—gave the designer a "terrific grilling," when he wasn't wisecracking or taking over the floor. His role was clearly to find the weaknesses in Norman's plan, to confuse or anger him, as necessary.

Drives, controls, chains, escalators, ascending and descending belts, and hundreds of thousands of model components, all to be fabricated, tested, adjusted, and installed in less than a year?

Do you appreciate the ramifications, should you fail?

Like Grant and Knudsen, Kettering was particularly adamant that the highly complex Polyrhetor sound system couldn't be completed in time, assuming it could be successfully built at all.

Other factors, spoken of only behind closed doors, also contributed to GM's hesitance. During the past seven months, GM's stock price had dropped some sixty percent, more than it had after the 1929 Crash.[376] It clearly wasn't a time for risk-taking. Or was it?

Unlike the majority of his colleagues, C. F. Kettering believed in the importance, and rarity, of imagination. The Wright brothers had, he once said, flown "right through the smoke screen of impossibility." And he respected what he called "the fighting heart," the ability to push for what one believed in, and to bounce back after being knocked down.

"When I couldn't answer technical questions, I usually managed to shift the emphasis, respond with a question, flatter my interrogator, or simply burst out laughing," Norman recalled years later. Still, it was rough going; he came close to losing his nerve. "They were disorganized, stronger in virulence than fact. If 'Boss' Kettering had stuck to the engineering aspects, my position would have been difficult. He was a great repartee fencer."

Did Geddes really believe that designing soda-pop dispensers and Broadway musicals qualified him to be responsible for the corporate image of one of the country's industrial giants?

Wasn't it all a bit . . . grandiose?

Perhaps instead of "the futurama" they should just call it "Geddesburg."

The wall clock read 12:30 p.m. Norman had been talking for three and a half hours.

"I know what's wrong with you, Geddes," Kettering finally determined, waving a long finger. "You're too goddamned Waldorf Astoria."

"Meaning what?"

"You're too high hat, too sophisticated to understand the average American. General Motors sells to the general public, *not to New Yorkers*. All of us here are from this part of the country. The average man has never set foot in the Waldorf. Me, I'm a cafeteria man," he added. "You ought to eat in a cafeteria now and then."

"Mr. Kettering," Norman replied, "I was born in a small town forty miles from this room and lived in Detroit before moving to New York, which I like, as do eight million other Americans. As for the Waldorf, Manhattan has a dozen infinitely better restaurants and hotels. My wife and I enjoy dancing and we sometimes go there for Eddie Duchin's fine music. That doesn't make me a Waldorf Astoria man. It's true, though, I don't like cafeterias. I was a busboy in one when I couldn't afford to eat anywhere else.[377]

"Tell me," Norman added, "when you're in New York, where do you stay?"

"At the Waldorf," Kettering admitted.

"Where do you eat?"

"Horn and Hardart."

Boss Kett, who'd grown up a poor Ohio farm boy, seemed oblivious to the irony. The "new" Waldorf, opened late in 1931, was a grand affair with 300 imported marble mantels, walnut paneling inlaid with ebony, and a staff of 1,600, the favorite destination of visiting kings, presidents, maharajas, and celebrities. Horn and Hardart was where those with a few Depression nickels to spare dined on macaroni and cheese or creamed spinach displayed behind automat windows. It was a ridiculous concession for a top GM executive.

"May I speak to you as a cafeteria man, then?" Norman's gray eyes betrayed a hint of mischief. "I suppose you're also familiar with the Waldorf's cafeteria?"

Kettering hesitated.

"You've eaten there?" Norman insisted, knowing full well that no such place existed.

"Of course."

"Where is it, exactly?"

There was a long pause. The room was completely quiet. Then Kettering stepped forward, extending his hand. "Why don't I take you to lunch, Geddes?"

That evening, Knudsen gave "the futurama" a green light.

•

WHILE BEL GEDDES'S staff had hastily substituted "General Motors" for "Goodyear" on all viable plans and sketches, his lawyer had drafted an eighteen-page contract, with riders, based on service rates established by the Fair's Design Board. But Norman never had a chance to remove it from the pocket of his suit jacket. On May 3, 1938, rather than wasting time ironing out codicils, William Knudsen personally wrote out his terms on a yellow legal pad as the designer watched. The overall budget was cited at $6 million (in excess of $91 million in 2017). The Dane would keep track of expenditures, leaving Bel Geddes free to concentrate on the task at hand. In addition, GM's pavilion would now be known as Highways & Horizons, an unmistakable tip of the hat to Norman's 1933 bestseller.

The building would cover seven acres. Though Norman's Map Lobby and ride would constitute only one-quarter of it, Knudsen's contract stipulated that Bel Geddes was responsible for designing the entire complex, from lighting and exterior landscaping down to public restrooms and the placement of vacuum cleaner outlets for the maintenance crews. He would confer with architects, engineers, contractors, and suppliers, and hire and train staff, supervising the fabrication and installation of the largest, most complex Fair venue ever made. Anywhere. It was an unprecedented commission for any single individual, least of all an "outsider."

There would be seven deadlines. Norman's $100,000 fee (the same as Grover Whalen's salary) would be paid in increments.[378] What was "barely a sketch on paper," to use Bel Geddes's self-effacing phrase, was to be completed and operation-ready by March 30, 1939. Eleven months from start to finish. The Fair officially opened on April 30.

Design-control drawings for everything were due in twenty-eight days. Norman left Knudsen's office with a large check that would allow work to begin immediately.

Assisted by some 150 architects, engineers, and draftsmen, Albert Kahn translated Bel Geddes's meticulous drawings and notes into working blueprints. In the end, the two men developed an amiable working relationship—Kahn received his regular fee; Bel Geddes got to retain his Baron Haussmann hat.

He purchased hundreds of geodesic maps, which he enlarged and used in reverse. The futurama's opening scene, a valley of rolling hills, was based on terrain now underwater at the Ashkodan reservoir in the Catskills. A composite of St. Louis, Council Bluffs, New Bedford, Concord, Oneida, and Colorado Springs provided the templates for 1960s cities and towns. Dockyards and a large port were added. The mountainous region was borrowed from Yosemite Valley and Yellowstone. California's coastline was juxtaposed with the Mississippi River, instead of the Pacific, then abutted with parts of Illinois. Norman pieced the whole thing together on his living room floor.

•

IN THE MIDST of this intensive preliminary planning, with Rosamond's suicide and Bel's death still fresh, Frances's health worsened. There'd been signs of trouble within months of their honeymoon when, stricken with double pneumonia, she'd been forced to spend fourteen weeks in bed. Three years into the marriage, about the time of their ten-week anniversary trip to Mexico, she was diagnosed with tuberculosis.[379] Fatigue and appetite loss hadn't stopped her from taking on increasing responsibilities within Norman's company. By now, she had a vested interest. In the wake of the Bel Geddes–Howe dissolution, she'd loaned her husband several thousand dollars to help meet payroll and other expenses and worked without salary. Her name graced the company letterhead just below Norman and Pax.

Norman had "taken on" his wife's condition with his usual attention to detail: making a study of tuberculosis, having long talks with Frances's doctor, making sure people knew when her birthday was, throwing an occasional dinner party and sending friends into the bedroom where she was resting, clocking them so as not to tire her. But now he felt obliged to place his young wife (thirty-four to his forty-five) in the Connecticut sanatorium where, he noted with encouragement, Eugene O'Neill had been cured.[380] Frances's father sent a large check, but Norman returned it. It was, he felt, his responsibility.

What began as a trial month would stretch into six.

Frances had barely settled in when Norman's youngest daughter, now fifteen, announced she was getting married. To her cousin. Or they might elope. She hadn't quite decided.

The Bel Geddes children were striking reflections of their mismatched parents. Joan shared her mother's dark coloring, religious bent, and bookishness. Barbs had inherited her father's fair features and mischievous temperament. She'd always been a dreamy kid, prone to inventing little plays and striking languorous poses. Her cloistered adolescence in the New Jersey suburbs (Bel didn't approve of children going to the movies) had been broken only by Norman's "rare, explosive visits"[381] or the chance to attend one of his rehearsals.

In the wake of Bel's death, Norman had taken his daughters on a three-week Caribbean cruise, after which Barbs had been enrolled at Putney's, a coed Vermont boarding school where she spent much of her time daydreaming, flirting, and sending her father missives. "I simply must have an evening dress immediately," one read. "Please send me two or three because I might not like the one *you* pick. Please hurry. Send them *special*. Make them *grown up*, for heaven's sake. Don't make me look like a child!!!"[382]

Bored, she auditioned for a production of Synge's *Riders to the Sea*. Her teachers were unimpressed. "She tried exceedingly hard," the headmistress wrote her father, but both she and Mr. Rogers, Putney's drama instructor, had concluded that Barbara "does not think deeply enough

[and] does not have dramatic talent."[383]

Come summer, Barbs had traveled to Ohio to visit relatives and gotten reacquainted with her first cousin, Buel,[384] rekindling a "romance" that had been percolating for years.

"This business of my becoming a mother as well as a father to an adolescent daughter doesn't fall on me too naturally," a distraught Norman wrote to the alleged groom's parents, family he ordinarily had little contact with. "Barbara disturbed me greatly when she returned . . . with a story that she and Buel were very much in love . . . and that . . . you had told them that although they were cousins, it was perfectly all right for them to marry."[385]

•

IN SEPTEMBER, NORMAN got in touch with his pal Eddie Rickenbacker, who'd purchased the fledgling Eastern Airlines from GM (a feat he managed despite a seventh grade education) for $3 million. Arrangements were made for a dozen of Bel Geddes's staff to fly over western Pennsylvania to observe, firsthand, exactly what could and could not be seen from a low-flying plane; special permission was obtained to fly over Philadelphia at between 1,000 and 15,000 feet and over other parts of the state as low as 500 feet.

Daughter Joan was working in Dad's office as an assistant, researcher, and publicist. A student at Barnard, she was engaged to jazz scholar Barry Ulanov (Miles Davis would later call Ulanov the only white critic who ever understood him or Charlie Parker.[386]) Designated the flight's official stenographer, braced against her fear of the enterprise with spirits of ammonia, she noted down everything from the color and clarity of lakes and rivers to reflections and shadows.

What Norman called "human interest" items, GM's building crew would call "nuisance architecture." These eventually included a meticulously crafted miniature clotheslines hung with tiny laundry, cow paddies, thousands of nails representing people sitting, standing, kneeling, and bicycling, a model of Notre-Dame de Paris, and a park sculpture of Brancusi's *Bird in Flight*, a pretzel-like sky-ride (which Walt Disney would "make real" at the 1964 New York World's Fair), plus animations—the spray of miniature waterfalls (created by combining tiny water and air jets), low clouds clinging to mountainsides (chemical vapors), moving shadows on the ground cast by miniature airplanes. Arguments that many of these details were too small for the futurama's visitors to notice fell on deaf ears.

Flo Ziegfeld had spent a fortune—$600 apiece!—on Irish lace petticoats, invisible beneath the bouffant skirts of his *Follies* girls, because, he said, *they* knew they were there, and it made a difference in the way they walked.[387]

The detail hound would lose his battle with Knudsen over installing tens of thousands of minuscule transparent windows in the buildings (with working shades for night scenes) and for traffic haze (smog), which was being touted by some as a sign of progress. Night scenes would, however, make use of fluorescent pigment activated by pulses of ultraviolet light (from 100-watt mercury vapor lamps with special purple glass filters) to create "moving" headlights and his highway lane separators.

There would be "nuisance" paint jobs, too. Thought to lack depth, the hand-painted vines on miniature houses would be reapplied through a stencil, then "blown" with flock while the paint was wet. To create the proper "distance" perspectives, shadows required differing color values. No expense was spared. (The finest artists' pigments, some costing as much as $150 per gallon, were used throughout the pavilion.[388]) Completing the overall effect would be theatrical lighting, something Bel Geddes was master of.

•

DESPITE A HERCULEAN workload and the occasional late-night out at 21, the Stork Club, El Morocco, the St. Regent, or Spivy's Roof, Norman wrote to his wife almost every evening, whiskey in hand, before falling into bed ("Don't get any ideas that I may change because you're sick. You are the niftiest girl I've ever known or ever hope to,"[389]) and he took the train to Connecticut most weekends. He would send her friends up when she was stronger and arrange for weekly drawing lessons from muralist Anton Refregier. He made sketches for a "Microvision" device to help her read in bed.

Meanwhile, he was thinking about moving his office to the lavish new Rockefeller Center, in which case he wanted Frances to help design and furnish it, from her bed if need be. Then again, maybe not. "If this war business continues," he confided, "I may lose my nerve about everything. There's a clause in GM's contract that they may cancel [the Fair exhibit] in the event of war, and everyone is so jittery."[390] The hope was that Hitler was bluffing.

COMMANDEERING ENOUGH DRAFTSMEN posed the most immediate problem. From New Jersey to Connecticut, most were already engaged with hundreds of other World's Fair jobs. Norman managed to muster an initial crew of twenty-eight. Each man was asked to venture out at least 100 miles and return within a week with two or three draftsmen they could vouch for. Following an hour's talk, the new recruits were then sent out on the same mission, while the original twenty-eight were assigned their first hands-on work, geared to Knudsen's schedule. The pattern was repeated until several hundred "mostly good" draftsmen had been rounded up. Meanwhile, GM still wanted their beloved automobile assembly line display and Norman was being pressured to find a place for it.

When no reputable contractor could be found to bid on the finished designs—at this late date they, like the draftsmen, had more work than they could manage—it was decided that Bel Geddes & Co. would handle the job. The work would be broken down into categories like woodwork, metalwork, and painting, each with myriad subdivisions.

For a general project manager to oversee construction, GM's top brass made what seemed a peculiar suggestion. Years before, an hour prior to the opening of his annual display in a Chicago hotel, Alfred Sloan had pronounced the immense ballroom's atmosphere "cold." George Wittbold's Florist Shop happened to be right around the corner; within half an hour, it was emptied of its inventory. Wittbold had gone on to supervise many GM showroom displays.

It's tempting to think that disgruntled in-house staff had a hand in the decision to have a florist oversee a complex, groundbreaking, multimillion-dollar project. Norman, however, was unfazed. He liked the affable florist with the cherubic smile, who, despite knowing nothing about models and finding Norman's "perfectionism" an ongoing irritation, proved a good choice. He told George Wittbold the same thing he'd told Knudsen and Grant. *The job is as simple as adding a column of numbers. It adds up correctly or it doesn't. If it does, you can call it perfectionism. If it doesn't, call it whatever you like.*

•

THE OLD COSMOPOLITAN Motion Picture Studio, once the site of numerous Marion Davies's shoots, occupied a large block at 126th Street and Second

Avenue. It was here, on three rented floors, that Futurama (now with a capital "F") and the rest of Highways & Horizons took shape.

Given the extremely tight deadline, exhibits were designed in order of importance and size. A twelve-foot-square wooden model of the entire site was constructed so that progress could be kept track of in three dimensions.

At seven o'clock each evening, the model department received the day's blueprints and an all-night shift set to work. By 10:00 a.m., new developments and corrections had been built in and painted yellow; previously approved work was white. As Norman made corrections, the affected parts were numbered in red. By noon, these had been added to the drawings and, by evening, to the model. All corrections were recorded with a Dictaphone. Nothing was left to chance.

Futurama's 408 panels were arranged on individual twenty-by-five-foot tables at four-foot intervals, so the thousands of technicians, draftsmen, carpenters, and model makers could move around them.[391] Each panel had a number and a work progress card of twenty itemized steps. After

Blueprint in hand, Norman stands atop a section of the unassembled Futurama model.
(Courtesy: Harry Ransom Center, The University of Texas at Austin, photo by Gjon Mili.)

one employee got lost amid the labyrinthian arrangement, Wittbold had red arrows painted on the walls that read THIS WAY OUT.[392]

"At the moment we have 200-and-some draughtsmen, between 400 and 500 carpenters and about 25 sculptors," Norman wrote to Frances, who'd headed up the Vegetation Department on the Shell model, "plus some 30 people experimenting with ways of making trees," more than 1 million of them. The exhibit called for eighteen different species, with three different colors for different perspectives. A special moss had been imported from Norway, then chemically treated and dyed. Mountaintop trees, in particular, had to be carefully watched; those more than one foot tall threw off the scale.

An additional 500 crew members were soon added to "locate" buildings and vegetation on the models.[393] Yet another group tackled geographical surfaces. The skeleton topography layers (including progressively terraced farm areas) were cut from fiberboard; some of the twelve-foot mountains required more than 288 stepped layers. Plowed fields were made of ribbed corduroy velvet, forest floors of green cork. Forty tones of quick-drying "dope plaster" were used for ground color. "The stuff used to stuff birds," presumably excelsior, provided the necessary surface texture. After asking a taxidermist for several tons of it, Norman ended up buying his shop.[394]

Though Norman had sold off most of his photography equipment with his reptiles and amphibians, he'd kept a camera or two in reserve. He used them now to document his brainchild being born. Men wearing ties, vests, and rolled-up shirtsleeves pouring plaster of paris into rubber molds to create miniature skyscrapers, an observatory, a refinery, a hydraulic dam, power plants. Others wiring, drilling support holes, calibrating, filing and spray-painting, soldering circuits, using magnifying glasses to paint tens of thousands of tiny skyscraper windows (the compromise to installing actual windows). A young man holding a cigarette in one hand, a brush in the other, as he works. The makers of "nuisance architecture"—beehives, mailboxes, hot dog stands, and woodpiles—are there, tweezers in hand. Others fashioning bridges, highways, an amusement park, and assembling the experimental orchards (rows of trees under individual glass domes) and experimental farms (vats of green "grow" chemicals gleaming in the sunlight).

If Norman's War Game (at 80 square feet) had been chess taken to infinity, Futurama (at 35,738 square feet) was an intricate jigsaw puzzle with similar aspirations. Every field, park, and snow-capped mountain, every orchard, farm, and vineyard, every companion city and satellite town ,had to be accurate to a fraction of an inch so that, when the immense panels were joined, their edges merged into a meticulous, unbroken whole.

THE PAVILION'S FACADE – massive, towering, windowless, its contents impossible to guess—rose from 40 feet to over 100, a dramatic, unbroken ten-story sweep ending in a vertical "hook." Future neo-futuristic architect Eero Saarinen, twenty-eight, worked on the project for forty-six days, receiving the crew's second-highest salary.[395] (At fourteen, Saarinen had seen Norman's *Miracle* staging. He would make it a career goal to bring architecture and lighting effects into greater harmony.) Still, noted a contemporary, the lack of ornamentation was, "as many people are aware, strictly Bel Geddes."[396]

One of the Fair's most difficult fabrication and assembly jobs, Highways & Horizons required, among other things, a 3,000-ton frame— more steel, Norman claimed, than the Empire State Building. Its concrete floors rested on a base of 2,400 35-to-90-foot wood pilings. A vast, sleek, metallic whale, it anticipated aspects of Frank Lloyd Wright's 1959 NYC Guggenheim Museum (which, like Futurama, brought visitors to the top of a spiraling ramp, from which they then descended), Saarinen's 1962 TWA terminal at JFK International (curved mezzanines and free-form walkways), and Frank Gehry's titanium-covered 1997 Guggenheim Museum Bilbao.

•

IN MID-OCTOBER, THE project fell behind schedule, precipitating the arrival of Jack Dineen, the director of GM's sales division, now in charge of Fair exhibits. A lean, hard-drinking Irishman, he'd announced his displeasure with the whole business months before. He disliked dioramas, and Norman's age—he was forty-five, Dineen was in his thirties—made his judgment suspect. Besides, Dineen insisted, Futurama's "1960" was too far away, though some had suggested targeting the year 2000. "I can't," Norman had replied, "because *there won't be* any gasoline cars by then. I'm no Jules Verne."[397]

Just off the train from Detroit, Jack Dineen began making threats. The work plans were impractical, he announced, impossible to achieve on deadline. Geddes's contract would be cancelled, Wittbold's too, and a lawsuit filed. Dineen was moving to New York to take over operations personally.

Coinciding with Dineen's rant was a threat of a different sort.

With a clever rewrite of a forty-year-old H. G. Wells novel and an actor who imitated FDR's voice (familiar to radio listeners from his Fireside Chats), twenty-three-year-old Orson Welles orchestrated a nationwide panic with his "War of the Worlds" broadcast. The Munich Crisis had occurred a month before; it wasn't a stretch to conflate a Martian attack with a German one.

Norman, himself a prankster, was much amused. Welles outdoes Wells. He sent Orson a telegram, one theater man to another. *Too bad such a large part of your audience walked out on you.*[398]

Two extremely tense weeks after Jack Dineen's arrival, "the battle of the century" was over. Grant and Knudsen had been quick to arrive in Dineen's wake and, seeing the model, were delighted. "We won 100 percent—all points," the relieved designer wrote his wife. "We've been

Ramps leading into the massive, mysterious building were always full.
(Courtesy: GM Media Archive)

given a completely free hand and the only GM checkup from now on is that Mr. Grant, personally, will once a month come out and make an inspection . . . no one could be more fair minded . . . what an appalling state everything was in last Friday as a result of Dineen's five days in New York."[399]

For Highways & Horizon's exterior, Bel Geddes chose a silver-gray lacquer suggestive of the aluminum facade of DC-3s, the same color as *Horizons'* book jacket, the color of his 50,000 miniature alien-like Futurama cars. The lacquer had the added virtue of absorbing changing light, and with sufficient gloss to reflect the colors of sky, surrounding trees, flowers, and adjacent buildings. After dark, projected cloud patterns would transform the towering facade into an immense, revolving planet.

"None of us are scared now," Norman wrote Frances in early November. "We feel we have the thing under control. I never was scared," he claimed, "but I've been the only one who wasn't for a year . . . Of course, we can't let up for a minute."[400]

Of all the compromises Norman had been forced to make (the dreaded auto assembly line had been jettisoned), those involving his "surprise finale" had been particularly frustrating. Determined to maintain Futurama's illusion to the very end, he'd wanted to create a vanishing perspective by finishing the ends of the full-scale streets with mirrors. But with an estimated additional cost of $50,000 to $60,000, Knudsen had put his executive foot down.

As for the cars on the full-scale intersection, Norman's proposal had called for chassis "dummies" shrouded to resemble Futurama's silver teardrop cars. Emerging from their plush armchairs, visitors would be able to "drive" them (automatically guided by hidden tracks and rollers) beneath the elevated sidewalks, accelerating and braking for stoplights. But both the intersection and the salon showroom, which occupied a quarter of that intersection, were slated to display GM's latest model Chevrolets, Pontiacs, Oldsmobiles, Buicks, La Salles, and Cadillacs. Fresh from the future, visitors would be confronted with boxy sedans and semi-boxy coupes. But the corporation wasn't about to forfeit sales for the integrity of Norman's vision.

Making the best of it, he and his team had designed an elegant, understated auto showroom. A rosewood column widened as it rose from the carpeted floor to a copper ceiling, which softly reflected an expansive, spiral-nebula-like light fixture, the world's largest, made up of 2,000 transparent bars.[401] The walls were covered with a metallic fabric previously used only for women's evening shoes and handbags.

Against the odds, it was all coming together.

"It was fun experimenting today, arranging the principal buildings in our metropolis," he wrote Francis later that month. "The photostatic plan of the city was stretched out over a 60 × 40 foot drafting table and half a dozen of us walked around with pockets and hands full of sky-scrapers" arranging them.[402]

Frances called it "playing God."

> *Fifteen [Futurama] panels went to the Fair this morning and another fifteen tomorrow, so it won't be long before we can begin to set the jigsaw together and see what it looks like.*
>
> *I am still distressed over the idea that you will have to live in sanatoriums for another six months and that it will probably take five years of careful living to make the cure complete, but . . . you will still be five years younger than I am now . . . I can't help but miss you terribly.*[403]

By the end of the month, 108 more panels had been transported to the fairgrounds and 40 of them installed (in all, the 408 panels filled 408 trucks), and Bel Geddes got a first look at his onsite office, a cubicle in the conveyor tunnel reputedly designed by someone in GM's purchasing department. A desk was nailed to the floor and a lamp nailed to the desk. There were bars on the window, and the walls were upholstered to resemble a padded cell—a reminder that many up and down the food chain, from "suits" to technicians and manual laborers, thought the designer was "nuts."

When Norman left town for a couple of weeks, a joke circulated among the construction crew that when he returned, he'd spend an hour or so eyeing what had been done in his absence, then say, "Let's take the whole thing down about a foot and a half."[404]

By February, things had progressed so well that even Jack Dineen had come around. "He's a changed individual," Norman wrote Frances. He'd even stopped drinking, had nothing but praise, and went so far as to say he thought the job would make Norman famous.

Frances, too, had progressed. "You're coming out in the clear," wrote Natasha Rambova, busy charting her friend's astrological vectors. "Both your Sun and Mercury are in the sign of Scorpio . . . I do think you could slowly increase your doses of violet light."[405] Mrs. Bel Geddes was finally coming home, though there was talk of a rib re-section if it turned out the tuberculosis had traveled to her lungs.

•

WILLIAM KNUDSEN HAD done everything in his power to make the project a success, but he hadn't made it easy. He was a large man, slow and thorough, who only subscribed to what he could *see*. Norman's "artistic" habits were the subject of particular annoyance. Like F. L. Wright, he was an insomniac who sometimes worked until four or five in the morning. Knudsen, whose employees were at their desks at 8:00 a.m. sharp, took to telephoning from the train station at the crack of dawn, inviting his charge to last-minute 7:30 a.m. breakfast meetings that required Norman to traverse seventy-five blocks of traffic between his apartment and GM's New York office.

His weekly inspections expanded now, from several hours to half a day, with a string of stenographers in tow. In the project's early days, department heads had made occasional appearances, but as the pavilion approached its final stages, GM higher-ups began swarming, reporting back to Detroit with long memos on color, "good taste," and styling, and generally getting underfoot. "It's a damned good thing you people didn't become so interested earlier," Bel Geddes confided to the Dane.

For six months or more, the two had traded punches. Knudsen's irritation over Norman's sleep schedule was compounded by the designer's insistence that GM employ the best theatrical illumination engineers, not just Westinghouse or GE technicians, to achieve the subtle effects he wanted. Ultimately, Futurama incorporated more electrical apparatuses than all Manhattan's theaters combined that year.

Then, about two months before the Fair opened, Knudsen's strict tone and cool demeanor, what Norman dubbed his "dry ice treatment," noticeably changed.

It happened one March morning during an inspection. The corporate president veered from protocol, suggesting the two of them take a look at how the other exhibits were coming along. But walking the grounds, Knudsen didn't pay them much attention. With tangible evidence of what

Futurama worker placing cars "downtown."
(Courtesy: Harry Ransom Center, The University of Texas at Austin, photo by Richard Garrison.)

Norman had been entertaining in his imagination all along, he'd come to recognize a fellow "production man," not just an "artist" filled with hot air and wishful thinking, not an easy thing for a man like Knudsen to admit.

Then, with a few weeks to go, the first test runs of the "silent" conveyor belt, a combination train-elevator-escalator, took place.

Knudsen, Richard Grant, and a small group of GM employees (Dineen among them) stood on a strip of floor moving at the same speed as the conveyor belt (102 feet per minute), then settled into seats heavily upholstered in top grade mohair for the seventeen-minute ride.

The extraordinary piece of machinery worked flawlessly. The belt extended for one-third of a mile, moved by ten motors (designed to handle overloads, in the event of a breakdown), so perfectly counterbalanced that it cost only 21 cents an hour, or $2.73 a day, to operate.

One thing still bothered Knudsen, though. The serpentine entrance ramps Norman had fought so hard for—why did he insist on making visitors climb up fifty feet—five stories—only to walk back down?[406] It was bad planning, bad crowd control, and expensive to build. Why not just let people enter on the straightaway at ground level, where the entrance was three times wider?

"It's what's called 'good showmanship.' No one's *required* to use the upper ramps."

"And they won't." The Dane shook his head. "Unless you're planning some kind of special inducement."

The inducement, Norman assured him, was already in place.

•

IN LATE MARCH, Field Marshal Hermann Goering was enjoying a weight-loss vacation in San Remo, his 250-pound bulk jiggling on the tennis court "like a huge hurried custard," noted *Life* magazine's photographer, a workout interrupted by Hitler's annexation of Bohemia and Moravia.

Norman was en route to Washington, D.C., with Pax for drinks, dinner, and a presentation at the White House about ideas for a new type of naval craft—covered with a rounded steel shell that deflected bombs, causing them to explode off the side—one of several top-secret projects he was working on. He'd prepared an eight-page speech and brought along his model for the world's first streamlined, ocean-going diesel yacht (the one designed for Axel Wenner-Gren), along with a baseball bat to demonstrate the shortcomings of current ship hulls. But he'd also been invited because Roosevelt was interested in worthwhile relief projects and believed none were more national in scope than road building.

Unofficially, Norman hoped to broach the idea of Roosevelt personally opening GM's pavilion. There were already plans afoot for the president to inaugurate the Fair remotely. (In 1935, he'd pressed a gold telegraph key in the Oval Office that set lights flashing and sirens screaming at the start of Manhattan's Industrial Arts Exposition.) What better way to arouse public interest in, and endorsement of, a project the president believed in: a transcontinental highway system with high-speed motor ways? And Futurama had a security advantage. It was the only exhibit the president could experience without having to walk around in full public view.

Alfred Sloan had been warmly cordial to Bel Geddes from the beginning, but the news that Roosevelt would be delighted to accept Norman's suggestion, a public relations coup of the first order, on condition of a written invitation, left him unmoved.

FDR's New Deal, with its encouragement of unions and collective bargaining, struck Alfred Sloan as so abhorrent that GM had secretly stockpiled more than $24,000 worth of tear gas. In 1936, it had come in handy when 140,000 workers at seventeen GM plants staged a "sit down" strike, bringing the corporation to a standstill. To Sloan's fury, Roosevelt had sent in the National Guard—not to put it down but to protect the striking workers from corporate goons.

No invitation was forthcoming.

•

NORMAN DECIDED, REGARDLESS of how Futurama was received, to abandon the shabby, East Thirty-Eighth Street firetrap. On April 1, 1939, Norman Bel Geddes & Co. began its move to Rockefeller Center's Associated Press Building, taking over the entire eighth floor. Like the huge gilded Prometheus that graced the main building's entrance, the ninth-grade midwestern dropout had stolen fire from the gods—his eponymous corporation ensconced in Manhattan's newest jewel in the crown.

The cost of converting the raw space—installing a ventilation system, fixtures, wiring, painting, carpeting, and more—ran to $19,329 (in excess of $312,660 today). Completed, the expansive headquarters featured ivory-colored walls and a sixty-three-by-ten-foot carpeted hallway inset with seating alcoves and display vitrines. There were twenty-two rooms in all, including a fully equipped and soundproofed machine shop for creating models, a spray booth, a color room, a conference room with ten chairs, a draftsman's room, and a small print shop (run by a man who'd learned his trade at Black Mountain College). There was even a Play Department, where scripts were read. Bel Geddes–designed furnishings were scattered throughout.

The choice of venue was also an homage to the memory of Raymond

Sitting areas of Norman Bel Geddes & Co., Rockefeller Center, 1940.
(Courtesy: Harry Ransom Center, The University of Texas at Austin, photo by Richard Garrison.)

Hood, Norman's "principal recompense" for having worked on the Chicago Fair. "The man who made modernism sexy"[407] had died five years before, at fifty-three, without having seen his vast Rockefeller Center complex completed.

On April 19, William Knudsen addressed 568 guests at a celebratory preview dinner. "It's only fair to say that Norman Bel Geddes had us guessing in the beginning . . . From the reactions I had today, I know [Futurama is] absolutely original, and its execution has shown he had a good deal more foresight than we had hindsight . . .

"It cost us $6,700,000. We hope you like it."[408]

Eleven days later, a hot, stormy Sunday, 200,000 turned out for the Fair's official opening. Miss Futurama was on hand for press photos, resplendent in a futuristic robe (semitransparent, with exaggerated collar, shoulders, and cuffs). Surrounding her was a bevy of female consorts (all GM employees) in various combinations of rubber, glass, acetate, and rayon. That evening, just prior to throwing the master switch that would inaugurate the ground's phantasmagorical fluorescent lighting system, Albert Einstein spoke to the crowd.

The Fair, he said, "projects the world of men like a wishful dream. Only the creative forces are on show, none of the sinister and destructive ones which today, more than ever, jeopardize the happiness, the very existence, of civilized harmony." It wasn't what the crowd, some of whom had scrimped and saved to be there, wanted to hear. As if to emphasize Einstein's remarks, the master switch triggered a system overload and the fairgrounds went completely dark. Grover Whalen rallied by ordering the fireworks set off.

Just before noon that first day, the telephone rang in Norman's new office.

"Geddes, why aren't you out here?!" It was Knudsen.

"Anything wrong?"

"Of course not. But you should see the crowd trying to get into the building."

"Happy to hear it."

"You were right," the Dane said. "The upper ramps filled up immediately, and the crowds keep swarming in. How could you have known? No one has used the ground floor entrance yet; I've been watching for over an hour. Why do you suppose that is?"

"Why don't you ask them?"

"I did," GM's president replied. "They don't know either. They just smile and shrug their shoulders. You've done a better job than you know, Geddes."

By the end of the first week, Futurama was being hailed as the Fair's most popular destination. By July, 2 million people—as many as 15,000 at a time—had snaked their way up the upper ramps, waiting two to five hours to get in, then exiting with cobalt-trimmed I HAVE SEEN THE FUTURE buttons pinned to their collars, even as world events lent the catchphrase an increasingly ominous edge. Aerial photos revealed a sea of straw-colored hats, the women in dresses and skirts, most of the men in tailored shirts and ties, despite the summer heat. "Even in youth, one could sense the central themes of mobility, speed and adventure," wrote one visitor, looking back from a distance of more than half a century. "Children of the Depression had travelled very little and led simple lives

Viewing the future in comfort, from a safe distance.
(Courtesy: GM Media Archive. Photo by Gjon Mili)

... The unforgettable sixteen minutes in the grip of Norman Bel Geddes ... surpassed all the other impressions of the Fair."[409]

Alfred Sloan—perhaps having forgiven the FDR faux pas—congratulated Bel Geddes on a corporate public relations triumph, the first time General Motors "had hit the bull's-eye."[410]

"The model landscape strikes the armchair audience dumb with amazement and admiration," applauded *The New York Daily News.* "Norman Bel Geddes built it in eight months, with 2,800 people helping him." *The Christian Science Monitor* noted that Futurama's plan would open up large sections of the country, something to consider if the United States found it "necessary to participate in that sad circumstance," the European war, with "great blocks" of refugees needing resettling.[411]

The New York Times' science editor Waldemar Kaempffert devoted an entire page to the circuitous "carry-go-round" ride, marveling at its 1,586 feet ("about seven NYC blocks") of track, its rubber tires and mats ("no clanging of shoes on steel plate"), and the convex-concave "wishbone" coupling of its cars that allowed sharp twists and turns as it rose as much as twenty-three feet, and all without jerks, jolts, screeching, or skidding, carrying some 2,200 visitors per hour, 28,000 daily. Four months of mathematical figuring had been required to create this perfectly counterbalanced magic carpet. "An error of a only an eighth of an inch," Kaempffert marveled, would have meant "a total error of forty inches in track and train."[412]

He was equally in awe of Futurama's narrative heart, the sound system that, like the circuitous conveyor belt, Grant and Knudsen had declared a pipe dream, unbuildable, eleven months before.

At twelve feet and twenty tons, the Polyrhetor was the largest precision instrument in the world. Employing the same photoelectric principle as motion-picture projection, its $250,000 revolving steel drum housed twenty-four bands of sound film pierced at 150 equally spaced points by brilliant needles of light. The voice currents, generated by light-sensitive cells, were magnified by individual amplifiers, then flowed through a unique system of trolleys to speakers embedded in the wings of each pair of chairs.

In other words, Kaempffert informed his readers, those twenty-four bands carried 150 versions of the same recorded script, each one of which was synchronized to what a visitor was seeing, regardless of where in the 7-block-long, 552-seat chain he sat, and regardless of the conveyor's speed.

This extraordinary machine never went out of sync, never varied more than a thousandth of an inch, a precision approaching that of the world's great telescopes. During the Fair's first six months, it repeated the narration script more than 5 million times. Housed in a customized room, its temperature was held constant within two degrees, its humidity within five.

And then there was the intimate narrative itself, voiced by actor Edgar Barrier, a member of Orson Welles's Mercury Theatre. One reviewer described it as a "disembodied angel" explaining Elysium.[413] According to one of Norman's acquaintants, the concept had been "borrowed" from *The Eternal Road.*

Equally important to the overall effect was Futurama's lighting. Five hundred stage-light projectors had been suspended ten to fifteen feet above the diorama. Nine different tinted filters replicated varying effects of sunlight, light amber and sea-green creating shadows on the tall buildings. With 108 different kinds of glowing bulbs, including

116 5,000-watters, the largest number ever burned continuously any-where,[414] the average temperature on the model floor was sufficient to melt the sprinkler heads, setting off fire alarms. Like the Polyrhetor, the floor model required its own separate air-conditioning system.

All told it was, wrote one reviewer, "a new phase in the study of architectural design . . . a new step in the integration of art with nature, with life, with the universe."[415] For many, experiencing Futurama was nothing less than a reality shift, a space-time warp, a kindergarten version of string theory.

The plush navy mohair chairs proved an important aspect of success. Wandering the Fair was a fatiguing business—"Flushing feet" was a common complaint. Most exhibits required additional walking. Like the cool, deeply carpeted, cathedral-like Map Lobby that greeted visitors who had stood, often for hours, in the bright hot sun, the chairs offered welcome comfort. Visitors had only to sit back and relax; the exhibit did the rest. The high-sided tandem seats proved particularly popular with sailors and their dates as a "necking" retreat.

•

THERE WERE CRITICS, of course.

An Illinois farmer wrote claiming that one of the farm buildings was a hay barn, "and hay barns went out of date in 1935."[416] Visiting Europeans, experiencing the war firsthand, had mixed feelings about all the relentless, exuberant optimism.

The brilliant but notoriously grouchy Lewis Mumford decried Fair promoters for being tied to the "completely tedious and unconvincing belief" in the triumph of modern industry. "The less said *about that* today, the better." He wondered what would happen if a blizzard struck all those high-speed highways. Still, he conceded that it was "an achievement of the first magnitude," so good that one felt the "airplane" should go up another thousand feet for safety's sake. Mr. Geddes was "a great magician, and he makes the carrot in the goldfish bowl look like real goldfish."[417]

While the general public may have found it all dazzlingly new, the cognoscenti knew better. Much of Futurama was a pastiche of existing theories and concepts that had appeared in everything from H. G. Wells's stories and Fritz Lang's *Metropolis* to sketches by F. L. Wright and Raymond Hood. And certainly Le Corbusier was in the mix. Success, as the proverb goes, has many fathers.

Granted, Bel Geddes had invented neither the carry-go-round nor the Polyrhetor. Those tasks had been farmed out to mechanical and electrical engineers. But the brash concepts were his, part of the complex vision that tenacity, stubbornness, and dogged faith had set in motion.

His genius was in fitting all the elements together, arranging and integrating the pieces into an accessible and interactive "experience," both entertaining and educational, building up from a base of existing parts to create a new whole—a process that sixty years later would be dubbed "cultural ratcheting."

And while his 50,000 instantly iconic futuristic silver teardrop cars were enamoring millions, Norman continued to drive his beloved 1928 Packard roadster.

(Courtesy: GM Media Archive)

CHAPTER 17

Crystal Lassies; or, The Future Will Be Topless
1938–1940

The Town of Tomorrow is morally pure,
in the realm that its pinnacles smile on,
Smile on;
For who would get tight in a Plaza of Light,
or sin in the shade of a Trylon?[418]

The Fair's Midway—traditional home of less-elevated (and more lucrative) entertainment than corporate pavilions—had a time-honored rule of three: peeks, freaks, and rides.

"One thing World's Fair veterans may find lacking is sex," reported *Time* magazine, somewhat prematurely, of the first. "Despite the appearance of such numbers as Della [Rose Dance] Carroll, who once lifted Adolf Hitler's brows several pegs, Grover Whalen last week insisted that there would be nothing at the Fair to shock anyone."[419] Billy Rose, the official brains behind the Midway, had gone to great lengths to convince Whalen that "girl shows" were more nightclub than burlesque, certainly less risqué than what had been playing at Minsky's since the mid-1920s.

Two months later, with the Fair gates open, *Time* noted that the Midway was home to more unadorned female pulchritude "than any place outside of Bali."[420] Not only the quantity but the palette had shifted. In the past, only dark-skinned "primitives" had appeared stripped to the waist.

Della was the tip of the iceberg. The Living Magazine Covers exhibit featured young ladies wearing only shoes and a discreet "patch." Exstasie was home to Salome's "Dance of the Seven Veils," a display described by NYPD plainclothesmen as exhibiting "intense lust and passion."[421] A "Miss Nude of 1939" beauty contest featured Fair employees willing to pose in a G-string. At the Crystal Palace, Rosita Royce performed with garment-untying parrots and doves. Frozen Alive offered more family friendly fare. For fifteen cents, spectators could converse, through a microphone, with Arctic Girl, clad in an abbreviated bathing suit, entombed in a 1,400-pound block of ice.

Voyeurism under the guise of history or high art was a classic Midway tactic that fooled no one. The Cuban Village offered "exotic Yanego rites dating back to the aborigines." At Salvador Dali's Dream of Venus, visitors could watch Lady Godivers, naked but for fishnet stockings, cavort underwater amid rubber telephones, exploding giraffes, and the painter's signature soft watches. One young performer, a former Girl Scout paid to lounge, covered with a few strategically placed branches, on a red satin couch, told *Life* magazine that the faces on the other side of the glass "looked like hungry wolves."[422] The Congress of Beauties consisted of professional strippers billed as ancient followers of Heliotherapy. (Hollywood choreographer Busby Berkeley was listed as an advisor.) In the Glamazon Village, female gladiators sparred with helmets, shields, and javelins, one breast bared per Amazon.

The Fair as a whole was rife with sexual innuendo, especially if one was looking for it, starting with what became the most recognizable symbols of any exposition since the Eiffel Tower—the priapic 700-foot Trylon obelisk and eighteen-story breast-buttock Perisphere.

Even Highways & Horizons wasn't exempt.

GM's immense building, cut through top to bottom with a vertical vermilion slit, struck some observers (and many feminist critics yet to come) as being more than a little suggestive of things best relegated to a gynecologist's office.

Upon entering Futurama, one descended on twilight-lit switchbacks. Thick carpet. Soft music. All was hushed, reverential, womb-like. The roadways of today and tomorrow appeared as spidery, vein-like filigrees across the immense wall. Then visitors, in chromosome-like pairs, were led to plush velour love seats that proceeded to glide into a dark tunnel. As "the future" appeared below, their carry-go-round seats transported them in a serpentine (sperm-like?) pattern, now high, now low. An authoritative but intimate voice narrated the journey, which ended with a "climax"— entrance into a bright, new world.

Second in the Midway's triumvirate were the freaks.

Marvello, the Fingerless Pianist. The Lion-Faced Chinaman. The Man with the Twelve-Inch Monkey Tail. Adonis, a bull with "transparent" human skin, was insured by Lloyd's of London for $15,000. Visitors could watch a man pull a girl in a cart with his eyelashes or a woman swallow her own nose. When possible, freaks and peeks were combined. Duckbill Ubangis, pink-eyed Icelandic albinos, giraffe-necked Burmese, Batwa pygmies, and Romanian giantess sisters all appeared in various stages of undress.

The Fair Committee had waded through hundreds of Midway proposals, including a booth of satirical contraptions (e.g., an Anti-Suicide Machine to prevent stockbrokers from jumping out windows), courtesy of Rube Goldberg.[423] Though Norman was busy tracking down a sponsor for what was still a "City of the Future" without a "ride," he'd submitted his unrealized Chicago Fair theater and restaurant blueprints, along with half a dozen Midway ideas ranging from a Nutshell Jockey Club game to a photo booth that (thanks to Dr. Seuss's patented two-lens Infantograph) merged a couple's features into a likeness of their future offspring.

One passed muster, a "highly scientific" nudie show, a "gazing palace" that would, Norman assured Grover Whalen, be "high class," though it could, he confided, "be as naughty as you might, at any time, decide."[424]

With everything else on his plate, Norman became responsible for raising $125,000 ($2 million today) to cover construction, air-conditioning, maintenance, and eventual demolition.

•

MORRIS GEST STILL sported his signature black fedora, but beneath it his cheeks were sunken. At sixty-three, his career had lapsed into predictability.

Time had called him the producer of "staggeringly magnificent extravaganzas,"[425] an initiator of comets across strange and daring skies. But there would be no comets now, no Ballet Russe or Eleanora Duse, *Chauve Souris* circus or Moscow Arts Theatre.

He was reprising his Midget Village from the 1933 Chicago Fair (the largest population of "little people" ever assembled), ratcheted up several notches with an elaborate 36,000-square-foot "postage-size city" (the exact square footage of Futurama). Fair visitors would be able to watch the mayor at work, dispatch telegrams (the couriers in miniature regulation uniforms supplied by Western Union), even have their hair cut in America's only midget barbershop.

The "city" sailed to New York on the *Ile de France* along with five of the world's smallest horses, a tiny church organ, and dozens of what Gest called "masterpieces"—trapeze artists, magicians, jugglers, trick cyclists, singers, comedians, musicians, and a palm reader, including five married couples, ranging in age from nineteen to fifty-six, in height from two feet two inches to four feet three inches—gleaned from across Europe. Also in the cast was the Doll family, fresh from Todd Browning's *Freaks*, and several Munchkins from *The Wizard of Oz*—a film for which Bel Geddes's costume and scenic design talents had been considered.[426] The *Wizard's* Manhattan premiere, slated for August 1939, was expected to be box office boffo; Gest was banking on it working in his favor.

"Midgets are gigantic!" he declared in a thirty-three-page illustrated brochure. "Little in body but great in mind and heart . . . I believe I'm setting a living example of how all Nations can live together in harmony."[427] He christened his latest production *Little Miracle Town*, referencing Reinhardt's *Big Miracle*, now fifteen years behind him.

Norman's Midway "gazing palace" owed a debt, of sorts, to owl-faced Alex Woollcott.

Since the demise of the Algonquin Round Table, Woollcott had been pining for a replacement venue, a meeting place for those he considered worthy. He'd even come up with a name, the Elbow Room—something small and comfortable, with soft lighting and plenty of elbow room—an invitation-only club where VIPs would never be bothered by news photographers or autograph hounds, and where Alex's culinary preferences would reign. He found an unused space conveniently adjacent to the Berkshire Hotel's kitchen on Madison and Fifty-Second, convinced Beatrice Kaufman, wife of the Broadway playwright, to become his business partner, got two realtor cronies to put up the funds, and commissioned Bel Geddes to create something suitable.

Only three years had passed since Norman's partnership with George Howe had been unceremoniously shut down. Despite the risk of putting Norman Bel Geddes & Co. in "a somewhat dangerous position"[428] should he be found out, he submitted his blueprints to the city's building department under a pseudonym.

The space was long and narrow. To open it up, Norman custom ordered blue-tinted mirrors for the side walls, the largest mirrors Libbey-Owens had ever made. So large that the soon-to-be-legendary glass had to be shipped via Canada and down through New England, as they didn't fit through a railroad tunnel on the direct route.[429]

Glass was very much in vogue, a material thought as futuristic as aluminum. "You . . . enter a skyscraper of colored glass, are whisked upward in a glass-enclosed elevator shaft, enter an office with glass floors, walls and ceiling," predicted *Popular Mechanics* of the world to come, "and perhaps doff a topcoat of fibrous glass before sitting down at a glass desk."[430]

Curved partitions and changes in the floor level added to the Elbow Room's illusion of spaciousness. Un-mirrored walls were painted chocolate brown. Everything, down to the porcelain plates and gold tableware, was customized, for a total design tab rumored at $200,000.

The result was an exercise in quiet luxury. Diners entered through a pair of sleek, unadorned bronze doors into a circular, pigskin-covered foyer topped with a thirty-five-foot-high copper ceiling. Concealed lighting around the upper periphery suggested a streamlined solar eclipse.

The narrow, twenty-foot dining room, home to the mirrored side walls, was set a few steps down. (Henry Dreyfuss claimed the mirrors were so heavy that the walls sagged.) There was seating for sixty-four in round-backed, voluptuously upholstered armchairs with automobile-like armrests that could be pulled down for added comfort. Tables (walnut on copper bases) adjusted to two different heights, accommodating both cocktails and meals. Automobile parking lights, discreetly embedded well below the dark blue, star-flecked ceiling, cast a diffused light, minimizing the room's eighty-foot height.

Using the automobile as a design motif wasn't new. The Chrysler Building's deco facade sported half a dozen winged radiator caps and a frieze of hubcaps and mudguards. But those were external and merely decorative. Bel Geddes's armrests and ceiling lights were both decorative and functional, their novelty and understated elegance incorporated into an interior space.

Opened in the spring of 1938, Woollcott's new playroom was immediately recognized as one of the city's most elegant and exclusive private clubs. But after operating at a loss for several months, the Elbow Room closed.

"A group of intelligentsia" wanted a place "safe from the eyes of the vulgar herd," wrote *Esquire* magazine, only to discover "the sad truth that egos can't live on other egos without replenishing the supply."[431] It seemed that VIPs were staying away because they *liked* being bothered by autograph hounds and the press. The press, meanwhile, ostracized by Woollcott (a former newspaperman, no less), had offered no publicity. Plus, in his role as newly minted nightclub entrepreneur, Alex had set the prices so high that even the upper crust hesitated to venture in, and his lambasts of the decor deterred even his own friends from visiting.

"You've been in my apartment often enough to know what I consider pleasurable," he spewed, venting his fury at Norman. "Your wildest dreams never committed such a palace for incubus as this Coney Island mirrored mockery where I see myself making fun of myself and, I presume, you're peeking through a carefully concealed hole, laughing your diabolical head off. I don't see why," he whined, "when I have always shouted your praises to the skies, you had to let me down in this fashion."[432]

Outraged sensibilities were a Woollcott trademark.

In January, following a few alterations, the venue reopened as the Barberry Room. With the public now cordially invited, the food French, and the wine list excellent, the place was descended upon "with squeals of delight" by those who could manage its still-rarified tariff.[433] Norman kept his hand in, gratis, advising on everything from the headwaiter ("he must buy himself a well-tailored dark suit . . . shave at five o'clock every afternoon . . . not smoke cigarettes in the dining room") to marketing strategies.[434]

The Barberry would remain a premier Manhattan destination for the next twenty years, its decor inviolate.

Lunching in the "Coney Island mirrored mockery" one afternoon, Norman noticed a "luscious, hip-swinging blonde"[435] strutting past his table, reflected and re-reflected in his enormous (and enormously expensive) blue mirrors. What if, he thought, she danced instead of walked? And wasn't so . . . thoroughly dressed?

Sixty-four mirrors—squares and rectangles ranging from six and a half to thirteen and a half feet—eventually lined his Crystal Gazing Palace walls at calculated angles. A fourteen-foot mirrored dais in the center served as the stage from which a sole dancer, lit by eight 2,000-watt spotlights, was transmogrified into "uncountable millions" who imitated her every move, "a whole chorus of World's Fairettes."[436] Some dancers gyrated to rumbas, others to Camille Saint-Saens.

"A polyscopic paradise for peeping toms," wrote the *New York Post*, "an illusory harem [with] enough angles to confuse an Einstein."[437]

Even given Norman's relentless energy and drive, the work load was extreme—organizing a corporation of some twenty Wall Street bankers, attorneys and friends to finance his kaleidoscopic mirror show, supervising

Futurama's pre- and onsite construction, overseeing the details of his new office, preparing to meet with Roosevelt, dealing with daughter Barbara's elopement plans, and commuting to Connecticut on weekends.

He was also trying to help Teddy Backer (whom he'd worked with on the Barberry Room) sell the idea of turning the top sixty floors of the Empire State Building into a hotel, something that could translate into a job on a par with the Highways & Horizons commission.[438]

Through it all, Norman's "fickle mistress" still held way. He and film producer Buddy De Sylva were planning to reprise *Lysistrata*, this time as a musical, if only Cole Porter could find time to compose the tunes. Mae West seemed a natural for the bawdy romp.

"I went home yesterday afternoon to dress for dinner," Norman wrote Frances in early January, "threw my clothes off in a great hurry and jumped in the tub for fifteen minutes to relax. After a few minutes, on reaching for the soap, I was aware that I hadn't taken off my underwear."[439]

•

BY MID-FEBRUARY 1939, the Gazing Palace's blueprints were well underway. Though the concession was remote from anything "intellectual," early sketches reference Norman's old friend Claude Bragdon,[440] a former architect who'd written, at length, about simultaneity, multiple perspectives, and the fourth dimension as represented by a prism-shaped tesseract, all theories that had informed Bel Geddes's design classes in the 1920s. The peep show's final structure—a twenty-six-sided octagon, thirty-two feet from plane to plane in all directions, supported by a steel frame—may also have been a nod to the Cirque d'Hiver, the twenty-sided oval, dating from the reign of Napoleon III, in which Norman had enjoyed many Parisian spectacles.

Alerted to this latest madness, helpful friends suggested trading in the Victorian-sounding Crystal Gazing Palace for Sexorama, Mir-O-Peep, 5,672 Ravishing Females, Prismatic Peaches, or Crystal Asses. In the end, it would become Crystal Lassies (the Scottish in Norman coming into play) though in the Fair's official guidebook it appeared as "Bel Geddes's Mirror Show," in which the designer "known the world over" for his "distinctly serious intent . . . takes a vacation and turns his talents to the lighter side of life."[441]

Spectator anonymity was a priority. As many as 600 visitors at a time would stand on "peeping platforms" on three levels provided with window slits. Thanks to carefully arranged strips of one-way mirror (also known as Chinese or "speakeasy" glass), each would have the illusion of being the sole onlooker, and dancers would be unable to observe their observers once the lights came up.

The dressing room, lodged beneath the stage, was an eerie, blue-lit stronghold. An elevator lifted performers, one by one, up to the mirrored dais. Twenty-four times an hour, a wardrobe mistress worked the lever of the lift's weighted trap door. Dressed in a harem outfit—a G-string, a gauzy chiffon skirt, the occasional veil—each girl danced for three minutes before being replaced by one of eight colleagues. Between dances, performers amused themselves knitting, swapping stories, or sending out for coffee.

Norman had worked out all the details. The talent worked two-hour shifts; they cost him $500 a week (compared to Billy Rose's Aquacade, 230 performers for a weekly total of $15,000). Admission was fifteen cents; after ten minutes, a customer theoretically got his money's worth.

Theatre Arts Monthly would call Crystal Lassies "a techno-soft porn extravaganza." Norman made no high claims. He did it for fun, designing the harem-esque costumes and personally auditioning the girls who'd wear them, one of whom was Eve Arden, soon to achieve radio and TV fame as *Our Miss Brooks*.

When, in anticipation of the Fair, *Vogue* commissioned nine industrial designers ("These men know tomorrow like their own streamlined pockets") to predict the Woman of Tomorrow's attire, see-through clothing reigned.[442] People in 2055 would "not wear costumes of cellophane illuminated by neon lights . . . not nightmare stuff, not jazz," H. G. Wells objected.[443] But the author of *The Time Machine* was of a minority opinion.

A similar *Vogue* article predicted a world of gene-manipulated Stepford Wives with perfect bodies and sympathetic dispositions, all "ordered like crackers from the grocer."[444] The eugenic ideals that were passing for wit in America's premier fashion magazine were, at that exact moment, serving as a death warrant rationale in Vienna. In Paris. Berlin. Prague. Belgrade.

.

"LET'S GIVE THESE poor bastards a cocktail party," Norman told his Crystal Lassies partners.[445]

Grover Whalen arranged a motorcycle escort for Norman and mistress of ceremonies Mary "My Heart Belongs to Daddy" Martin. Two days before the Fair's official opening, 200 investors (some prominent New York attorneys), journalists, and Park Avenue blue bloods watched the performance, then sipped champagne and fraternized with the dancers.

Unbeknownst to all but a handful, Norman was grappling with the knowledge that Frances, who'd begun to continuously hemorrhage, was undergoing a trio of operations, the last to remove the remainder of the ribs on her right side in order to bring about the collapse of her right lung.[446]

"Don't think for an instant I don't know everything you're going through," he wrote her. "I know it so well my own head nearly blows off . . . Days and nights, nights and days, you're all I care about, all I think of. But I can't tell you [in person] what I'm telling you now—I'd go to pieces. And you would, too."[447]

Trouble wasn't long in coming. Lassies was forced to close for several hours on Wednesdays when the Boy Scouts and other underage groups came to visit. Worse, patrons under the age of sixteen were forbidden entry. There was nothing potentially offensive, Norman protested to Whalen. Women would enjoy the show as much as men, what higher praise could there be? The whole thing was ridiculous, and it was cutting into revenue.

One notoriously shrill journalist attacked the Midway as "un-American," a proliferation of "lewd, lascivious, indecent and vicious exhibitions" created by foreigners (European, Jewish theater producers like Billy Rose and Morris Gest) "who just won't get assimilated."[448] One outraged citizen implored Whalen to exchange the nudes for "clean shows like the Rodeo and Midgets and other fine things."[449]

Whalen changed the Midway's name to the Amusement Area.

A Queens pastor, having spent several hours personally inspecting the "vile resorts," wrote the license commissioner, asking *what he thought his oath of office meant?* Public Safety officers described one of Norman's dancers as being naked but for two bright feathers, "illuminating paint" on her nipples, and a tiny lower "shield." Performers' buttocks were bare, and the mirrors beneath their feet "entirely exposed" their "private organs."

Mayor La Guardia issued orders to make arrests.

Some of Crystal Lassies' dancers protested the harassment—after all, their sins were no more egregious than their neighbors'—by painting "Amusement Area" in large letters across their exposed midriffs.

In June, less than two months after opening day, plainclothesmen took Miss Kay Frop (Frozen Alive) and Miss Joan Vickers (Congress of Beauties) into custody, along with their managers.[450] Fair management responded by ordering "full uppers" for all the girlie shows.[451]

Norman would have none of it.

"Mr. Geddes got tired of the police, the censors, and everybody else coming in here," Lassies' manager Nick Holde told the press. "They told us what was art and what was not art until we were sick of it. When they made the girls wear big brassieres, we heard people say, 'If that's what Norman Bel Geddes thinks is art, he's goofy.' And then he decided to act."[452]

It wasn't the *show* that was vulgar, Norman determined, but the "uppers" that were "obscene and unspeakable." Would underwear improve the Venus de Milo? He posted an emphatic letter in the subterranean dressing room:

> *You will, beginning immediately, restore your costumes to the identical status as worn by you at the opening performance on May 26.*
>   *This eliminates the use of brassieres.*
>
>   *Norman Bel Geddes*
>   *Producer and General Manager*

If Lassies would remain topless, other concessionaires had no choice but to ignore management's edict or go broke. Who would pay to see "something" when they could see "everything" next door? Norman's revolt was gleefully picked up, far afield, by the press. A Letter to the Editor in Kentucky's *Louisville Times* stated that the only thing more obscene than brassieres were girdles. As for Norman's reference to the Venus de Milo, the writer continued, she didn't really enter into the argument, not having been created "to caper for gate money," shop for groceries, or attend church.[453]

•

IN ANTICIPATION OF the Fair's second season, which opened in April 1940, the grounds were refurbished with 100 tons of new paint, the Midway-cum-Amusement Area was renamed the Great White Way; Elektro, the seven-foot robot, was given a robot dog, Sparko; and the entire enterprise was reinvented as a county fair where common folk wouldn't be intimidated by "high silk hat" types. The once-dazzling array of souvenirs (clocks, radios, cocktail glasses, shower curtains, wallpaper) was reduced to armbands and zipper pulls, the gardeniaed Grover Whalen was replaced with a banker, and the man in charge of selling bonds to pay for the Fair, a former president of the U.S. Stock Exchange, was sent to Sing Sing for fraud.

Europe's escalating war was evident in the disappearance of nearly a dozen "lost" nations. The pavilions of Albania, Germany, the U.S.S.R., Poland, and Yugoslavia, had been torn down. Czechoslovakia's handsome edifice—with its many-thousand-jeweled coat of arms and a picture panel of 50,000 rare woods—now represented a country that, technically, no longer existed. Romania's all-marble pavilion, with its rock salt ceiling, managed for the duration. Hitler wouldn't invade his ally until the Fair's final weeks.

The privilege of a Great White Way renewal was contingent on a concession having been "credibly operated"—that is, sufficiently lucrative. (The Fair received 25 percent of gross receipts.) Crystal Lassies didn't make the cut. The problem, according to Frances, was that Lassies was "too refined, not vulgar enough, and the girls too beautiful."[454]

Still, the second season didn't lack female pulchritude. A new addition, The Hot and Cold Show, featured ice-encased girls (Arctic Girl competition) and "hot as sun-kissed asphalt" exotic dancers.[455] Dali's Dream of Venus lovelies, Living Magazine Covers (its eroticism "softened" by trick lighting), and Strange As It Seems ("genuine savage natives" from around the globe) all remained to compete for the crowd's nickels, dimes and quarters. Rosita Royce returned, briefly, with her garment-untying doves, though she would leave, citing exhaustion from seventy-four shows a week. The trade papers claimed her departure had more to do with rival stripper Tirza, who allegedly attacked Royce's doves with a BB gun.[456]

Futurama remained the Fair's number one attraction. The previous season's daily average of 28,000 (2,150 visitors per hour) would nearly double. It wasn't uncommon for people to revisit multiple times.[457]

Some critics had decried its absence of slums, others praised its absence of billboards. (Norman had fought numerous battles to keep the latter out.) Still others asked: *Are there no institutes of higher learning in the future? No babies? No toll stations? No religion? And where do all those cars go for fuel?* There was, in fact, a monastery and a replica of Notre-Dame cathedral (included for scale). Now, in a $15,000 advisory role, Norman dutifully added 600 churches, a university (with stadium and tiny score boards), hundreds of gas stations, a golf course, new miniature amusement area rides, a river showboat, and 30,000 new trees ranging from seedlings to forest regions. Animation was increased, with several thousand more little silver cars, streamlined trains, and tiny garments on backyard clotheslines that now "flutter[ed] in a Norman Bel Geddes breeze."[458]

During the six-month break between Fair seasons, Futurama's surface was cleaned and repainted, with many parts replaced. Railings and mammoth serpentine awnings were added to the exterior ramps, along with eight water fountains "cooled by Frigidaire," a GM affiliate. Air-conditioning was installed throughout the building, as were micro-blitzkrieg-like toilets that zapped each seat with ultraviolet rays—the same basic technology being used to reduce food spoilage and odors in refrigerators and for tenderizing meat.

Within ten minutes of the reopening, a long queue had already filled the switchback ramps and beyond. Futurama now remained open an additional four hours; still, at day's end, thousands were turned away. Even corporate guests and high-ranking business associates had a hard time getting in. It was the first time an industrial exhibit had out-pulled the Midway (by whatever name) at a world's fair or expo.

In a ludicrous effort to level the playing field, Henry Ford commissioned an eighteen-minute ballet in which Old Dobbin returns from the past to report on his displacement by the automobile, an idea credited to Walter Dorwin Teague.[459] Forty dancers, including two boys who played the horse, performed "A Thousand Times Neigh" every hour on the hour, twelve times daily.

New York Park Commissioner Robert Moses, whose single-minded ferocity had already reshaped much of New York's urban landscape, with more to come, told the American Society of Civil Engineers that Futurama was "interesting but hardly scientific," in some cases "plain bunk," a "work of the imagination," and that transcontinental speedways were, in any case, unnecessary.[460]

Under the present system, new roads are outmoded before they're even completed, Norman shot back. "It's not only a vicious circle but a silly one." He, for one, couldn't be as calm as Mr. Moses about the annual highway death toll—some 32,000—not to mention property loss. As for being "hardly scientific," Futurama had been based on seven years of careful research, most of which hadn't been previously compiled, never mind weighed and evaluated. And as for being unnecessary, that was a "defeatist view . . . strangely incongruous" with the building of the railroads and the work of the Wright Brothers and Henry Ford.

Moses, meanwhile, was already looking into obtaining permits to run bridges over streets so that pedestrians could cross above the traffic.

It didn't help that, several years before, the building czar had incurred Bel Geddes's wrath by razing the Central Park Casino, one of Norman and Frances's favorite dancing and dining venues.[461] Influence was brought to bear. Fifteen years later, Norman would still be railing: the Casino made money, had a great staff, had never been connected to scandal. "Commissioner Moses, who perhaps has never danced in his life, doesn't drink and can't eat because of his ulcers or something, who I think the world of but who is also a horse's ass, tore it down."

Moses's opinion of Futurama was further compromised by the fact that, despite being hailed as "America's greatest road builder," he'd never learned to drive, relying, like Raymond Loewy, on a bevy of chauffeurs. "I have great admiration for the Park Commissioner," Norman told *The New York Times.* "No one excels him in resetting bushes in a park and finding play spaces for underprivileged voters."[462]

It's estimated that Highways & Horizons ultimately attracted more than 27 million people, three-quarters of the Fair's total attendance, more "than saw all [Bel Geddes's] theatrical productions put together," noted theater critic Morton Eustis.[463] More, Norman added, "than ever saw any motion picture, more people than attended all the baseball games in the country those same two summers."

Mrs. Bel Geddes was among the millions. Norman made special arrangements, in the autumn of the second season, for Frances to visit in a wheelchair. Following the ribs removal, she'd been declared free of TB, "a little battered up" but on the mend.

On a more pragmatic note, Futurama helped boost GM's annual sales for cars and trucks in the U.S. and Canada by 916,000, an increased net profit of more than $190 million,[464] a victory the corporation would tap into for decades to come. The "genius with his head in the clouds" had been vindicated.

But even the enormous success of Futurama, the Parachute Jump (more than half a million riders), and Billy Rose's Aquacade (grossing in the millions),[465] augmented with souvenir royalties and food concession revenue—16 million hot dogs, half as many hamburgers, 14 million cups of coffee, and 294,000 doses of bicarbonate of soda—weren't enough to tip the financial scale.

The Fair's planners had estimated a draw of 50 million visitors, 2.5 times the attendance at the 1933 Chicago Fair, thirteen times the population of the United States the year George Washington was inaugurated.[466] They hadn't anticipated that the same citizens who stood in awe at the Fair's scientific and electronic wonders believed that the Depression, and its concomitant joblessness, which worsened under the planners' watch, was the *consequence* of that technology.[467] In theory, world's fairs were self-supporting. In truth, they rarely broke even. The New York Fair had cost $157 million, compared with the Chicago Fair's $47 million, and it closed $119 million in the red.

On closing day, October 27, 1941, Fair attendance soared past the half-million mark. It seemed impossible that such an elaborate enterprise could just end, literally overnight, struck down like the stage set that it was.

Futurama's technical crew had the honor of the final ride. After exiting the plush upholstered seats, a tear-stained Robert Murray, the chief maintenance engineer, led his team onto the model floor, where the men stuffed miniature trees, cars, and other bits that weren't nailed or glued down into their pockets.

Shortly after midnight, a bugler played "Taps."

•

"IT WAS ALL built in eight months with a crew of 1,800 for somewhere between five and seven million," noted a *New York Sunday News* editorial. "We should think the Futurama would be worth either of these sums in advertising, good will and legitimate propaganda for more and better highways. And we hate like poison to think of this miraculous exhibit being dismantled."[468]

Encouraged by letters from hundreds of "starry-eyed visitors," one of whom compared the exhibit to a Rembrandt or a play by Shakespeare,[469] Norman fought hard for Futurama's preservation. GM was interested enough to pay him $7,500 ($121,000 today) to draft up a proposal.[470]

A caravan of trucks could transport it around the country, circus-style. It could be permanently displayed in downtown Manhattan, with moving sidewalks, in a yet-to-be-built GM Public Relations Center (another NBG project). It could be installed on Atlantic City's Boardwalk, or be rebuilt to fit on a Mississippi showboat.

Better yet, a lightweight Futurama reproduction could be installed in the hull of a giant zeppelin (with moving chairs to circulate around it), "an object of awe and wonder" flown worldwide on a goodwill tour. (Taking into consideration fuel, personnel, and crowd control, Norman estimated a cost of $4 million (the Graf Zeppelin had reputedly cost $1 million), plus another $400,000 for operating expenses.

The dirigible plan was no more impractical than Futurama had been, he insisted. It was, in fact, the only imaginative idea that outdid it. But in the end, the trouble and expense of taking apart and reconstructing the vast "jigsaw puzzle" would, engineers determined, cost as much as building a new one.[471] A few sections of the 100,000-ton conveyor belt and the 20-ton Polyrhetor were salvaged for display at Rockefeller Center's Museum of Science and Industry, but lacking Norman's magic carpet ride, they seemed sadly anticlimactic. The Army "borrowed" a large segment of the floor model for use devising wartime camouflage, and another was featured in the St. Louis Art Museum's urban planning exhibit. Some of the meticulous tree miniatures ended up at a Seventh Avenue hobby shop, sold to model train enthusiasts.[472]

Few exhibits survived. The Parachute Jump was dismantled and reassembled at Coney Island. The Aquacade's enormous pool was turned over to the citizenry of the Bronx, and the House of Jewels, once home to the finest offerings from Tiffany and Cartier, was left standing for use as a public restroom. The lion's share of salvageable metal, some 40,000 tons, went to the war effort. The Perisphere and Trylon were converted into battleships, shell casings, and gun forgings, just in time, as it turned out, for Pearl Harbor. The remainder of the World of Tomorrow headed for the junk heap.

On October 28, as the bulldozers were beginning their work, GM ran a three-quarter-page clarion call in *The New York Times* under the heading: *We Hope We Set a Boy to Dreaming.*

"Our Futurama has come and gone," it read, "its lights are darkened, its sound chairs silenced," but perhaps among the crowds there were "youngsters" (though presumably not dreaming girls) whose imaginations had been stirred, who had "caught the glimmering of a vision" they would pursue "by turning to their home chemistry sets, their construction toys, their tool kits and work benches," thus setting out "on the road to usefulness and service to themselves, their country and their fellow man."[473] It had the distinct sound of Charles F. Kettering, who liked to say that if one kept on going, the chances were he'd stumble onto something, perhaps when he least expected. No one ever stumbled while sitting down.

CHAPTER 18

Elephants in Tutus

1940–1942

Elephants are no harder to teach than ballerinas.
  —GEORGE BALANCHINE[474]

My music is best understood by children and animals.
  —IGOR STRAVINSKY[475]

For a long time, John Ringling North lacked any aspiration to rule his uncles' eponymous empire.

John Nicholas Ringling, the longest lived of the founding quintet, had parlayed the family business into a ninety-car Greatest Show on Earth by buying out every other circus in the country, including Barnum & Bailey. His assets, at their height, were estimated at $200 million; he traveled in a Rolls-Royce designed for a Russian czarina, wore alligator shoes and the finest custom-made suits, and collected Old Masters (he had a soft spot for Rubens) by the shipload. His home, a fifty-six-room pink marble confection inspired by the Doge's palace, with Florida's Sarasota Bay standing in for the Grand Canal, was an elaborate exercise in bad taste of which he was exceedingly proud.

Childless, Uncle John took an interest in young Johnny North, his only sister's wayward boy, who as a Yale undergrad had squandered his inheritance (roughly a quarter of a million today) on horse racing, roulette, fancy accoutrements (raccoon coat, roadster, silver hip flask), and women. Taking the young dandy under his wing, he brought him along on summer act-scouting trips to Europe.

But the Ringling empire was soon floundering, thanks to an expensive divorce settlement, a government crackdown on millions in back taxes (Uncle John considered taxation unconstitutional), and a loan that coincided with Black Thursday. The first of a series of heart attacks followed.

In 1936, when Uncle John died, The Greatest Show on Earth was plummeting toward bankruptcy. Johnny North—now thirty-three, having spent the better part of ten years in a New York brokerage firm—obtained power of attorney. He would, he promised, find a way to get the family enterprise out of hock, on the condition that, should he succeed, his contentious aunts and cousins would vote him president. When he managed to secure a million-dollar loan, they reluctantly handed him control for five years.

Unlike the rest of the Ringling clan, all staunchly averse to change, North recognized that a radical overhaul was crucial. The family's fifty-five-year-old enterprise had become shabby and unsafe, frozen in time, a money-leaking hodgepodge—operating costs averaged $17,000 a day, in Depression-era dollars—and meager competition against Hollywood's ever-expanding shadow. The war in Europe was also taking a toll. Some performers had joined the military; a few had been interned for having German or Japanese roots.

By the time Johnny North phoned Bel Geddes, he'd already created a fury by selling off hundreds of Percherons, Belgians, and Clydes used to haul the show, in exchange for a fleet of tractors (an annual savings of $500,000), and by painting the Big Top's traditionally pristine canvas—dark blue at the peak, then paling down as it sloped (to prevent lighting effects being bleached out during sun-blazing matinees). He'd invested in a motor-driven blower system that took three extra flatcars to transport, fifty more men to run, and 7,328 tons of ice per season, creating what everyone said was impossible: an air-conditioned tent. Other blasphemies included hiring a "lady" clown and spraying lilac, honeysuckle, and tea rose aromas into the air, at sawdust level, to counteract more earthy odors.

Most important, he'd rushed to import a windfall of foreign acts "before the door closed" on an increasingly volatile Europe. Among them was rapid-fire juggler Massimilliano Truzzi, equestrian Roberto de Vasconcellos, the acrobatic Akimoto and Cristini families, and Alfred Court, who handled every kind of wild cat and four kinds of bears, all without the usual whips, guns, and chairs.

Ringling's Midway offered Fat and Tattooed Ladies, a snake charmer, a Rubber-Necked Man (former railway clerk), a Living Skeleton, a Human Pincushion (who drove six-inch spikes into his forehead), and a Fire Eater. Miss Patricia swallowed swords and the Great Waldo swallowed live mice (then coughed them up). Alfred Langevin could smoke a pipe or cigarette, play the recorder, or blow up a balloon with his eye. Popeye, the facial contortionist (he'd had all his teeth pulled in order to make his faces), died in 1937, and Omni the Zebra Man (a former British Army officer who'd paid thousands to have himself tattooed from head to toe in bold geometrics) had left to try his luck in Maori country.

The circus had once been America's premier popular entertainment. There had to be ways, North reasoned, to keep the best of the old traditions while upping the ante. More showmanship. More continuity. And more intimacy, a sense of connection between the crowds and the performers.

•

NORMAN WAS ONLY too happy to bring his talents on board. He retained vivid childhood memories of the circus's arrival, the animals emerging from their box cars like emissaries from the pastel-colored maps in his schoolbooks—Africa, India, and Asia. The inside of the Big Top glowing like a dream castle. The carnation-red curtains.

Visiting Ringling's setup for the first time, he was impressed with how well things were managed, given the heavy, outmoded wagons and all the various riggings that had to be set up, broken down, packed, loaded, and moved to the next venue, sometimes every twenty-four hours during its annual eight-month run. Modernizing the operation would be an enormous undertaking and a continuing source of controversy. "If this gloomy interpretation of plans underway at the Ringling winter headquarters is borne out," ran a *New York Times* editorial, "then the world as we know it is really tottering . . . they have hired a Big Name to further their conspiracy . . . This is clever, but we see through it, of course." The streamlining plan was clearly a scam to get gullible older folk to rush out for a last chance to see what was left of the *genuine* circus before it irrevocably disappeared, creating "a war-size boom in tickets."[476]

By the end of 1940, some 240 editorials nationwide had addressed the impending "face lift." Rumors that Bel Geddes was bringing back the original Ringling "strident strain" calliope, expanding the number of clowns, and generally brightening and unifying the extravaganza did little to stem the mounting indignation.

"If anybody can be counted on to break with tradition," observed *The New York Herald Tribune*, "Mr. Geddes is the man. His Futurama . . . will be long remembered by those who were, and are, convinced of its improbability."[477] "It may be some comfort to the nostalgic," wrote *The Minneapolis Star Journal*, "that even Norman Bel Geddes can't streamline an elephant."[478]

"I had not realized" he told *The New Republic*, "that redesigning the circus could be, as the political writers say, 'significant.'" Many of the changes would happen completely out of the public's view. It was, he explained, a job as essential as adding fire escapes and exits to a beloved theater. The Machine Age required concessions all around, but the fat lady would still be fat, the snake charmer charming.[479]

During that first circus season, Bel Geddes replaced the wood-burning cook stoves (responsible for 4,600 daily meals) with electric models of his own design. Gasoline motors were traded for diesel generators, creating more electricity for less, and a machine for guying out the tent ropes replaced gangs of bare-chested men chanting "Heave it! Heavy! Hold!"

He changed the menagerie from a ragtag line of cages into an "environment" suggestive of the animals' natural habitats, with more air and light, Rousseau-like paintings for background, and built-in pedestrian traffic flow. The giraffes were given their own elevated platform, high above the milling crowds ("makes 'em look taller than ever"), the elephants were grouped together rather than strung apart ("makes 'em look bigger that way"), and the monkeys got a glassed-in mountain playground. On the production side, he took charge of the elaborate Mother Goose–themed "spec."

It would be, the designer announced, "a circus for a plastic world," with softly illuminated plastic handrails and a new bandstand with special sounding boards; plastic for the aerialists' horizontals and bars (simplified and safer) for clown equipment and props; and translucent plastic for the center rings. The latter, embedded with colored lights, could now be lit from both within and above, allowing for a range of dramatic effects.

Traditionally, performers supplied their own outfits. In order to create an overall integrated "look," Norman applied his talents to redesigning thousands of wardrobe pieces for everyone from trapeze artists, stilt walkers, clowns, and elephant wranglers to giants, midgets, and showgirls, from band members and ushers to sartorial monkeys.

The challenges of wardrobe give some idea of the circus redesign's overall complexity, and scope. Besides having to coordinate with new stag-

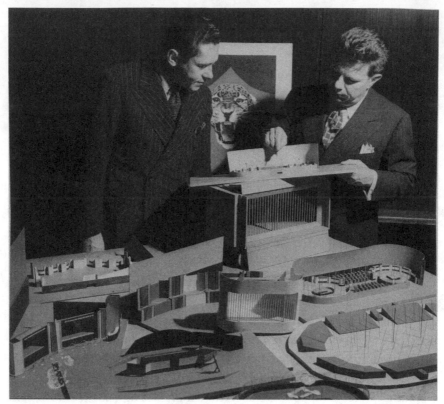

Caption: Johnny North and Norman Bel Geddes with "streamlined" circus model, 1940.
(Courtesy: Harry Ransom Center, The University of Texas at Austin, photographer unknown.)

ing, choreography, and lighting, costumes and uniforms had to be water repellant, the dye and paintwork guaranteed to not rub or sweat off. And they had to last; they were too expensive to allow for "doubles" in the event that something wore out. (The recent introduction of air-conditioning in the Big Top helped.) Costumes worn by animal riders had to have leather-reinforced seats and thighs. Animal costumes worn by "little people" required mask adjustments and tricky padding. Tradition dictated that ticket takers uniforms be made without pockets. Given the international political situation, German and Japanese fabrics were to be avoided. Last but not least, performers' personal preferences had to be accommodated.

Revealing, eye-catching costumes lent the circus an edge over the carefully monitored geometries of flesh on view in movie houses (thanks to the vigilance of the seven-year-old Hays Office, busy salvaging the nation's morality by cracking down on revealing necklines and "'sweater shots' in which the breasts were clearly outlined")[480]. There would be lots of feathers, plumes, and sequin work, all outsourced to Paris; Johnny North was big on sequins. Also made in Paris, by Hermès, were some of the show's saddles, bridles, and harnesses. There were also wigs, props, and floats to be designed and built, and everything had to be ready in five months.

Norman's most talked-about design that first season was an enormous $20,000 glass-enclosed "honeymoon suite" for Ringling headliner Gargantua the Great. (Leaning on his Yale education, North had named the 6-foot-6, 525-pound gorilla after the giant in François Rabelais's novel.) The suite's thermostatic controls and humidifier, designed to replicate the Congo, were the work of the air-conditioner company that had sponsored Arctic Girl in her Tomb of Ice.

The "Vehemently Vicious, Frightfully Fiendish Man-Like Monster" had been an expensive gamble. Lions and tigers could be had for $800, camels for $1,000 a pair, elephants for $4,000. The only Ringling animal pricier than Gargantua's $10,000 tag was the rhinoceros at a cool $12,000. The circus hadn't had a major animal headliner since Jumbo, the massive African elephant whose celebrity had added a new word to the English language. But when it came to gorillas, the Ringlings had had very bad luck. Prone to respiratory ailments, young apes fared poorly on the road.

Having learned of an eight-year-old "domesticated" candidate—raised by an eccentric Brooklyn housewife, indulged with lollipops and ice cream, dressed up for afternoon drives, and taught to "take tea" in the garden—Johnny North had jumped at the opportunity, in part to capitalize on the lingering popularity of *King Kong*. The future pet had been captured in Africa as a baby and brought to the States by a ship captain, only to have a disgruntled crew member spray it with a fire extinguisher. The creature's scarred face gave it a ferocious look. "That acid," Johnny North would write, "was worth a million to us."

His calculated recklessness paid off. Gargantua had quickly garnered *Life* and *Newsweek* covers, a guest appearance in London, and a cameo in the Marx Brothers' film *At the Circus* as "Gibraltar" (played by an actor in a gorilla suit). North let it be known that his new star—a vegetarian—had an insatiable appetite for human flesh.

When it came to animals (or midgets), "romance" translated into box office boffo. North was set on marrying the big brute off. As luck would have it, word came of a young domesticated female, M'Toto, who'd been raised "at home" in Havana by a wealthy socialite and indulged with liberal rations of apple pie and Coca-Cola. That her name (Swahili for "Little One") suggested Dorothy's black terrier, Toto, in *The Wizard of Oz* was fortuitous. North, who was about to marry French actress Germaine Aussey, made a point of honeymooning in Cuba to clinch the deal.[481]

"Science expects to learn a great deal from this union," Norman informed the press (meaning the gorillas' union, not their owners', though both would go badly), "facts that may or may not be pleasing to all the followers of Darwin and Huxley." He was hoping to use Plexiglas bars for both the honeymoon suite and the menagerie cages strong enough to do the job but more or less invisible. *But how will Gargantua react?* one journalist objected. *Already frenzied by a dark, malevolent hatred of the human race, would such an idea not drive him to even greater excesses?*[482] Johnny North himself couldn't have written a better publicity hook.

Another Bel Geddes design that caught the media's attention was his Pole-Less Tent, a gigantic, ultra-modern "streamline" auditorium slated to debut the following season. *Popular Science* published a schemata.[483] The Big Top, with its 130,000 square feet of canvas—weighing 18 tons (considerably heavier when it rained)—required dozens of side and quarter poles for support, obstructing audience sight lines. Norman eliminated the problem by suspending the canvas on cables attached to a 160-foot external steel scaffolding. More external supports facilitated air-conditioning the space and created an unprecedented 540-by-220-foot interior that allowed for a fourth ring, thus doing away with the venerable "three-ring circus."

It would, Norman determined, require 50 percent less labor and half the time to raise, and have national defense applications as a portable airplane hangar.

Asked by reporters how much he was paying Mr. Geddes for all this modernity, Johnny North replied, "Plenty."[484]

It's hard to say which project had been more complex, the revamped circus or the made-from-whole-cloth Futurama, so different were their components, but the long-term risk for the Ringling family far outweighed any General Motors' considerations. The latter would have survived a modest turnout; the former would not.

When the first Bel Geddes season finally opened at Madison Square Garden, a venue ordinarily home to hockey and boxing matches, "grown men were startled," reported *Time*. The circus looked "as fabulously beautiful as they imagined it had when they were children," and if the effect on adults was "powerful, on the less mature it was staggering."[485] The new version was "awash in fiery hues that would have blanched a rainbow," noted *The New York Journal-American*. The only commodity to retain its original tint was the pink lemonade, and Bel Geddes was "expect[ed] to get around to that any day now."[486]

The gate swelled to an all-time high during the Garden's twenty-eight-day, fifty-five-performance run, followed by Ringling's best tour in years. The North–Bel Geddes collaboration played to "straw houses" for every show;[487] thousands more showed up to watch the unloading and loading of the transport trains. By the end of the 8-month, 20,030-mile, 137-city tour, nearly 4 million ticketshad been sold. Fifty thousand military personnel got in free.

The magic, as promised, had remained inviolate, with one exception: the fat lady—Mrs. Ruth Pontico, also known as "Baby Ruth"—missed the beginning of the season. While hospitalized to have some excess "streamlined" (she weighed between 685 and 800 pounds), her bed had collapsed, forcing the postponement of her surgery.[488]

In August 1941, Norman signed a second Ringling season contract. His core creative team got raises and their names advanced to the front of the program. They were joined by *Ziegfeld Follies* producer and former antiques dealer John Murray Anderson.

Another new addition was Barbette, a female impersonator and trapeze artist fresh from performing in full drag at the Moulin Rouge and Folies Bergère. (Jean Cocteau was an ardent admirer.) Barbette's mandate was to choreograph a thirty-six-girl aerial ensemble to back up Elly Ardelty in the circus's most dangerous act. Bedecked in "fine Bel Geddes feathers," the petite Russian blonde would "ascend into a Bel Geddes sky"[489] in a gilded cage and proceed to balance—on her head—on a trapeze bar, arms and legs outstretched as she swung in forty-foot arcs above the ring without a net.

Norman's role, as before, was to oversee the design of costumes, uniforms, props, floats, and lighting, make studies for safety features and cost reductions, and produce the new season's "spec," a holiday theme this time, with elephants sporting reindeer antlers and pulling a forty-foot sleigh.[490] A decade after Thornton Wilder described Bel Geddes as "a brilliant lion tamer in spangles," the designer was coming remarkably close to inhabiting the role.

•

WHAT BETTER WAY to help pull in the crowds than to commission a big-name composer? In the era of radio, some of the old tunes had become, well, hackneyed.

Cole Porter claimed to be "toed up for at least three months."[491] Norman told North that the man who'd rhymed camembert with Fred Astaire was unhappy about the modesty of the fee. (Porter was, in fact, working on tunes for a Fred Astaire–Rita Hayworth movie called, ironically, *You'll Never Get Rich*.)

Irving Berlin also had prior commitments.

"If you still want me for the complete score next year and I am free, I would love to do it," Berlin wrote. "I really believe you exaggerate the importance of a so-called 'name' writer . . . Any one of half a dozen competent men will give you what you wan."[492]

Norman, meanwhile, had his own priors, including a half dozen top-secret military projects, a line of postwar radio cabinets (RCA), a parade float (U.S. Steel), and a cigarette vending machine (U-Need-A-Pak). And he was about to take on yet another, at the behest of "Bob" Moses.

Allowing himself a much-needed break, Norman had visited the Hayden Planetarium. "Just a magic lantern show," he wrote, expressing his great disappointment in a long letter to the director of the Museum of Natural History. "In all sincerity, it is an insult to average intelligence—not even equal to the brand of science perpetrated in the comic strips."[493] Surely astronomy contained more than enough dramatic material to seize the popular imagination?

Six days later, unaware of the letter, Robert Moses dictated a lengthy one of his own, addressed to the designer, about the moribund Hayden's precarious status:

> At its dedication, I said that Charlie Hayden had purchased immortality very cheaply because he gave nothing but the instrument which was made in Germany, [forcing] the Trustees to sell more bonds . . . Immortality appears to have been too strong a word.[494]

Unless some other millionaire could be convinced to ante up, in exchange for having his name replace Hayden's over the entryway, the Planetarium's days were, likely, numbered.

Moses and Bel Geddes had run into each other the previous Thanksgiving, a blustery, rain-thick Manhattan night during which, aggravated by their children's bad behavior, both had walked out on family dinners. Wet and cold, wandering the mostly empty streets, the two exiles ended up having a drink together. Norman's irritation with Moses over Futurama, a ping-pong match conducted on *The New York Times* editorial pages ten months before, had been forgiven, though the Central Park Casino's demise would remain a sore spot.

Would Norman be interested in a commission, making recommendations to increase the Hayden's ability to be self-supporting? And could he do it in seven weeks? One idea might be to install telescopes (preferably donated) on the roof, so visitors could go from watching the show to observing the real thing.

The designer's response was immediate. "Your letter quite took my breath away . . . it was mental telepathy."[495]

His subsequent thirty-page "strictly confidential" report, with nine pages of diagrams and cross sections, recommended the banishment of "gadgety atmosphere," "mediocre tableaux," and the lecturer's dry drone. As for the Hayden's "gloomy" edifice (with visible joints in the dome!) for what was meant to be a theater of the Heavens . . . visitors should be rewarded with drama and mystery, awe and excitement, a sense of infinite space.

He was particularly keen on developing *A Trip on a Comet* show.

It would begin with a speck (a cloud of billowing gases) that enlarged as the spectator "got closer." Soon it would be shining above the viewers' heads in all its arresting detail, a nebula with a central whirling vortex. Then, circling in at close range, it would reveal itself at startling angles.[496]

•

THE MOST UNUSUAL and talked about aspect of the 1942 season would be created with the help of two brilliant Russian émigrés, a Norwegian prima ballerina, and more pachyderms than Hannibal conscripted for his famous march across the Alps.

To thousands of fans, elephants were the essence of the circus; they'd been stock players since Roman times. A circus without elephants, wrote veteran clown Robert Sherwood, "would be like *Hamlet* without Hamlet."[497] Ballet was still a novelty to American audiences, something exotic, a high-toned European oddity. Combined, they would create a new way to feature the unique dancing talents of Old Modoc, a Ringling veteran, the largest Indian elephant this side of the Atlantic.

An elephant ballet. It's unclear whose initial idea it was; Bel Geddes's notes suggest it was his. True, Disney's *Fantasia*, which had had a limited release the previous November, included a segment of ballet-dancing hippos and elephants, but few saw the original, pedantic, 220-minute "film symphony." And there was a world of difference between quadrupeds prancing on celluloid and an elaborate performance piece starring live ones.

In an audacious stroke, Johnny North contacted choreographer George Balanchine, master of the idealized and the elevated. Having recently completed his masterwork Ballet Imperiale, the Russian was on the lookout for projects, and he had a soft spot for Broadway and Hollywood dazzle. (His *Romeo and Juliet* ballet in the 1938 Goldwyn Follies had included tap-dancing Montagues.)

Balanchine, in turn, telephoned fellow ex-patriot Igor Stravinsky in Los Angeles. (The genius behind Fokine's *Firebird* ballet, Stravinsky had also composed works for Woody Herman and Paul Whiteman and turned down several "serious" commissions to work on a Billy Rose production.) Abstract and intellectual as his music was accused of being, his infamous *Rite of Spring* had accompanied *Fantasia*'s dinosaurs.

The conversation, in Russian, went something like this:

"Igor, I wonder if you'd like to do a little ballet for me."

"For whom?"

"For some elephants."

"It would be difficult," the composer replied. "I'm preparing a rather important piece, twice the size of the Dumbarton Oaks Concerto. Twenty-five instruments. It premieres in February and I'm conducting."

"Just a polka. A short one," Balanchine assured him. "Something a circus band can play."

"Elephants, you say?"

"*Dah.*"

"How old?"

"Oh . . . very young."

"How many?"

"Fifty. And fifty ballerinas. Also very young."

Stravinsky paused.

"All right," he agreed. "If they are very young elephants, and you can wait until March, perhaps I can do it."[498]

Balanchine began spending time at Ringling's winter headquarters in Florida to collaborate with elephant trainer Walter McClain, who explained, among other things, all the ills of the flesh his charges were prone to, from indigestion and in-grown toenails to colds, blisters, bunions, and tangled eyelashes (they grew to four inches long). The choreographer quickly learned to recognize each pachyderm and what each could do, and he took to calling them by name. For the uninitiated, these powerful herbivores—averaging eight feet high at the shoulder and weighing between 6,000 and 8,000 pounds—inspired trepidation, if not terror. But according to one of the fifty young women hired for the act (part Ringling starlets, part professional ballerinas imported from New York), all would have "walked into a cage full of hungry lions" for Balanchine, whose "romantic good looks and soft-spoken courtesy" were in marked contrast to the "cigar-filled faces" of circus men and John Murray Anderson's signature sarcasm.[499] The choreographer inspired similar affection in the elephants, all females, who were soon following him around the grounds.

Also in the cast was Balanchine's wife, Vera Zorina. The Norwegian beauty (née Eva Brigitta Hartwig) would make a guest appearance opening night, a benefit fund-raiser, as Modoc's partner.

Maintaining the Corps des Elephants was a serious financial enterprise. Each of the fifty pachyderms consumed 125 pounds of hay, half a bushel of oats, and 7 pounds of bran every day, and together, by season's end, enough water to fill an Olympic-sized swimming pool. Once the show was on tour, they stuck their heads into the dancing girls' trailers, opening drawers (and pulling everything out). Refrigerators and the popcorn concession were also fair game. Grooming included "shaving" their bristles every few days with an acetylene torch and applying top-quality neat's-foot oil (more than 150 gallons for the season) to keep their thick hides from cracking. Then there were the unforeseeable hazards. The previous season, in Atlanta, eleven elephants had been poisoned.

On top of everything else, it was inevitable that, within the confines of Ringling's winter quarters, a contingent of fifty exceptionally large elephants would come in contact with the horses, whom they loathed. Other hazards included zebras (mean and likely to kick at anything nearby), camels (deadly accurate spitters), and performers' pampered pets (a trained pig, a goose). One clown had a Chihuahua he'd incorporated into his act. Dressed up as an elephant, it terrified the real ones.[500]

Still, they'd prove their worth above and beyond a terpsichorean mandate. In Lexington, they'd be called on, as in former days, to haul up the Big Top when the tractors got hopelessly mired in deep mud.

The initial version of Stravinsky's *Circus Polka*—a witty mix of French ballet music circa 1910–1920, a Charles Ives–like American vernacular, and allusions to Schubert's *Marche Militaire*—called for clarinets, saxophones, cornets, horns, trombones, tubas, drums, cymbals, a xylophone, and a booming Hammond organ. The first time the band played this "Harvard music," as veteran bandmaster Merle Evans called it (his boss was a Yale man), the animals, patient by nature, began to trumpet and flap their enormous ears. Mrs. Balanchine described the scene as "unadulterated hilarity."[501]

Evans was less amused. Experience had taught him that good elephant music was something with a solid beat—a march, a one-step, a fox-trot, a cakewalk, or a soft, dreamy waltz. Stravinsky's music confused and frightened them. The band, the ballerinas, and the bull men weren't having an easy time of it, either. The cornet part was too high, the clarinet had too many notes. There were six meter changes in the first fourteen bars alone.

In spite of the stunts they were made to perform, the quadrupeds were dignified animals and deserved better, the bandleader lamented. "It would have taken very little during the many performances . . . to cause a stampede."[502] Instead, the trumpeting became part of the act.

Various distinguished persons, and what North called "society razorbacks," traveled to Sarasota to watch the new season getting into shape. Poet Edith Sitwell sailed from blitz-ravaged London, and from Paris, renegade couturier Elsa Schiaparelli. (Her 1938 Spring line had made headlines with its circus-inspired dresses.) Mining heiress Evelyn Walsh McLean entered the Big Top with the Hope Diamond resting celestially above her cleavage. One toothless, ill-kempt roustabout who happened to be sitting nearby when the perfumed Mrs. McLean walked past, stunned the famously unflappable Murray by asking, "Pardon, Mr. Anderson, but isn't that Lanvin's *Arpege*?" It was.[503]

•

IN THE WAKE of Pearl Harbor, the production adopted a patriotic theme. Norman covered the main arena, where the ballet would take place, with blue sawdust, the end rings with red and white. He decided on pink tulle tutu skirts and jeweled, feather-crested headbands for the quadrupeds, with matching pink skirts and huge satin hair bows for their human partners. The pachyderm contingent would require 6,000 yards of fabric. His dreamy twilight lighting may have been informed by his ideas for the foundering Hayden Planetarium.

By the time the 1942 program was ready to roll, 12,000 pounds of white lead paint, 1,000 gallons of blue enamel, 750 of red and 500 of silver, 500 packs of gold leaf, and 1,000 gallons of varnish had been deployed. The Bel-Geddified Ringling Bros. and Barnum & Bailey Circus, season two—feathered and heavily sequined, layered and lit, "streamlined anew in keeping with the modern trend"[504]—was about to meet its public.

Early on the morning of April first, the 79,000-ton production set off on the two-day and two-night trip north, but not before its ninety double-length steel railroad cars were blessed by the circus chaplain. Roosevelt had arranged special rail travel dispensation, countermanding restrictions imposed by the war. On board were, among other things, 41 tents, 4 miles of electrical cable, 1,600 people (representing 29 nationalities), and 500 wild animals, including Mr. and Mrs. Gargantua the Great.

Undeterred by the couple's lack of wedded bliss (Gargantua seemed intimidated; M'Toto wasn't interested), North would continue to exhibit "the most publicized primates on earth" in adjacent cages, milking the "will they or won't they" question for the remainder of the simians' lives—a "sucker bait" strategy, *The New York Times* felt obliged to point out, in the best P. T. Barnum tradition.[505]

The first week at Madison Square Garden was given over to rehearsals. Introduced to their costumes and Norman's lighting scheme for the first time, the elephants, apparently excited, raced through their routine ahead of the music. Toward the end, pirouetting on their backsides, they "let go" in unison, releasing steaming piles of excrement.[506] When they rose, the under-portions of their tutus were smeared brown, rubbed in by four tons of pressure. The elephants seemed to know something was wrong; their heads hung low as they exited to laughter and applause. Clearly, the sitting-and-pirouetting finale had to go. But a taught routine was not easily altered. Duplicates sets of the 130-yard costumes were rushed to completion and the strongest available cleaning chemicals stockpiled, just in case.

But ultimately, the 300-legged "original choreographic tour de force"[507] proved worth all the trouble.

It began with an inter-species duet. Vera Zorina entered the arena perched atop Modoc's head to a vast roar of applause, as the band played a rendition of Von Weber's "Invitation to the Dance."

We were alone in the blinding spotlight. I felt the incredible grandeur of riding on that noble beast, who knew exactly what to do . . . She walked majestically into the deafening fanfare . . . and entered the middle ring. She knelt down and gently let me off . . . [Her] "dance" . . . consisted of lifting up one front foot (which took eight bars), then slowly lifting the other, then doing a turn.[508]

The duet complete, the two knelt in a reverential bow, touching foreheads in the sawdust. When Modoc "offered" her trunk, Zorina lay back on it, grasping the elephant's jeweled headband as she was lifted up and gently carried offstage.

Then the rest of the troupe entered the stadium's heart, lit as an ethereal blue dusk. First the fifty all-pink dancers—holding garlands of flowers above their heads, their tutus standing out in crisp net folds, the ribbons of their ballet slippers crisscrossing their calves. Then their fifty enormous partners, sporting Norman's pink skirts around their ample middles, feathered aigrettes perched coyly between their eyes. A blue spotlight followed them through the blue dusk, their gigantic scale making the girls look tiny.

Vera Zorina and Modoc. "Like riding in a gondola."
(Courtesy: Life *magazine/Getty Images. Photo by Herbert Gehr.*)

The ballerinas danced over and between the three circus rings, performing fifty *pas de deux* as their towering partners nodded rhythmically and gravely swayed.

The arc of sway widened and the stomping picked up with the music. In the central ring, Modoc the Elephant danced with amazing grace.[509]

The elephants linked trunks to tails in an endless chain, the ground shaking with their measured steps.

They balanced atop individual stools while the dancers arranged themselves in formations.

They rose up on their hind legs to create an immense oval, each pair of front legs resting on the skirt of the beast before it, trunks elegantly curled—a great gray necklace—while the dancers posed in fifty arabesques atop fifty enormous heads.

Through it all the bull men, like Bunraku puppeteers, kept things moving, there but invisible.

At the very end, the smallest Ringling midget, in page boy attire, presented Modoc with an enormous bunch of American Beauty roses, the stems twice as tall as the fellow himself. One blossom dropped from the bunch. Modoc lifted it with his trunk and handed it to Zorina, who'd returned to share the final bow.

It was an absurd and extraordinary event—a mix of classical ballet and circus acrobatics, designed by the twentieth century's greatest choreographer, set to music by the era's greatest living composer, costumed and lit by the much-lauded designer of *The Miracle* and Futurama.

Variety found the Stravinsky score "weird" and the tutus "ridiculous."[510] Norman's perennial rival also chimed in. "A pseudo-sophisticated mess halfway between stage and tanbark. All *veddy, veddy distingué*," wrote Raymond Loewy, a man who wore silk ascots to go pheasant hunting and boasted that he put Chanel No. 5 in his scuba gear to mitigate the smell of rubber. "All frightfully dull."[511] He'd apparently been curious enough to buy a ticket.

But those were minority opinions. The ballet would ultimately be performed 425 times, and most of the millions who saw it agreed with the initially skeptical *New York Times*. It was "breathtaking."[512]

Seated beneath Madison Square Garden's roof that early April evening, transported, it was impossible to imagine the absurd and extraordinary events of a very different kind underway at that same hour. Starving Parisians picking through garbage as their greatest paintings and sculpture were shipped, by the trainload, to Germany. Londoners coping with the aftermath of the Blitz. Shanghai and Nanking reeling from the massacre of hundreds of thousands by the Japanese. The opening of Treblinka and Sobibor in occupied Poland, and the introduction of Zyklon B at Auschwitz-Birkenau. The prelude to the Bataan Death March in the Philippines. Preparations for the U.S. bombing of Tokyo.

"...that stinking, dirty, filthy piece about me in *The New Yorker*"

1941

Gotham is one of New York's many names, and Gotham was a village in England whose people were proverbial for their follies.
—BARRY ULANOV, *A HISTORY OF JAZZ IN AMERICA*, 1954

W hen we laugh at the grandeur of some of the conceptions in Flushing," wrote longtime *New Yorker* contributor Wolcott Gibbs, referring to the 1939 Fair, "it is because we ourselves have never dared to dream of anything more spectacular than getting a fifteen-cent magazine out on time with most of its pages right-side up.

"Life will probably always be like that," he added. "The man of vision creating, the little men carping, with terror and amazement in their hearts."[513]

Harold Ross, Gibbs's employer, had been a member of the Nutshell Jockey Club. As Norman's star rose, Ross twice broached the idea of running a profile on him. Ross could write anything he wanted, he was told, as long as, on his word of honor, he'd treat Norman's professional dealings, his bread and butter, with respect. Nothing ever came of it.

In the wake of Futurama's extraordinary success, Ross tried again.

"You know my condition," Norman said.

A staff researcher spent three weeks in Norman's office with free access to all the files.[514] Norman made time, on a dozen occasions, to meet with Geoffrey Hellman, the writer assigned to the piece; Hellman also interviewed more than a dozen of his subject's friends, acquaintances, clients, and employees. The final profile ran to nineteen pages spread over three issues, the longest in the magazine's history.[515]

The first installment appeared in the February 8, 1941, issue. The morning it hit the stands, Norman's attorney, Carl Austrian, phoned. The profile was biased, even libelous; Norman would be hard-pressed to find anything complimentary in it. Austrian suggested taking action.

Next came a call from *Time* magazine. They were preparing a story and wanted to double-check a few things. Did he completely rebuild the Century Theatre for *The Miracle* and the Manhattan Opera House for *Eternal Road* because the stages were too small for his scenery? Did Standard Gas Company scrap an entire factory to produce his stove design?

Where in the world did you get that information? Norman asked.

Why, Mr. Geddes, the reporter replied. In *The New Yorker.*

•

WITH HIS FIRST Ringling Brothers season about to open, his office (now with a staff of seventy) was preoccupied with dozens of last minute details, plus a Montgomery Ward catalog overhaul, several hundred oilcloth and wallpaper patterns ($20,000, plus expenses), a Patriot Radio for Emerson (with built-in "super loop" antenna and illuminated dial), a Kelvinator refrigerator, and costume jewelry for Trifari, Krussman, and Fishel of Fifth Avenue that incorporated cabochons ("jelly bellies") made from discarded military aircraft windshields. There was a major redesign of the exclusive La Rue Restaurant on East Fifty-Eighth Street, at a reputed tab of $40,000 ($680,000 today), with Norman getting $5,000 (in excess of $80,000).

And then there was Sonja Henie.

It Happens on Ice had been seen by 1.5 million at Radio City, but the three-time Olympic skater had bigger plans. Norman was hired to reimagine the sets and costumes and supervise the lighting. But first, he was charged with creating "America's first ice theater" in what had originally been an art deco movie palace with 75-foot African mahogany walls. The result would be a gigantic 7,000-square-foot rink with graceful curves extending almost into the spectators' laps. His 80-foot, semicircular white curtain, woven with 315,720,000 individual fibers of spun glass, would be noted in the press as a typical—if magical—Bel Geddes extravagance. As was often the case, it was a pragmatic solution to a complex problem. The shimmering fabric conformed to fire department regulations and had the added benefit of not absorbing moisture and mold from the ice.

Meanwhile, the designer's name was appearing in social columns nationwide: spotted at a Ritz-Carlton cocktail party for Mr. and Mrs. Cole Porter, who were off to the South Seas. Seen at Spivy's in the company of Gypsy Rose Lee and Jock Whitney. Renting a summer house with a slip for his sailboat. Sketching on restaurant tablecloths. In May, *The New Yorker* ran a cartoon of a couple seated on a park bench. The caption read: "*It beats me why they didn't have Geddes do something about these fireplugs.*"[516]

Another cartoon appeared in the the the *New York Post*:

"*$500 to re-decorate his room!*" a Princeton student's mom complains, perusing the dorm. "*Is Bel Geddes doin' it?*"[517]

The redesigned La Rue opened in September, *Ice* in October. In November, Random House released his second book, *Magic Motorways*, further elaborations of his transportation theories, just in time for the opening of the Pennsylvania Turnpike, "the nearest thing in America to those super-super highways depicted in the Futurama."[518] In December, he joined Mrs. Astor, Mrs. Hearst, Dorothy Draper, the editors of *Vogue* and *Harper's Bazaar*, Loewy, and Teague to judge a Fifth Avenue Christmas window competition.

"Where do your interests end, Mr. Geddes?" one interviewer asked.

"They'll end," he grinned, "in a perfection I'll never know."[519]

•

NOT SURPRISINGLY, HE'D forgotten when, exactly, *The New Yorker* profile was scheduled to appear.

The errors began in the first paragraph. "Among his accomplishments have been a four-ton section of a steamship in a play called *La Nave*," read one line. Aside from the fact that the steamship was a galley and the play was an opera, Norman noted dryly, everything in the sentence was correct.

"The dividing line between fact and fantasy" was "a difficult one to distinguish," the profile read. "Geddes states that he designed the first streamlined ocean liner [but] no streamline ocean liner has been built as yet. A number of Geddes' greatest successes . . . have never existed at all."

Whether or not *The Divine Comedy* and *King Lear* had been staged, or the liner or triple-deck Aerial Restaurant built, all had been worked out in complete detail, production-ready. The advent of the Depression hadn't helped, and he certainly wasn't the only architect-designer for whom this was true. (F. L. Wright would ultimately design 1,150 buildings of which only 552 were built.) It didn't mean, as an unnamed "friend" was quoted as saying, "his head is in the clouds but his feet are certainly not on the ground." Unlike Loewy, Teague, and Dreyfuss, Bel Geddes was described as "impractical and visionary"—which left open the question of what a *practical nonvisionary* was likely to achieve. Not one quote was attributed. Like the Inquisition, you didn't get to confront your accusers.

"The amount of scoffing that has been hurled at me for getting my head so high into the clouds that my feet could no longer maintain contact with the ground is something I'm proud of, rather than embarrassed by," Norman would write, a decade later. "I'm only sorry I didn't recognize its value earlier."

Hellman claimed that the changes had been imposed from above; Norman believed him. Despite surrounding himself with some of the most interesting writers of his generation, Harold Ross was notorious for his rewrites, what Norman called "Rossifection." He was also profoundly suspicious of "anything smacking of scholarship."[520]

Norman had told Hellman that profits on his occasional hits had been greater than the losses on his failures. (He'd succeeded at least 50 percent of the time, which in Broadway terms was pretty damned good.) Distilled through Rossifection, this read that Norman's success depended on whether one adopted "an aesthetic or a bookkeeping point of view." A master of turning "small, concrete" projects into "gigantic, theoretical" ones, the text ran, he'd coerced veteran theater producers and corporate titans alike into relinquishing vast sums. In which case, Norman wryly observed, Toledo Scale, Zalmon Simmons, Standard Gas Company, and General Motors would all have gone belly up long ago. As for the more than 200 stage productions he'd designed, he'd exceeded his budget perhaps fifteen times.[521]

"He always improves what he sets out to do," banker business partner George Woods had explained to Hellman. "Then it costs more. So, a lot of misunderstandings arise."[522] But Woods's comment didn't appear in print.

Concerned and upset, Norman telephoned Alfred Sloan, Henry Luce, and Walter Winchell for their impressions.

When he rang up Ross's office, the publisher was out. He's having a hectic day, his secretary noted, getting ready for a winter break in Palm Beach. "Please tell him," Norman said, "that before boarding the train, he'll be served with an injunction that his friend and my attorney, Carl Austrian, has already procured. He'll be responsible for gathering up and accounting for every distributed copy of the current issue and destroying the rest. It may prove rather expensive. Other than that, I wish him a relaxing vacation."

Ten minutes later, Norman's phone rang. It was Ross. "What's all the fuss about?" he asked.

"You might want your attorney to call Carl."

"What the hell?"

"I gave your people full access to me *and* my files, Harold. After badgering me for years to do this, you must have had a field day making me look like a fool. Did you ever consider how my business clients might react? I counted twelve intentional lies and half-truths, all of them easily disproved. Quotes taken out of context. And *it's only the first installment.* The thing is nasty and dishonest."

"We're a humor magazine, at heart," Ross explained. "You're taking it all too personally." This from a man who took *everything* personally.

"My hat is off to you, Harold. It's a rare thing you've done—setting out to create something unique and succeeding at it, against the odds. Then managing to *keep* doing it, year after year."

"That's great, Norman. But why—"

"After all, there's a great difference between *designing* something and actually having it *produced.*"

"Sure, but—"

"I tracked down a copy of your prospectus. You know, the one you wrote in '22 to raise money to start the magazine? Interesting document. *The New Yorker* 'will print facts that it will have to go behind the scenes to get, but it will not deal in scandal or sensation for the sake of sensation. Its integrity will be above suspicion.' All very noble. When was the last time you read it?"

"We exaggerate," Ross allowed. "The profiles don't pretend to be *serious.*"

"You mean, Harold, that you intentionally distort the truth about people who've accomplished something in order to liven things up, make their stories more entertaining?"

Ross hesitated. "I suppose you could say that. Jesus, Norman—"

"Maybe knock them down a few notches in the process? Destroy the public's illusions about celebrities while showing them how smart you are?"

"Fuck, Geddes—"

Finally, the famous Ross profanity.[523]

"Hold on a minute," Norman interrupted.

A confirmed Dictaphonist, Norman kept a machine in his office, another in his bedroom, and took one on his travels. Having recorded the conversation, he played it back into the receiver. Hearing his own voice, Ross hung up. Twenty minutes later, he burst past Bel Geddes's secretary and arrived "raging and sputtering" at Norman's office door.

•

SINCE HIS TWENTIES, the word "genius" had been increasingly linked to Bel Geddes's name; the more famous he became, the more it was used as a euphemism for his perceived eccentricities and excesses. Geniuses, it seemed, were talented misfits to be admired from a distance but approached with caution. It was interesting how the media chose its targets. Gershwin and Chaplin weren't singled out for ridicule, but Einstein was—so brilliant and high-minded he couldn't tie his own shoelaces!

Norman appreciated that a genius spendthrift who realigned rainbows and blueprinted castles in the air made good copy, and he believed that writers had a right to try to sway readers—but with honest argument, not manipulated "facts" and manufactured "truths." It struck him as supremely ironic that his Rossified profile accused him of trying to rewrite and reconstruct every production he was hired for.

A few days later, Austrian called to say that an agreement had been signed. It was a bit late in the day to round up unsold copies. The second installment had been "toned down" but was already at the press, so revisions were limited. The third installment would be overhauled completely. Still, in the weeks that followed, he was portrayed as "flail[ing] his arms about like a semaphore," firing employees who had the temerity to say "good morning," and greeting dinner guests clad "in practically nothing but moccasins" (a misguided reference to his teenage fascination with Indians).

His office printing press, readers were told, was used to create clothing lists, under complex headings, that resulted in Norman's butler packing him formal wear and boiled wool shirts for casual summer weekends. He enjoyed thinking he was "surrounded by spies," an allusion to Loewy's corporate mischief when the Electrolux was being designed, but offered in the profile without a context. Then there was the "gigantic code" he'd devised so his office could save money on cables when the boss was out of town, a code he subsequently forgot about, complaining when the non-sense cables arrived. Included, too, was a short list of Norman's more memorable jokes, like the Easter he sent a live duck and a pregnant rabbit alternately every hour to his business partner's wife.

By the third installment, Hellman managed to include a few lines that could be interpreted as apologies: "His ideas, while *sometimes* not immediately marketable, have *possibly* stimulated the profession more than those of any other industrial designer."

"He has retained a high regard for Wall Street and for big business. In turn, brokers [consider him] not only a genius but a congenial nightclub companion and a somewhat elegant bohemian." Norman was credited with an income that allowed for a $50,000-a-year lifestyle ($800,000 today), suggesting that, despite rampant rumors to the contrary, his toes remained on terra firma.

Nearly a decade after his "Design for Living" profile was released, the designer would still be smarting, comparing Ross's mythical "halo of integrity" to a "barbed wire wreath." Though ultimately, he admitted, "that stinking, dirty, filthy piece about me in *The New Yorker*" had done no real harm.

•

THE LAST OF the three installments ran in the February 22, 1941, issue. Four days after its release, Richard (Ryszard) Ordynski sat down at a table in his bare-bones rented room on East Fifty-Eighth Street to write his old colleague a note.

My dear Norman,

Just to tell you that I found the article on you in The New Yorker *very interesting. I was sorry that so far I did not see any mentioning of our mutual collaboration, a thing I always recollect with great pleasure and pride.*

As ever,
Richard Ordynski[524]

His last missive, regarding Norman's call for "forward thinking" work for the Chicago Fair, had been twelve years before.

Since then, Ordy had directed a dozen films (sometimes doing triple duty as director, writer, and art director) and judged four Venice Film Festivals. He had to have been aware of Norman's highly publicized second collaboration with Reinhardt.

Twenty-five years before, at age thirty-seven, Ryszard Ordynski had fled the escalating mess of Eastern Europe and with it a successful career ("Ordynski was a big man over there," wrote Eugene O'Neill), ending up in the oasis of 1920s Los Angeles. Now, with Hitler's invasion of his native Poland, the only wise thing had been to pull up roots again.[525] And so, at sixty-two, he'd returned to New York.

He would almost certainly have contacted Max, but the financial disaster of *The Eternal Road*, followed by a lackluster production of *Faust* at L.A.'s open-air Pilgrimage Theatre (Goethe's poem translated badly), had dampened Reinhardt's prospects. For Max, art was everything and money a means to an end, a credo that made him a heretic in Hollywood, where he'd been living for the last six years.

Europe's great director had also recently lost his most valuable and sentimental asset; after annexing Austria, Hitler lost no time confiscating Leopoldskron, turning it into a retreat for high-ranking military personnel under the mandate that all Jewish-owned property belonged to the Reich. Then within months, the Fuhrer "gifted" Max's beloved fixer-upper to Princess Stephanie Hohenlohe-Waldenburg-Schillingsfurst (née Steffie Richter), a red-haired, Vienna-born Jew, for ongoing services (which included sleeping with Hitler's former superior officer, Fritz Wiedemann) to the Reich.[526]

Leopoldskron was "a pure love affair, an irreplaceable treasure," Max wrote his son upon hearing the news. An eighteen-year love affair. "Yet one doesn't die when love is lost. Certainly one can exist without marble halls and 80 rooms."[527] But no stoicism could prevent him imagining SS commandants swilling brandy in his library, cavorting in his bedroom, perhaps wearing his dressing gown. It was, he wrote his wife, "the harvest of my life's work."[528]

Ordynski had managed through the Thirties, but these days he was barely scraping by, translating the occasional Polish play to the accompaniment of an increasingly growling stomach.[529] He'd once been quoted as saying he'd "rather live on crackers and cheese in Los Angeles and do as I please than take five thousand dollars a month trying to please David Belasco." Twenty-four years later, Ordy had his wish, in that twisted way wishes come true in fables. Or fever dreams. Belasco, the impresario who dressed like a priest, was gone, and Ordynski was struggling, not in sunny Hollywood but in the heart of a wretched New York February.

Seven months after his initial note, Ryszard Ordynski's restraint dissolved.

Labor Day, September 1941

Dear Norman,

Have you any real recollection of our days at the Little Theatre in Los Angeles on Figueroa Street? I wish I could plunge you for a moment, in on our interesting days when Nju or Papa was rehearsed—can you? And then when you emerge, you would ask or rather wonder how it is possible that I have been here now over a year, that we had so many memories in common, so many views—and that all this should only produce a most casual and rather hasty few words you said to me or some other meaningless phrase of sympathy?

I am writing you this not as a reproach, to which I have no right whatsoever, but as a question whether or not you did, for a second, emerge from that plunge and see me with your eyes as it was years back that we were linked with a kind of work which has certainly left a mark in your career—and all our so frequent talks we had?

Here I came to this country suddenly struck with the lot of a refugee, having left all behind me, and there was not one moment where I would have seen in your words any real desire to help me—what other people much less related to me have done. You know too well that there is never money that I would ask you for, nor accept, only good will.

If I am writing this it is only that I am now asking myself, that I must have been guilty of some crime that you could treat me as you do. All I can say, and all I can swear, is that: Were you in a similar position, here or on the other side, and I the one who is having his daily bread and butter—I would have acted entirely different. The only answer is that I am, and probably was all my life—a fool. Please don't take this sentimentally, it may just be an outburst of a man who lived many

*years in this country and thought that he would be entitled to some
place here, in order not to starve literally?*

*I always liked you very much, I have talked about you and I do it con-
stantly whenever I can praise your extraordinary talents and power of
conception—and that is probably why I feel so badly being deserted.*[530]

Norman didn't require "plunging." He hadn't forgotten any of it—the
long walks amid palm trees and scented orange groves and Ordy's patience with his many, many questions. The dinners with him, Aline, and
Bel, mapping out the future of American theater. The afternoon Ordy
introduced him to Nijinsky and the three of them had attended a Chaplin
picture show. How afterward, Nijinsky had perfectly imitated Chaplin's
walk. Nor had he forgotten that Ordynski had ultimately been responsible for his first New York job, a contract with the Metropolitan Opera, no
less. And later, how he'd listened patiently as Norman mapped out his
vision for Dante's "comedy," the very thing that led him to Reinhardt and
The Miracle.

Whatever his thoughts about Ordynski's role in the demise of Aline's
dream, the Players Producing Company was half a lifetime away, and
Norman's ties with the heiress had long since broken. And though he'd
once pegged him as a cad, Norman might have learned, given the opportunity, that Ordy had subsequently lived with Aline for a time in New
York. And for at least the first nine years of Sugartop's life, he'd made a
point of spending time with his daughter and her mother on holidays.[531]

Both men had changed, as the world had. And surely Norman no
longer worried that Ordy's choices would override his own on a given
project. Norman's reticence was out of character, given all the people he
had helped, none of whom he'd been indebted to—former students like
Henry Dreyfuss and Aline Bernstein, muralist Anton Refrieger, Isamu
Noguchi, and the dozen or so others he'd recommended for Guggenheim
fellowships, actors Raymond Massey, Sidney Greenstreet, Miriam Hopkins, and more. And he seemed to have no such qualms about his other
older mentors, Europeans all—Otto Kahn, Caruso, Max.

Or Frank Lloyd Wright, with his bohemian capes and canes. Privately,
Bel Geddes blamed the architect—who knew "next to nothing" about
how theater functioned but "loved it as a child does"—for having also
contributed to the "disaster" of Aline's dream, which might have had a

significant impact on America's Little Theatre Movement, not to mention the course of Bel Geddes's entire life, and Barnsdall's. Even so, the blame hadn't interfered with Norman's abiding affection for Wright.

A self-made New Yorker, Bel Geddes would always be a midwesterner at his core. Something remained unspoken, unacknowledged, but unshakeable. The theater world, for all its geographic spread, was a relatively small place. His intransigence may have had to do with rumors or suspicions that his handsome, once-trusted, blue-eyed friend was bisexual.[532]

The "most casual and rather hasty few words," the first they'd exchanged in twenty years, might have transpired on the phone (likely a Dreyfuss model). Or perhaps Ordy's curiosity got the better of him and he'd risked a visit to Rockefeller Center, finding his way through the imposing glass-and-chrome doors, past the lobby's majestic marble floor mosaic and gleaming brass turnstiles, and up the elevator to Norman's bustling eighth-floor office.

The letter would remain unanswered. As with Ordynski's previous correspondence, Norman had his secretary type up a duplicate for his files.

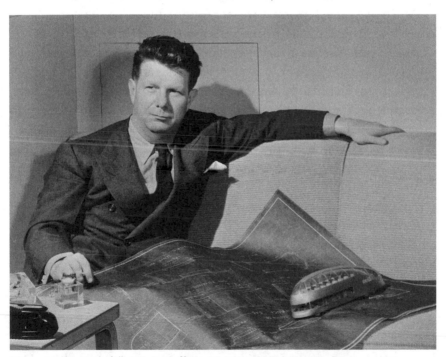

Norman in his Rockefeller Center office. *(Courtesy: Bettman/Getty Images, photographer unknown)*

CHAPTER 20

All the Wonders That Would Be

1940–1945

For I dip't into the Future, far as human eye could see;
Saw the vision of the World and all the wonder that would be . . .
    —ALFRED LORD TENNYSON, 1842

O stensibly created as a way to help people live well (affordable mass consumption thanks to mass production), industrial design was now embraced as a way to kill them more efficiently. And at a comfortable distance. Some 65 million combatants would die by war's end, the majority never having looked their enemy in the eye.

Much of Europe's military weaponry remained basically Victorian—rifles, single-shot pistols, an unwieldy, water-cooled World War I–era machine gun. Rescue came in the guise of a portable, automatic carbine capable of firing 600 bullets per minute, created by a British factory draftsman at his kitchen table using spare parts—tailpipe tubing, a bed spring, standard-issue screws. By early 1941, the Sten Gun was in production—in a toy factory. In short order, the factory's owner, an expert on doll houses and tricycles, sat down at *his* kitchen table, reduced the number of parts, and configured the Sten to fire the same 9-millimeter ammunition the German's were using, a handy "plus" when foraging in enemy territory. The new improved Sten quickly became the icon of Allied resistance.

Across the Atlantic, a twenty-nine-year-old Bell Labs engineer was working on a machine that used a pen to record voltages on strips of paper. Attached to the pen was a small device (the "potentiometer") that controlled the high-speed magnetic gyrations. One night, he dreamt that the potentiometer, mounted on a gun barrel, brought down an airplane with every shot. In other words, an antiaircraft gun, a device he knew nothing about. In December 1941, a working model of the M-9 was delivered to the U.S. Army for testing. By the summer of 1944, radar-guided 90-millimeter guns controlled by M-9s were shooting down the vast majority of all German buzz bombs launched over London and Antwerp.

•

BY THE END of his second Ringling season, Norman was weary of his constant association with streamlining, now an acknowledged and ubiquitous style on both sides of the Atlantic. Though he more than anyone had helped promote it, he was hoping it would die a quiet death. "I detest popularized terms like 'streamlining,'" he told an interviewer in 1942. "Thank God it's almost finished now, like jazz, or something that gets overdone."[533]

The writing had been on a wall from the beginning. His 1934 treatise in *The Atlantic Monthly* had admonished ad copywriters for playing fast and loose with the term that was already being applied to anything simplified, modern, up-to-the-minute, or ultra-advanced, used anywhere sharp angles had been turned into curves or levers replaced with buttons. Before long there were streamlined radios, typewriters, and Chippewa potatoes (the "absence of deep eyes reduces waste in peeling and also speeds up the job for the housewife"[534]), streamlined financial cutbacks, weight loss programs, inkwells, and coffins.

When redesign costs were deemed prohibitive, things were *decorated* to suggest friction-less virtues. Chromium ribs, indentations, and bulbous curves appeared on staplers, cameras, and cocktail glasses. Hairdryers, irons, and lawnmowers now resembled automobile grilles, airplanes, and the occasional spaceship. Aluminum roasting pans proved particularly ripe canvases; one was modeled after a streamlined ferry, another after the Burlington Zephyr locomotive.

Even Fifth Avenue's reigning beauty salons succumbed. The Ballets Russes–inspired decor of cosmetics giant Helena Rubinstein began to look overblown beside the "unabashedly mechano-morphic" sheen of rival Elizabeth Arden's venues.[535] The savvy Rubinstein "upgraded" accordingly, going so far as to special-order the world's first transparent Lucite sleigh bed, with edges that changed color via concealed fluorescent tubes.[536] Crude petroleum alchemized into the stuff of fairy tales.

Edgar Kaufmann, Jr., director of the Museum of Modern Art's Industrial Design Department, found streamlining's pervasiveness as irritating as it was surprising, given industry's general apathy to modernism.[537]

Streamlining, he quipped (borrowing Norman's metaphor from six years before), was "the Jazz of the drawing board"—popular in its appeal (sufficient grounds in and of itself for condemnation), far removed from its characterizing source (aerodynamics, "Negro" music), and highly commercialized.[538]

Kaufmann's British counterparts were similarly elitist. Streamlining, most agreed, was vulgar, a shorthand for American commercialism. "Design" (good)—as opposed to "style" (bad)—was European in origin and grounded in history, something too sophisticated and refined to be embraced (even if they could afford it) by the masses.

The teardrop shape exemplified by Norman's Futurama cars had, complained the cognoscenti, "swelled, divided and multiplied, become garnished with ribbons and chrome and . . . elevated on an altar of sales, while statistical Magnificats were sung in its honor."[539] The once-elegant scientific principle had morphed into a parody of itself.

The following appeared in *Yank*, a U.S. Army weekly written by enlisted men:

A PLEA TO THE POSTWAR PLANNERS

OR

DON'T STREAMLINE YOUR MOTHER WHILE I'M GONE

> *Sitting here in my foxhole muddy*
> *please don't think me fuddy-duddy*
> *if I say that I'm not fighting*
> *for plywood pants and neon lighting.*
>
> . . .
>
> *I've little need for breakfast toasters*
> *built like shiny roller coasters*
>
> . . .
>
> *I need no girdle wove of plastic*
> *nor baseball ball of glass elastic*
>
> . . .
>
> *or trips to Mars in Roman candles*
> *or caskets trimmed with Lucite handles.*[540]

The disillusionment with streamlining was partly a disillusionment with technology, which had been sold to the public as the key to "progress," which went hand in hand with if not a "frictionless" future then at least a world less troubled, mercurial, and violent.

But at decade's end, engineers would still be touting its virtues:

- Rounded molds allowed for the smooth, even flow of molten material, and its subsequent removal.
- Rounded shapes could be cut and polished by machine; sharp-edged corners required hand finishing.
- Curves distributed weight and stress more evenly, making breakage less likely.
- Curves brought out the reflective beauty of glossy materials.

Henry Dreyfuss coined the term "cleanlining" in an attempt to better describe the virtues of jettisoning "useless protuberances and ugly corners."[541] Not to be outdone, Loewy came up with "tumblehome," a nautical term referring to inward-sloping sides of a ship that would, at an imaginary point, converge.[542] *His* streamline ethos was something developed "far beyond the ideas of its originators."[543]

Design arbiters, soldiers, even cultural historians were beginning to insist that enough was enough. Roger Burlingame devoted an entire chapter of his weighty *Engines of Democracy* (a study of "mature" America) to what he considered a "hoax" perpetuated by "adroit propagandists." Streamlining, he wrote, is about speed, and speed is ultimately about saving time. But to what end, beyond haste, restlessness, and corporate profit?

•

Men standing in a bar, "killing" time. Women and children fretting over a game, a puzzle, a contest, to kill time . . . the subway, the stove, the can of condensed, cooked food all worked so fast . . . Even the radio is slow. We might go out in the car.

Millions of people killing time in cars . . . What does that snail think he's doing? . . . Now we've missed the light. A cigarette, please, to keep us from doing nothing . . . It's time to turn the old bus in, anyway . . . The new model has a hydromatic drive, there'll be nothing to do with your hands. What, then, shall we do with our hands?[544]

The 1932 *New Yorker* cartoon of the businessman who wanted Norman to style his new biscuit had been prescient. A decade later, hoping to capitalize on Bel Geddes's escalating fame, Loose-Wiles Biscuit Company contracted him to design packaging for a dozen products ranging from Cheez-Its and Oyster Crackerettes to Sunshine Fig Bars (a Fig Newton competitor), Hydrox (an Oreo clone), and Chocolate Queens (a Chocolate Mallomar upstart).

His detractors may have seen this as a new low for the once-lauded designer of *The Miracle*, but Norman had Alphonse Mucha and Paul Cezanne for role models. At about the same time, Loewy took on packaging for LU French butter biscuits, Wrigley's chewing gum, Heinz soup, Four Square Tobacco, and Nabisco; Dreyfuss would design packaging for Higgins Vegetable Glue and Whitman's chocolates; and in the late 1950s, retired "car architect" Harley Earl would be hired to create an "aerodynamic" new shape for the aforementioned Fig Newtons.

The public's fantasies about artistic purity remained inviolate. And many hardcore businessmen still thought of art, even when combined with mechanical ingenuity, as "smack[ing] of afternoon tea and Greenwich Village."[545] Imagination, wrote the authors of a 1946 design textbook, "comes from the clear-eyed examination of facts and alert attention to possibilities. Most people are so half-awake that such processes seem magical. But one proceeds from the known to the unknown in *any* problem."[546]

•

HOPING FOR SALES in the $50,000 range, Valley Upholstery Corp., New York, commissioned Bel Geddes to create four quasi-Finnish living room furniture groups (divans, sofas, chairs, and "occasionals"). "Sales have reached $300,000," Valley's president subsequently informed his designer, "and had we been able to meet the demand, we could have sold $1,500,000 more."[547] The Englander Co. of Chicago signed him on (a $1,000 fee plus monthly costs) as spokesman for their "streamlined Mattress of Tomorrow." Too bad Old Man Simmons wasn't around to see it. "You're on the Mat (tress), Mr. Geddes!" read one ad. "Shake Hands with our PROFIT Prophet!" read another.

At the end of 1941, Englander was forced to suspend the contract "until such time as Peace is installed."[548] Norman, who had more work than he could handle (including a "landmark" Bacardi rum distillery in Puerto Rico[549]), responded by sending beautifully dressed Christmas pheasants to three of Englander's top brass and seventeen other business associates and friends across the country. With the "Buddy, Can You Spare a Dime?" Depression still in progress, the gesture cost him today's equivalent of nearly $5,000.

•

"BILL" KNUDSEN AND Bel Geddes had become warm friends. Though not much of a theater man, Knudsen went so far as to present Norman with notes for a play he had in mind.

When Knudsen's native Denmark fell to the Nazis, Roosevelt appointed him a lieutenant general, the only civilian ever to join the U.S. Army at such an initially high rank. Alfred Sloan, who disliked losing quality staff almost as much as he disliked Roosevelt, had no choice but to let him go. Knudsen's new role was as consultant and troubleshooter, overseeing industrial war mobilization. Sometime after his appointment, he sent off the following note:

Dear Geddes,

What are you doing about camouflage? Very truly yours.

Dear General, Norman wrote back. *Keeping as invisible as possible. Very truly yours.*

Can you do it for industry? read the next missive. *Very truly yours.*

Yes, with pleasure.

Norman's "invisibility" projects for the military, all classified top secret, would include:

For the Navy, Floating Submarine Bases that appeared as icebergs or islands (and also harbored supply ships), as well a design for self-inflating airborne targets. Also, a camouflaged factory for a company contracted by the Navy to produce flying boats.

For the U.S. Air Force and Army Corps of Engineers, portable, self-camouflaged suspension hangars large enough to house two American Flying Fortresses and sturdy enough to survive severe weather.[550]

For the Army, camouflage patterns for military vehicles and aircraft.

He also advised on blackout techniques for "civilian" Manhattan and its greater metropolitan area. Given the ongoing problem (particularly in Britain) of pedestrians being hit by cars, he came up with an adjustable, luminous "Blackout Belt" compact enough to fit into a pocket, a design, he noted, that could also be manufactured for dogs.

Bel Geddes's BATLRAMA, a foldout relief map of northwestern Europe, had been published about the time the 1939 Fair gates opened. A two-dimensional outgrowth of his three-dimensional War Game, it was designed to aid public understanding of the reasons behind high commands' strategies. For ten cents, anyone could purchase its depiction of mountains, rivers, bridges, railroads, and other influences on military maneuvers, tack it to a wall, and keep track, with pins, of their son's, brother's, cousin's, and father's units.

Now he set his sights on "exploding" that map, creating a series of vividly realistic, three-dimensional reconstructions of naval battles and aerial attacks. Few actual photographs of military confrontations existed. Soldiers concentrated on shooting artillery, not film. Images they did capture were mostly obscured by smoke; available literature carried conflicting accounts.

Perhaps he might be able to help, as he put it, "clear things up."

After seven years as Norman's right-hand man, Pax had moved on to become an art director at *Life* magazine. He knew his old colleague had been giving plenty of thought to how his Lilliputian flotilla, much expanded since his War Game, could aid the war effort. Working together, the two raced to see if the necessary replicas could be amassed, and their collisions re-created, photographed (using a stop-action system dubbed the Bel Geddes Method), and disseminated before their "real life" counterparts clashed again.

Working overtime, Geddes and his machine-shop crew produced everything from smoke, lighting, burning planes, anti-aircraft puffs, and smoke screens to the fleets themselves. In the end his flotilla, built to precise scale (1 inch to 100 feet) with the help of several professional jewelers and an array of dental tools, expanded to nearly 1,700 battleships, cruisers, destroyers, submarines, submarine tenders, aircraft carriers, hospital supply-and-relief ships, torpedoes, motor cruisers, and mine layers—in all, the complete navies of England, the United States, France, Germany, Italy, and Japan, plus the best known ships of Russia, Greece, Turkey, and the Scandinavian countries, and twenty typical merchant ships of each country.[551]

In November 1941, *Life* ran photos of Bel Geddes's topographical model of Guadalcanal in the Solomon Islands, shot from above. A week later, the magazine ran a nine-page spread showing the Allied landing at Dieppe, complete with scouting planes, tanks, the surprise attack, and men "mopping up" the enemy.[552] Norman personally covered the $1,500 tab (in excess of $24,000 today) for the custom-built, twenty-foot-square cement "table" irregularly raised to simulate mid-ocean conditions.[553]

That same month, *Popular Science Monthly* ran a Geddes reconstruction of Japanese ship formations that caught the eye of military higher-ups, resulting in the first of several U.S. Navy contracts. Shortly after the bombing of Pearl Harbor, he was commissioned to photograph fifty views of specific ship models, at precise horizontal and vertical angles. It was a rush order. He had seven weeks. The prize was a whopping $25,500.[554] A rented room adjacent to Bel Geddes's office became Ground Central, its windows taped over, its access restricted. A vault, opened only under Navy supervision, was installed to secure the footage. By installing a small 16-millimeter film projector at the top or bottom of a periscope, direct comparisons could be made between the images in the eyepiece (on a screen, adjacent to the periscope sight) and photos of Norman's models.[555] The result—the Mark IV Submarine Range Finder—provided quick and accurate identification of any vessel within periscope range, both aiding invasion strategy and reducing the number of depth charges mistakenly dropped on whales.

On his own, Bel Geddes was experimenting with a TV-controlled aerial torpedo—the Television Bombing Plane. A transmitter would be placed *within* the torpedo to assist with direction and control. In the summer of 1940, Norman telegrammed RCA's David Sarnoff about it; it turned out that a similar idea was already in the works. Undeterred, he submitted ideas for "psychological" weapons: a "palm-of-hand" mapping technique for infantry scouting, a "rubber map process" for briefing landing forces (presumably in wet environments), a demountable pill box for jeeps and four-wheelers, and a military tank with a streamlined, ergonomic interior.

•

IN THE SPRING of 1942, *Life* published a five-page reenactment of the Battle of the Coral Sea, just seventeen days after the actual event took place, with Norman's name given pride of place in the headline.[556] As the conflict had been fought almost entirely in the air, Japan's first important naval defeat was re-created as if viewed through a cockpit dome. "There has never," the text began, "been a clear and complete photograph of a naval battle." Now, thanks to Bel Geddes's "amazing technique," shells splashed, bombs burst, anti-aircraft fire gave off black smoke that revealed flashes of ship gunfire from below, and targeted planes fell to the water in flames.

The smoke was cotton, carefully built up on wire armatures. Slanting wires gave the effect of distant rain; oil spills were simulated with paint. The lighting was carefully designed to match actual conditions. The greatest problem was simulating ship wakes. After experimenting with granulated sugar (obtained on Bel Geddes's ration card), the answer eventually emerged: bicarbonate of soda. Formulas were developed for camera placement and angles that best conveyed both scale and geographic elevations.

In July, *Life* ran Bel Geddes's "Battle of the Midway" sequence, a visual "retelling" of what had happened the month before.

Three years later, with the war finally over, the Navy would commission an *official* historical photo record of the battle, a seven-month task requiring a rented warehouse space at Manhattan's South Ferry, a crew of thirty, and $65,000 ($855,000 today) in military funding. The photo sequences would include believable depictions of everything from depth charge explosions and torpedo tracks to diving planes and shifting clouds. The Bel Geddes Method, wrote one twenty-first-century academic, anticipated a marriage between photography and fantasy unrealized prior to the development of digital image manipulation.[557] Various military higher-ups in both the United States and Canada mistook the reconstruction photos for the real thing.

On the theory that one should be prepared for the worst, *Life* also commissioned an elaborate model for the Battle of Gibraltar (which didn't occur) and pre-invasion projections of Norway and Germany's northern coast. "Norman Bel Geddes's models do not, as yet, include the entire surface of the earth but progress in that direction should not go unnoticed," quipped *Architectural Forum*.[558] Several of the models were installed at the MOMA.

•

LOEWY'S OFFICE, MEANWHILE, was modifying a glider to be used as a field hospital, Teague's office was redesigning Navy artillery controls, and

Dreyfuss's, mocking up interiors for a massive transport plane. Together, the three collaborated on the White House Strategy Room, a job that included phone systems, motion picture and slide projectors, complicated lighting, and the creation of several enormous rotating globes, one of which, at 100 feet in diameter, was large enough so that an aircraft carrier, made to correct scale (a vacuum cleaner was used to test its "airflow"), could be seen with the naked eye.

Additional Loewy wartime contributions leaned toward the quotidian—cigarettes and lip rouge.

A decade before, American Tobacco Co. president George Washington Hill had broken with tradition by eschewing imported Turkish blends and introducing the idea of smoking as a weight loss aid ("Reach for a Lucky instead of a sweet"), even persuading a bevy of debutants to march down Fifth Avenue, on Easter Sunday, puffing on "torches of freedom."

With 100 percent Virginia tobacco and a name that evoked the California Gold Rush, Lucky Strikes were 100 percent American—Loewy's preferred "target." He dropped some well-placed hints about his dislike of the iconic green and red bull's-eye package. Hill, he claimed, had to be persuaded. Like Zalmon Simmons before him, he wasn't hurting for sales.

Closer to the truth is that the U.S. military was in the process of requisitioning all copper and chromium based dyes and inks; the green had to go, regardless. Loewy exchanged it for a white background, moved the federal regulations to the side, and featured the red target logo on both sides of the package, so that it was always on view. A decade later, he would boast that "over 50 billion packs" sold as a result.[559]

Thanks, in large part, to the advent of Technicolor, lip rouge—once the province of prostitutes—was coming of age. Lip-*stick* was, like Luckys, a stateside invention. American females, so youthful and fresh-faced, risked becoming "highly vulnerable to gloom and dejection," the Frenchman reasoned. "Men would follow." A commission to remedy the problem came, he claimed, directly from Washington, D.C.[560] With metals and many plastics restricted, Loewy and his minions came up with a cardboard swivel tube.

"Hundreds of millions" sold.[561]

•

ANTICIPATING THE POST-WAR market, Dreyfuss designed a round thermostat; he'd noticed that rectangular wall models often didn't mount "squarely." A removable, clear plastic cover allowed buyers to paint the device to match the room. Released in 1953, the Honeywell Round T-86, like Dreyfuss's Model 302 phone, would become an instant classic. Donald Deskey worked on bowling alleys, newly popular thanks to automatic pin setters. Walter Gropius, now teaching at Harvard, was signed up by a New York ad agency, whose clients included Revlon, to handle their product design accounts. "When the echoes of this aesthetic earthquake die down," noted *Architectural Forum*, "it will be interesting to gauge the precise influence of the Bauhaus on the sale of nail enamel."[562]

Norman's postwar passions leaned heavily toward prefabricated housing.

By war's end, some 12 million military personnel, armed with GI Bill benefits, would be demobilized. Looking ahead to a shrinking demand for M-4 Sherman tanks (assembled in GM's factories at the rate of 2,000 a month[563]), General Motors had a vested interest in the "housing problem." The burgeoning decentralization of America would help sell more cars; the country needed standardized housing as efficiently produced as a Chevy or a Ford, a warship or toaster. "Why," asked Alfred Sloan, "doesn't somebody *do* something?"[564]

Affordable, factory-made housing wasn't a new idea. Edison, Wright, and Bucky Fuller had all taken on the challenge. But most architects saw pre-fabs as a threat to the integrity of their craft. Houses rolling off a production line? Instant slums, they argued, shelters for migrant workers and bums.

Norman's "House of Tomorrow," circa 1931, resurfaced early in the war as Expand-a-House, a dwelling that could, like his Oriole stove, be configured in various combinations. "Imagine a house with only 27 basic parts that can be delivered . . . in the morning and assembled into a finished home ready for you to move in by dinnertime!"[565] (Cookie-cutter Levittowns, assembled at up to thirty houses a day, wouldn't appear until 1947.)

His two-bedroom construct featured a flat roof, corner windows, air-conditioning, indirect overhead lighting, and a revolutionary heating system that interacted with body heat. Construction materials were yet to be determined, though brass sheets—welded onto each side of a honeycomb grid, filled with insulation and covered with plywood—had proven effective. Bel Geddes also devised a system for selling them, through the Housing Corporation of America, a cooperative venture run by former WPA (Works Program Administration) workers.

His detractors were often (not surprisingly) competitors. Case in point, George Nelson, who'd anonymously coined Norman's oft-quoted sobriquet "the P. T. Barnum of industrial design." Nelson's book, *Tomorrow's House*, released the same year as his "Storage Wall" (a modular alternative to closets), reassured readers that "design involv[ing] flat roofs and corner windows . . . is one kind of nonsense this book aims to expose."[566]

·

SOMETIME TOWARD THE end of his work on Futurama, Bel Geddes had met Libbey-Owens-Ford president John C. Biggers at a party. Norman was, arguably, the glass company's single best individual customer, given his custom mirror orders for the Barberry Room and Crystal Lassies.

Biggers made an offer. For years General Motors, a top customer, had been buying less and less. With designers trying to outdo each other, car windows were getting lower and lower, practically horizontal slits. Plus, bending them increased cost considerably. Biggers would hire Norman to design new bodies for every GM car, from Cadillacs to Chevys, paying the cost of all services and presenting the designs to GM, with no strings attached.

It was, Norman replied, one of the best jobs imaginable. Except for one thing. It had no chance of succeeding given the residual resentment toward him, especially in GM's Design and Styling Department, for having been awarded the Highways & Horizons contract.

It was the *results* that would count, Biggers argued.

Bel Geddes had been thinking about Plexiglas (Lucite) for windshields ever since his days designing five-years-into-the-future cars for Graham-Paige. Its advantages over glass seemed obvious. Along with being moldable, transparent, and less expensive, it was shatter-resistant, weather-resistant, light, strong, and as hard as copper or aluminum. It was, in fact, the "new plastic" Norman and Zalmon Simmons had observed in development, in Kenosha, in 1927.

Thanks to the war, Plexiglas had literally taking flight, used for cockpits, aviator goggles, and bombardier enclosures. Other plastics had found their way into everything from parachutes and gun turret seats to the tens of thousands of crucifixes and toothbrushes distributed to United States troops.[567]

Later, as Biggers was about to leave the party, Norman approached him. "Listen, Jack, you're not going to like this any more than you liked the last thing I said. *You're in the wrong business.* The idea of bent glass is only getting started. I know the mention of plastics to a glass man is like waving a red handkerchief in front of a bull, but that's what Libbey-Owens should be talking about. The two are identical, except that glass is brittle. Once the war's over, plastics will be ubiquitous. You're based in Toledo, where some of the best are already being developed."

Biggers left "in a positive gloom."

A year later, the financial page of *The New York Times* announced that Libbey-Owens-Ford had purchased Plaskon Inc., a Bakelite rival that boasted "1,001 Uses, 12,000 Colors," for "a good many millions." A major glass company investing in plastics was unprecedented, but perhaps not any more so than a cast-iron weighing scale company investing in "synthetics" research. It was Bel Geddes who'd convinced Toledo Scale Co. to establish Toledo Synthetics, for research and development, a decade before. Toledo Synthetics had eventually changed its name—to Plaskon.

Biggers never acknowledged Bel Geddes's lead.

"That plastics bug of yours panned out okay," Hubert Bennett, Toledo Scale Co.'s president, would tell Norman a few years later, over dinner in the Bennett home. Along with selling Plaskon to Libbey-Owens, he'd sold his personal Plaskon stock, resulting in "a pretty five-million profit of my own."

A few days after the dinner, Norman got to thinking over Bennett's "windfall." It had been Plaskon's rainbow of molded forms that had helped push plastics into the limelight, a postwar, multibillion-dollar limelight. It occurred to him that, had he been a businessman "instead of a creative type," he might have had the foresight to purchase an interest in Toledo Synthetics himself, early on. Ten years later he too would have had something substantial to sell to Jack Biggers.

•

THE WINTER OF 1943 proved particularly brutal, even beyond the weather.

On January 17, Frances's body finally gave out. Her doctor assured Norman that it was inevitable and likely for the best, as the TB had returned and her condition would only deteriorate. She was thirty-nine. Just the night before, Johnny North had thrown her a dinner party at the Stork Club, where she'd sat up until 2:00 a.m., happily drinking champagne.

Hidden among Bel Geddes's voluminous papers is a pencil sketch, lovingly rendered, of Frances, eyes closed, her hair disarrayed on a pillow. Seven years of treatments, surgeries, fatigue, and pain had aged her more than a decade beyond her years. The caption, in Norman's handwriting, reads: "Death Mask of F. Waite."

Less than a week later, Alexander Humphreys Woollcott, fifty-six, suffered a massive heart attack. Not too long before, he'd played himself on Broadway (as the selfish, autocratic Sheridan Whiteside) in *The Man Who Came to Dinner*. Two months after that, Rudolph Kommer, Reinhardt's longtime right hand, was found dead in his hotel room. Bald, rotund, and spectacularly unattractive, he'd spent his life surrounded by a coterie of exquisite, adoring women. Max wept openly at the funeral.

A wife and two very different friends gone, all within two short months. Bel Geddes did his best to bury his considerable grief in work.

A commission from the new School of Applied Tactics (AAFSAT), an Army Air Force appendage in Orlando, called for the design and construction of an automated, three-dimensional, spot-lit "tactical trainer." The result was a kind of surgical theater, a sixty-foot-square concrete pit representing 150 miles of rough terrain. Up to 200 soldiers and officers at a time, divided into "red" and "blue" teams, could observe from a pine balcony, looking down through glass windows, à la Futurama, placing commands for troop and tank movements and ordering bomb drops (acknowledged by light flashes) from Lucite airplanes (detailed down to the figures in the cockpit). A complex series of wires connected the control panel to a chain-and-pulley system run by 216 basement motors. The first-ever mechanically operated battle simulation room, it was a War Game offspring with workings reminiscent of the Nutshell Jockey Club.

Norman also produced *Civilian Defense in Action*, a film screened at Madison Square Garden; made a model-photography map of Tunis for the Army's Signal Corps; and was appointed to an Army advisory board (along with polar explorer Admiral Richard E. Byrd), to help solve feeding, clothing, and equipment problems. The Office of Strategic Services commissioned his "creative ability and unhampered imagination" for a secret weapons project that involved "front line sections in Europe and the Pacific," meant "hundreds in lives and millions in dollars," and required the subleasing of an apartment on East Forty-Ninth for six months.[568]

And he was collaborating with Reinhardt again, on an Irwin Shaw three-act at the Morosco,[569] and seriously talking about forming a corporation. Max's New York production of *Die Fledermaus* had just completed a 500-performance run.

For his next project, Max set his sights on a Broadway revival of Offenbach's *La Belle Helene* if only he could raise $20,000.[570] He went off to Fire Island to think.

It was late September. Almost everything was shut down. As he walked along the windy beach with his Scottish terrier, Mickey, a menacing, pug-faced boxer appeared. The terrier began barking and lunging toward the much larger dog. Max, now a portly man of seventy, dragged his furious pet into the only respite in sight, a phone booth. His decision to rescue Mickey was his undoing.

Once behind the booth's folded glass door, with the boxer growling and pawing outside, Mickey turned on his owner, biting Max's shoes, legs and arms, chest and sides. Struggling to defend himself in the narrow space, Reinhardt suffered a stroke, badly biting his tongue in the process. The boxer eventually wandered off. Reinhardt somehow managed to drag himself back to where he was staying.

It was a fate as cruel as anything Hitler's storm troopers might have dreamt up had Max remained in Europe, attempting to uphold his beloved Leopoldskron and the integrity of German theater. Additional strokes followed. A month later, he was dead.

"It must have been a shock to you," Max's son, Gottfried, wrote Norman a few weeks later. "Until all hope was gone, we tried to keep it absolutely quiet . . .

"You have been his best friend in America. And you have been one of his best friends in life . . . But what is even more important . . . is the fact that you were almost the only American who understood him for what he really was. He knew that always . . . He understood you, too, for what you really are . . . You really did not even have to speak."[571]

•

THE FIRST OPERATING machine to automatically and reliably execute long computations, the IBM Automatic Sequence Controlled Calculator, the Mark 1, initiated the Computer Age.

It was a monster: fifty-five feet long and eight feet high, weighing in at five tons. Its 765,000 components included 3,000 rotating counter wheels, 1,400 rotary-dial switches, assorted shafts, clutches, electromagnetic relays, interpolator knobs and plugs, a card feeder and card punch, "with parts sticking out every which way,"[572] plus input and output type-writers, all linked together with 500 miles of wire.

Like any self-respecting monster, it was ugly. Though still unclear as to its practical uses (like creating tables for aiming artillery shells and bombs), IBM chairman Thomas Watson was aware of the public's increasing interest in "thinking machines" and of the Mark 1's potential to boost IBM's image. On December 29, 1943, Bel Geddes was contracted to design a "skin" to keep out dust and muffle the noise (its clickety-clack sounded like a roomful of ladies knitting), then supervise construction and installation. The price tag for design, labor, and materials came to some $94,000 in today's dollars.

Along with dust and noise control, Norman's Plexiglas and metal "skin" strengthened the outer case. Thanks to carefully placed illumination, it also highlighted the machine's innovative works while downplaying its stock IBM parts. The final result—"like nothing before it . . . like a visitor from a future age"—put the Mark 1 in stark contrast to all other first-generation computers.[573]

Grace Murray Hopper, a U.S. Naval Reserve mathematician, was one of the people charged with keeping the Mark 1 running. (She was also the first person to literally debug a computer when she removed a moth that had fouled its innards.) Interviewed years later, she recalled that the case was "beautiful," that few knew Bel Geddes designed it, and that at some point it "got lost."[574]

•

IN APRIL 1944, Norman's old friend Paul Poiret died, franc-less and forgotten at sixty-five. Janet Flanner had called him a man "who helped change the modern retina," reintroducing color even beyond what his friends Bakst and Matisse had managed, a pirate who stole ideas from the future.[575]

But the future had changed. As the rotund Poiret's star was plummeting, the petite Coco Chanel's had risen, meteor-like, replacing his lush, orientalist fantasies with simplicity and understated elegance. The story went that upon meeting Chanel, dressed in her signature black, Poiret had asked who she was in mourning for. "For you, Monsieur Poiret," came the chilly reply.

July saw one of the worst disasters in United States history. When the Ringling clan wrested control away from Johnny North, Norman had asked, in protest, to be relieved from his third season contract, to the family's great relief. Despite the fact that upgrades had contributed to profits in excess of $2 million, the North–Bel Geddes's innovations—like replacing the tents' highly flammable paraffin-and-gasoline waterproofing mix with state-of-the-art chemicals—had, in the family's view, besmirched tradition. The old method was reinstated.

On July 6, a blazing hot afternoon in Hartford, Connecticut, a fire broke out in the Big Top. The nineteen-ton, paraffin-and-gasoline-soaked canvas burned so hot that it was gone in minutes. At 7,000, the audience was relatively small, but 168 died, most of them women and children, some trampled over by frantic adult males rushing to flee. Many others were hospitalized. The big cats, the only animals in the main tent, were herded out, unharmed.

On December 20, 1944, nearly two years after Frances's death, Norman, fifty-one, married Anne Howe Hilliard, twenty-nine, the Paris-educated, Social Register–savvy, recently divorced daughter of his former partner George Howe. Fiorello La Guardia did the honors at City Hall. Johnny North was best man. Two days later, the newlyweds left for a six-week Palm Beach honeymoon.

•

RAYMOND LOEWY WAS basically a stylist, but what was *designed* was his ascension to the rank of "most famous" industrial designer of the era.[576]

His staff included a private secretary (Helen "Pussycat" Peters), known for her exceedingly long eyelashes, and Elizabeth "Betty" Reese, a slender blonde whose talents would prove invaluable. During her first interview, "he told me right away that he wanted the cover of *Time*. I was a little taken aback."[577] She said it would take ten years.

Within three, Reese managed to place a lengthy feature in *Collier's*, public relations skillfully disguised as journalism, declaring Loewy's essential place in postwar expectations.[578] Under her guidance, *The New Yorker* described him as a twice-wounded war veteran who, "according to a court ruling some years back," was "the man who invented the profession."[579]

In a promotional film sequence she orchestrated, her boss wanders through a gaudy "room of horrors" (an idea borrowed from a Chamber of Horrors, staged a century before, at London's Victoria & Albert Museum). Standing beside a cherub-encrusted clock, Reese asks, "How do you dust this thing?" Blowing on it, she coats Loewy's hand-tailored suit with a heavy layer of dust. The public got the point.

She made sure interview questions were always available in advance. Unlike Bel Geddes, who thrived on spontaneity, Loewy "absolutely hated to be . . . surprised"; he'd spend hours practicing every word, every joke, for a public appearance.[580]

With her help, he would endorse everything from cookies and cellophane to—jewel in the crown—Rolex. In the full-page print ad, he wore a $1,075, eighteen-karat-gold Oyster Perpetual on his wrist,[581] posed beside J. S. Inskip, whose company imported Rolls-Royces. The ad implied he'd had a hand in designing both the watch and the automobile. Neither Rolex nor Rolls-Royce were Loewy clients.

Far better to be envied than pitied. It was a phrase Loewy claimed his mother had drummed into him as a child.

Betty Reese's father, a veteran newspaper man, had taught her some tricks and Loewy proved an apt pupil. When photos are taken with "some bigwig," always squeeze in to the *right* of the guy; that way, your name will always come first in the caption. Evert Endt, who worked in Loewy's Paris office for sixteen years, observed the pupil in action when they shared a flight to Germany. Upon landing, passengers were asked to remain seated so the press could photograph Miss America, also on board. Hearing this, Loewy "jumped up and made straight for the cabin door so that he would be in the picture at the beauty queen's side!" Squeezing in to her right.[582]

Loewy "was a great manipulator," Reese observed. "He'd decided at the age of about twelve or thirteen to create this character named Raymond Loewy and that's what he perfected."[583]

Betty Reese organized the first of many elaborate office Christmas parties, attended by clients and high-powered executives. On December 22, 1944, the "Raymond Loewy Players" performed a spoof of *Oklahoma*, Broadway's current smash hit, complete with costumes, wigs, makeup, sets, and a pianist. The entr'acte, to the tune of "Poor Judd Is Dead," was sung by "Bel Geddes" (his real-life counterpart en route to Palm Beach), "Teague," and "Dreyfuss."

I'm Norman Bel,
Yes, Geddes am I,

A fellow much respected by the press—by the press.
My Futurama's gone, and the other things I've drawn
Are pretty—but they can't be built, I guess—built, I guess.

My name is Teague—
It's "Walter" to you,
I'm known among designers as The Dean—as The Dean,
I've written several books, about function, form, and looks—
And, between us, I don't know just what they mean—what they mean!

Dreyfuss I am,
Initial is "H"—
I kick around a bathtub and a train—and a train;
I used to do some pens, with which Loewy now contends,
And my sympathy for him will never wane—never wane!
. . .
Dreyfuss and Teague
And Bel Geddes too . . .

we hope your ventures fizzle out—BUT GOOD![584]

Like his short-lived partnership with George Howe, Norman's marriage to daughter Anne would do just that.

•

IN JULY 1945, the U.S. War Department issued an oddly lyrical report describing an "atomic" bomb test, conducted in New Mexico's desert, as "magnificent . . . golden, purple, violet, gray and blue . . . the beauty that great poets dream about."[585] A century before, dreaming of the future's many wonders, poet Alfred Lord Tennyson had forecast, among other things, "pilots . . . dropping down with costly bales" and a rain of "ghastly dew."

On the morning of August 6, "Little Boy"—designed by some of the West's greatest scientific minds—was loaded into the belly of a B-29 christened *Enola Gay*, after pilot Paul Tibbet's mother, and transported to the skies above Hiroshima. "When I level out," Tibbets would recall, fifty-seven years later to the day, "I look up there, the whole sky is lit in the prettiest blues and pinks I've ever seen in my life."[586]

Three days later, "Fat Man" was released over Nagasaki, its impact the equivalent of 21,000 pounds of TNT. All told, a costly ($5 billion) bale of ghastly dew that burnt or maimed 200,000, invisible far below.

"My greatest disappointment," Bel Geddes would write, "has been the almost continuous warfare."

My imagination and interests would have developed much further and entirely differently if either, or both, of the two world wars had not taken place during my lifetime . . .

Technologically there was progress. Aesthetically and humanly, the difference will never be known, but my guess is that we were held back a full century.

CHAPTER 21

Quantity Trumps Quality

1945–1950

In America, which is the future, broadly speaking . . .
— MARTIN AMIS, *THE PREGNANT WIDOW*

T
here is going to be a new world when this is over," Natacha Rambova wrote to Norman in the fall of 1941, "even the dullest man in the street realizes [that]." That world depended on those with "the vision and the foresight to lay its foundations *now*," unlike the people she saw all around her, "burying their heads in nightclub sands like the proverbial ostrich while waiting for their backsides to be blown off."[587]

Predicting that postwar world became something of a national pastime. *National Geographic* reported the imminent arrival of run-less stockings (*hubba-hubba*) and permanently creased pants.[588] Henry Dreyfuss predicted mail dispatched by guided rockets that spanned the continent in half an hour, and face-to-face long-distance phone conversations. Waldemar Kaempffert, who'd waxed poetic about the Polyrhetor, envisioned wall-size TV screens (think plasma panels) and a movie subscription service (anticipating Netflix by fifty-six years).[589] "Living Unlimited," a GM promotional film, looked ahead to disposable bedsheets, ultrasonic dishwashers, and, for overburdened moms, an automatic spanking machine. *Science Digest* put its money on frozen foods precooked by world-famous chefs; zapped with sixty seconds' worth of high-frequency radio waves (microwave ovens), they'd "pop up like a piece of toast."[590] A "conservative" list commissioned by the U.S. Army prophesied electronic watchmen to protect children from polio; factories that run for a week on a single lump of coal; and a new science of "chemotronics" that "will knock the 'im' out of impossible." Bel Geddes's flexible housing and gigantic airliner were mentioned, as were bubble-shaped rear-engine cars à la Futurama.[591]

Real-time advances included cruise control for cars (invented by a blind man) and the world's smallest bathing suit (named after the Bikini Atoll, where the U.S. tested a post-Hiroshima bomb). Shortly after entrepreneur Milton Reynolds made a fortune marketing ballpoint pens ("It Writes Underwater!"), the ballpoint's delivery system was adopted by a deodorant company. Enter Ban Roll-On—an overnight success that quickly went international.

There was considerable talk about "personal flying machines." Dreyfuss came up with the Convair, a plane perched stiffly atop a sedan. Bel Geddes's "roadable" airplane, dating back to his Graham-Paige days, featured hinged, telescoping birdlike wings that retracted into the chassis.

"You drive out of your garage, get it out on the road, and press a button," he told an interviewer (not without a touch of mischief). "The wings come out and you take off."

"Is that just an experiment?" he was asked.

"Anything that's two or three years ahead is an experiment."[592]

Privately, he may have been more interested in developing a lightweight, folding bicycle.

Europe's factories had been paralyzed by the war. America's, the world's largest and most modern, had not. But many, if not the majority, were busy turning out poor designs made from low-grade surplus.

By ignoring what an increasingly savvy public wanted, industry was playing a perilous game, Bel Geddes had told a national forum of refrigeration engineers back in 1943. They were wrong to assume that the coming postwar demand for practically everything would be a free pass for cranking out shoddy merchandise. "The public never expects the impossible"; they knew the market wouldn't flip on a dime come Armistice. But they also knew that if industry could perform miracles in the name of destruction, they could perform miracles for peace.[593]

It would be a while before Bel Geddes's public had much choice.

Plastic—what *Forbes* had once described as "glamorous" because it was manmade—was now ubiquitous, replacing auto bumpers, milk bottles, and children's toys. Lucite cockpits gave way to "easy-wipe" sofas. Thanks to the birth of injection molding, "unbreakables" were promoted in lieu of heirlooms. Tupperware, the "Material of the Future," provided housewives with airtight, liquid-proof containers that sealed like paint can lids in reverse. Earl Silas Tupper, a poor farm boy who'd previously invented a fish-powered boat, customized cigarettes, and a better way to remove a burst appendix, eventually made enough to retire on an island he bought off the coast of Central America.

But the "new" seemed increasingly more perishable than the "old." It was becoming obvious that plastics aged badly, especially when compared to the patinas natural materials (wood, ivory, leather, copper, bronze) acquired over time; even things exposed to the brutality of seawater— driftwood, sea glass, encrusted pots from sunken ships—could be seen as having been enhanced. Plastic and other synthetics now struck many as best relegated to shower curtains and jukeboxes.[594]

But the villain wasn't simply plastics.

In the Twenties, Americans were perceived as having transformed goods into roses. Two decades later, those roses had devolved into thistles.

Arthur Miller's *Death of a Salesman* captured the zeitgeist perfectly. The American Dream as poisonous mirage. "They time those things," Willy Loman wails. "They time them so when you've finally paid for them, they're used up." Is he just talking about his refrigerator, his car— and himself—or also about the merchandise he'd devoted his life to sell- ing? Which was what, exactly? We're never told. When the curtain fell on the play's first performance, there was no applause. The audience was stunned into silence, followed by a tidal wave of weeping and sobbing.

•

THOUGH HE'D MADE the decision to focus on clients with $10,000 mini- mum billings, Bel Geddes took on a (largely self-funded) model of the Panama Canal's construction for the *Encyclopedia Britannica*, followed by another *Life* magazine sequence, a six-pager on Egypt's ancient pyr- amids. Using a combination of "fact, good probability and not so good possibility," he speculated on how levers, ropes, sledges, plumb lines, and a limitless supply of muscle had assembled nearly 6 million tons of stone, "only to surround the bones of a pharaoh."[595] (The Empire State Building had required a modest 305,000 tons.)

Far more ambitious was "Toledo Tomorrow," a sixty-foot-long, $250,000 model commissioned by a nonprofit Ohio newspaper, a fifty-year master plan designed to rouse civic support. Viewed from raised circular plat- forms, it incorporated many Shell Oil and Futurama concepts. Six 5,700- foot runways anticipated commercial plane travel as commonplace. *Life* obligingly ran a six-page photo spread.[596]

•

THE POSTWAR RADIO market was considered the hot ticket, a market "wide open for plucking."[597] In a few short years, NBC would take stewardship of the air, its coast-to-coast network of stations allowing snowbound New Yorkers to follow California Rose Bowl games in "real time," its Teletype machines connecting cities 24/7, making NBC "the only newspaper that's perpetually going to press." Shortly after D-Day, the Federal Telephone & Radio Corp. offered Bel Geddes a contract for a new line of radio cabinets.[598]

Far more interesting was talk of "a really new concept," the portable pocket device. Transistors would come of age in the 1950s.[599] But as early as 1938, a range of miniature Raytheon tubes, intended for hearing aids, were available; ruggedized versions, used in proximity fuses, had revolutionized WWII artillery. Bel Geddes offered Philco a design for a compact, lightweight, battery-operated "slim-line" radio that hung from a shoulder strap across the chest ("vertical suspension"). It featured recessed controls, a circular speaker (emphasized in white as a decorative element), a detachable top half that worked with earplugs, and a handsome leather carrying case.

Norman's shoulder-strap, ear-plug design predated Sony's first portable transistor radio, the TR55, by a decade. Cultural historian Christopher Innis cites it as an early (thirty-three years early) prototype of Sony's Walkman (1979), which was, in turn, father to the iPod (2001).[600]

•

RUNNING A BUSINESS had turned out to mean spending most of his time procuring jobs and administrating, rather than engaging in creative work. And Norman was growing increasingly unhappy about what he saw as his partners' intransigence on policies like giving new clients ownership of in-house-generated designs. Added to that, his annual overhead (reported at $884,000, in excess of $8.5 million today[601]) was reaching critical mass. Along with covering salaries, rent (an entire floor), and the wining and dining of clients, he'd personally financed the elaborate, labor-intensive models created for *Life* magazine. Expenses on the *Britannica* models and extensive research on his prefab housing ideas had far outweighed any recompense. He'd been forced to delegate more and more. Rumors of financial mismanagement followed. Combined with high blood pressure, it was all taking a toll.

His sprawling Rockefeller Center office was his Leopoldskron, his

Taliesen, proof to the world that, at forty-six, the midwestern high school dropout had mastered the dazzling pinball machine that was New York City. By late 1946, just six years after moving in, he saw no option but to let it go. He personally undertook to pay 90 percent of Norman Bel Geddes & Co.'s outstanding obligations and debts, in excess of $200,000.[602]

Otto Kahn and John Nicholas Ringling had died with their granite and marble fiefdoms intact. Norman had spent thirty years building his, with few resources other than his unique, turbo-charged, insomniac determination. Like Max, with his beloved Schloss, he would never get over the loss.

In short order, he was conducting business from his Park Avenue apartment (to the dismay of building management, who eventually got wind of it), employing a typist, a secretary, and one draftsman, and subcontracting out for specialized needs. There was talk of re-joining forces with Johnny North, back at Ringling's helm. Given that the circus still owed $2 million in fire damages, Norman offered to do away with his yearly retainer, working solely for expenses, "with some kind of understanding for the future."

Undeterred by Bel Geddes's at-home headquarters, Thomas J. Watson, the man who'd ordered the Mark 1 casing, green-lighted twenty-four "distinctly new" assignments, including a redesign of the ground floor, interiors, and exteriors of three IBM buildings at Madison and Fifty-Seventh; a "spectacular" rotating globe roof sign with time-telling features; designs for a large delivery truck ("with airplane pilot treatment of the driver space"); a series of "ergo-dynamic" desks; a card-sorting machine; a redesigned card-punch; and a redesigned trademark. Monthly billing was anticipated at $15,000 ($182,000 today) for at least six months.[603]

Other commissions followed—from Valley Upholstery (a second spate of furniture pieces) and A. S. Beck (window displays for 123 stores) to a consultancy on theaters being planned for the UN Center. The press, meanwhile, having quickly forgotten their storefront-less Leonardo, was busy touting the merits of "The Big Three."

Another postwar design, a collaboration with employee Eliot Noyes, was for a not-quite-streamlined typewriter casing to replace IBM's Electromatic, a clunky affair resembling Darth Vader's helmet. It debuted in 1947 as IBM's Executive Electric, featuring hidden under-mechanisms, a lower, more comfortably sloped keyboard, and untraditionally square, marshmallow-like keys (a prelude to computer keyboards).[604]

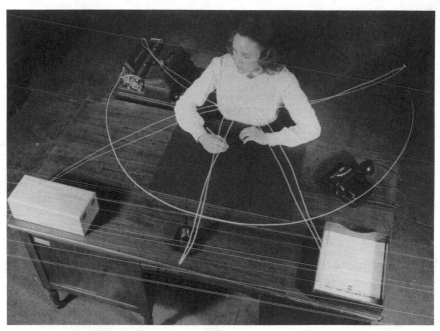

Ergonomic motion study conducted for IBM.
(Courtesy: Harry Ransom Center, The University of Texas at Austin, photo by Norman Bel Geddes & Co.)

Norman's personal IBM "Executive Electric," co-patented with Eliot Noyes.
(Courtesy: Harry Ransom Center, The University of Texas at Austin)

Even before the Rockefeller Center office shut down, Noyes hung out a shingle with Marcel Breuer, taking the typewriter account with him, with Norman's blessing. A subsequent seven-year collaboration between Noyes and IBM would result in the Selectric (stationary carriage; moving, interchangeable "golf ball" print head), which quickly dominated 75 percent of the market,[605] usurping the Underwood 5's half-century reign.

The Selectric's insides were revolutionary, but its body design varied little from the original Bel Geddes–Noyes patent.

•

"DO YOU THINK that luck only lasts so long and then lets a person down?" Amelia Earhart once responded to a reporter's comment, not long before she disappeared.[606] In March 1947, Norman fell "over cold" and took to his bed, under doctor's orders. The man whose only previous bed stay had been a bout of the flu, decades before, would remain there for the next eight months.

Dreyfuss, meanwhile, was helping Lockheed convert military planes for civilian transport and overhauling his own 1938 Twentieth Century Ltd. into a dazzling new Streamliner. He'd opened a second office—a tiresome, expensive commute—but California was conducive to his wife's allergies, and he hired a public relations pro, for a hefty $25,000, to quell rumors (in the wake of Bel Geddes & Co.'s demise) that his New York office was closing.

By the end of the decade, Raymond Loewy had opened branch offices in Chicago, Los Angeles, South Bend, Indiana, and London.

In February 1949, *Look* magazine ran a five-page "personality spread" in which he was seen relaxing at the wheel of his cabin cruiser (jaunty scarf at the neck), smoking in his "glittering" Park Avenue penthouse (mirrored fireplace, thick amethyst carpet), and pulling champagne bottles from a snow stack with his new bride, a Benson & Hedges public relations consultant described simply as a "young . . . ex-model."[607]

In May, *Life* published "The Great Packager," a six-pager.[608] It opened with an arrangement of objects; a Coca-Cola bottle and a Lucky Strike pack held pride of place atop a chair. Bearing a slight resemblance to Prince Rainer Grimaldi, who had, that same month, ascended Monaco's throne, Loewy stood holding an egg inside a brandy glass—two shapes he deemed beyond improvement.

"At 55, his appearance is controlled, as the professional phrase goes, by

Turkish baths, ultraviolet rays, calisthenics . . . He stands 5 feet 11 inches and weighs in at 162 pounds. The faintest tremor of the bathroom scale causes him to diet violently."

The text, presumably overseen by Betty Reese, veered between adulation and character assassination. "Though an unchallengingly masculine specimen with an impressive war record, Loewy does not hesitate to titillate his senses with feminine fripperies. He is one of the country's most aromatic males, given to burning incense in his office and at home, and consuming caseloads of perfume, scented soap, bath oil and skin lotions." (A 1935 article credited him with creating pencils "that give off a subtle Chanel fragrance in mid-office."[609])

His real estate holdings, listed in the *Look* profile, were expanded in vivid detail: joining the penthouse was a 200-year-old Long Island "mansion" (where he raced cars and motorcycles); a Loewy-designed St. Tropez villa (where he enjoyed deep-sea diving and his cook's unsurpassed bouillabaisse); a Paris apartment on the Quai d'Orsay; a rural sixteenth-century French manor (once home to King Henri IV's mistress) surrounded by peacocks, wild pheasant, and 100 acres of forest; and a modernist villa near Palm Springs (with heated spring-water pool that "meandered" into the living room).[610]

It's unclear what America's D-Day veterans, recently returned from risking life and limb to liberate France, made of this "determined hedonist" in their midst. Asked by a journalist, years later, if Loewy was a happy man, Betty Reese replied, "He was a *successful* man." A materialist. "Happiness is a word he wouldn't really have understood."[611]

Speaking "in a ripe French accent," he declared built-in obsolescence "unspeakably immoral."

In his autobiography, published the following year, he would adopt a very different stance. If Americans followed the European inclination to repair things instead of replacing them, "production would take a nosedive and the whole economy would suffer." And that, he observed, went a long way toward explaining "American prosperity" versus "European penury."[612]

He denied having invented the profession at which he excelled (another stance reversed in his forthcoming book), placing the indictment on a par with the quip that he and his "leading competitors"—followed by a short list in which Bel Geddes's name was obvious by omission—had "fomented more social unrest than Karl Marx." Walter Dorwin Teague, among those short-listed, went to the trouble of registering his objection. "I like Raymond Loewy personally," he wrote to *Life*'s editor, "but I want no part of his publicity, even by association . . . will you please print this explicit denial of any implication that I share [his] foibles, way of life, or attitudes toward my professional work?"[613]

Loewy found all the attention "heartwarming." In his own letter to *Life*'s editor, he wrote that he was pleased to see that his adopted country appreciated "the beauty and taste which all Frenchmen prize. Calvet, Braque and I must be grateful."[614] Corinne Calvet was a French actress. Braque was, of course, Braque.

Teague's wish for disassociation was not to be. Loewy was fond of comparing "Walter, Henry and I" to Boy Scouts, their good deed of the day having been that they'd "helped a lady across Style Street." The punch line was that it took three of them because "the old lady did not want to cross."[615] It's interesting to imagine the dynamics of Loewy's friendship with the kind-hearted Dreyfuss, whom he claimed as one of his closest friends. Henry Dreyfuss, who considered Bel Geddes one of *his* closest friends, referred to Norman as "Genius Number One in my life . . . the only authentic genius" among the first generation of U.S. industrial designers.[616]

"In less than a quarter century [industrial design] has mushroomed," observed *Time* magazine in October, "from a groping, uncertain experiment into a major phenomenon of U.S. business."[617] Later that month the much-coveted Loewy *Time* cover—the Holy Grail, the Golden Fleece—hit the stands with the caption "He streamlines the sales curve."[618] It had taken Betty Reese eight years, but she'd done it. The article predicted that U.S. business would spend some $500 million in the next twelve months, with Loewy's firm collecting $3 million of it.

Loewy's face was described as being "as inscrutable as that of a Monte Carlo croupier." The first of half a dozen photos showed him in a silk dressing gown, being served by a liveried butler. Observers "get the strange feeling that [Loewy] too is a design—by Loewy, of course."

There were hosannas to Bel Geddes ("industrial design's greatest prophet and visionary"), a Bel Geddes photo, and nods to Teague, Van Doren, and Dreyfuss.

The profile went on to mention a query from *Glamour* magazine that had asked Loewy for an article on theater design. "Wonderful," the Frenchman replied. "I've been waiting for a chance to tell everyone what's wrong with theaters."[619] Perhaps he hoped Norman would read it.

"I don't want to be publicized the way Raymond Loewy is, all that flashy stuff," Henry Dreyfuss was heard to remark, "but I really would like a *Time* cover story."[620]

Loewy's association with Coca-Cola began with a countertop dispenser, a 1939 Fair commission. A decade later, he designed a streamlined standing cooler and a Hobbs delivery truck body. In 1955, he'd suggest slimming down Coke's Mae West–like silhouette; it had been altered and slimmed before. John Epstein, Loewy's chief of staff (later firm vice president) drew the slightly refined curve.[621] Loewy may or may not have suggested the white embossing applied to the scripted name in 1957. In 1960, his firm would design Coke's first steel can.

But for Loewy, it wasn't enough.

American culture "is full of affirmations of what it means to be American," wrote British design educator Adrian Forty, "a characteristic that seems odd to non-Americans."[622] Like Miss America, with whom Loewy "created" a photo opportunity, like the Lucky Strike package, like Loewy's Airforce One and NASA projects, yet to come, Coke's fluted sea-green bottle—already iconic, its shape recognizable even in the dark—was an artifact that shouted "America," even in the farthest reaches of the planet.[623] If Walter Dorwin Teague hadn't had Kodak "sewn up," it would, in all likelihood, have been on Loewy's hit list. Or perhaps it was. Loewy wanted it both ways—to remain essentially French (provenance, refinement) while shaping the very notion of America (brash, new, innovative) more profoundly than any native-born peer.

But the drink, with its famously secret formula,[624] wasn't the issue. Loewy wanted people to believe that the Coke bottle was his brainchild.[625]

Though he never overtly lied about his role in its design (Alex Samuelson, a Coke company manager, is credited on the 1915 patent), Loewy "virtually monop-

Raymond Loewy, 1939. "Veddy, veddy distingué."

(Courtesy: Gemma-Keystone/Getty Images)

olized the subject of Coca-Cola packaging in the press during the postwar years,"[626] wrote design critic and historian Stephen Bayley, carefully orchestrating "misunderstandings," going on, often at length—during interviews, speeches, university and business school lectures, symposiums, conferences, and packaging competitions, both in the States and abroad—about its virtues. ("The Coke bottle is a masterpiece of scientific, functional planning . . . The most perfect 'fluid wrapper' of the day and one of the classics in packaging history" was a typical hosanna.) French design journalist Philippe Tretiack, Loewy's on-again, off-again booster, referred to "a certain well-maintained confusion" that reigned over many of his productions.[627] He talked about the bottle at every opportunity, concurred John Kobler, author of the 1949 *Life* profile, "brood[ing] a good deal" about its "aggressively female" shape, "a quality," the Frenchman noted "that in merchandise, as in life, sometimes transcends functionalism."[628]

He would still be pushing his agenda in the early 1960s, referring to the back end of his Studebaker Avanti design as "the Coke bottle look." Repetition and nuance eventually became "fact."[629]

"Raymond Loewy . . . the man who conceived the Coca-Cola bottle (and started industrial design)," wrote *The Philadelphia Inquirer*'s art critic in 2002.[630] "Perhaps his best-known design was that of the Coca-Cola bottle," reads his entry in the *Encyclopedia Britannica*.[631]

So pervasive was the myth, and its appearance in print, that in 2012, twenty-six years after the Frenchman's death, the Coca-Cola Company would finally post a disclaimer. "Many people have mistakenly attributed the [famous contour] bottle to Loewy," it read. "He did not play a role in the original 1915 bottle."[632] That the mistaken attribution had been a calculated one was left unsaid.

But in fact, Loewy *had* designed an original bottle, and a logo, for the beverage corporation—for the much lower-profile Fanta[633]—something he rarely, if ever, mentioned. Created specifically for a European market (patented in Germany and all Germany-occupied countries when WWII denied access to the famous Atlanta syrup concentrate), the orange-colored, orange-flavored drink didn't serve to promote Loewy as an architect of the American Dream.[634]

•

IN NOVEMBER 1947, when Norman, now fifty-four, finally "got up," as *Time* magazine put it, he began again with "bigger ideas than ever before."[635]

One of the first was securing a $15,000 "note" from Webb & Knapp, Inc., of Madison Avenue (while convalescing, he'd redesigned their lobby), repayable without interest, to help finance the Norman Bel Geddes Corporation of Connecticut, a new out-of-state office, in Stamford, to help his "tax situation." Meanwhile, still working out of his apartment, he expanded his staff to twelve and leased (and renovated) an East Fifty-Fourth Street penthouse so his draftsmen would have a work space.

The following June, he was scheduled to meet with Earl Carroll. Once owner of an eponymous Broadway theater, Carroll had gone on to open a highly successful supper club on Sunset Boulevard; a twenty-foot neon portrait of Carroll's girlfriend, Beryl, was one of Hollywood's most famous landmarks. Carroll was thinking about a mammoth new club and cabaret, but en route to their appointment, he and Beryl died in a plane crash.

Taking a break from a brutally hot Manhattan summer, Norman headed west to meet with J. Meyer Schine, who, happy with his purchase of L.A.'s 500-room Ambassador Hotel (he'd quickly recouped his $1.6 million investment), had invited Norman to "look the place over." What Schine (another former New York theater owner) considered plush, Bel Geddes declared a paean to bad taste, its marquee reminiscent of hot dog stands. As for its famed Cocoanut Grove cabaret, were cocoanut trees really necessary? "Does the Pump Room"—the Ambassador East's high-end watering hole—"have a pump?" he asked. Schine was willing to spend upwards of $10 million to get the job done.

Norman began commuting to California every two months for a week or two at a time, which gave him an opportunity to visit with the Dreyfusses when Henry was in Pasadena. Working with an architect (NBG remained unlicensed), he proposed doubling the hotel's size. Six city blocks' worth of surrounding land would accommodate 500 bungalows, each with a garden and a television (a hotel rarity at the time), all connected by a tunnel system to the main hotel (roof-level swimming pool, with a beach of washable, rubber-composition sand), so that no guest would be deprived of room service.[636] Completed, the pricey renovation revived the Ambassador's profile and bookings.

At about the same time, Norman was approached by the owner of what had been Miami Beach's famed Copacabana. Four years before, it had gone up in smoke. Standing amid the rubble, owner Murray Weinger, an expat New Yorker, had told a radio interviewer, "If I could get a man like Bel Geddes to come down here . . . it would be the talk of the country!" Alerted by one of Schine's associates, Norman lost no time getting Weinger on the phone.

Bel Geddes reinvented Copacabana as Copa City for a reported $1,125,000.[637] Thanks to three shifts of workers, the job was completed in a record four months. The result was a flesh-colored grand piano of a building—sinuously curving walls, inside and out. There were no right angles, just curving glass doors and serpentine, floor-to-ceiling glass walls. More than a nightclub, Copa City was a multiplex entertainment center, with TV and radio studios and shops. The "freewheeling," ever-curving interior led one journalist to say he was "beginning to understand what Professor Einstein meant by time without beginning or end."[638] *Variety* found the lush, vast, transparent illusion "overwhelming."[639]

Each of Copa City's five stories featured cantilevered, reconfigurable walls (think Expand-a-House) that could be raised, disappearing through ceiling slits, or lowered via slender overhead interlocking trusses. Everything hung suspended from the roof (as in *The Miracle*), with most of the weight carried by the exterior walls (like the Pole-less Tent). The cabaret room, with seating for 300, could, for example, be expanded to accommodate 1,300, all seated at tables. All above-ground floors were suspended (on one-eighth-inch diameter monel wires) for easy removal.

On December 28, 1948—twenty-six years after the Bel Geddes–designed Palais Royale debut—Copa City's sold-out opening was broadcast live, coast to coast. ("If this is how the future world is going to look, try and get a ticket!"[640]) Milton Berle headlined the floor show. Newsreel footage of the event was televised over the major networks and photo spreads appeared in national magazines.

It was the grandest of Miami Beach's nightclubs while it lasted, its biomorphic curves laying the groundwork for the Fountainebleau, Eden Roc, and other hotels destined to give Miami its distinctively "modern" look.

When Schine acquired the swank Boca Raton Club near Fort Lauderdale, he brought his Ambassador designer back on board. Boca remained a small farm town; Norman was charmed by its luxuriant vegetation, soft tropical moonlight, and enormous nocturnal moths. Bulldozers were soon reshaping the landscape into a gently rolling, four-square-mile plateau to be topped with a resort colony of deluxe single-level Bel

Geddes–designed homes—"stage scenery for a lifestyle"—slated to sell for $20,000 to $50,000 apiece.[641]

Eschewing the local pseudo-sixteenth-century Spanish "look," he opted for flat, cantilevered roofs (no snow in Florida), high ceilings, open floor plans, terraces (for outdoor living), and sinuously curved glass walls à la Copa City. Colony residents would also have the benefit of a Bel Geddes–designed shopping center, sports arena, restaurant, and a hotel-slash-yacht club. Half of his sixteen house designs would be built.[642]

While he was at it, he expanded his sights to the city of Boca Raton itself, drafting up a new town hall, a hospital, and a postwar street plan. But his plans hit a snag when he took local town leaders on a tour of Copa City, with its "up-in-the-air platform stages." All that mid-century modernism!

Schine's Boca commission got Norman thinking . . . what about a "tropical niche" stretching from South Carolina to São Paulo on the Atlantic side, from Monterey to Lima on the Pacific? A palette for a mass-production domicile that could make the most of sun and breezes? The only thing pre-fab about it, he wrote Henry Dreyfuss, were the sectional walls, complete with three-foot windows.[643] In good weather, they'd roll up in ten minutes or less; in storm season, they'd drop into place, weather-tight and hurricane-proof.

The roof would be supported by slender columns. Independent "cubes" would provide unlimited storage, "the first house to have *too much* closet space." Designed to allow for different configurations, depending on owner preference and lot size, these homes would be more affordable, per square foot, than conventional housing. When Hollywood released *Mr. Blandings Builds His Dream House* (Cary Grant, Myrna Loy), Norman saw it three times.

•

HAROLD VAN DOREN had predicted a postwar world in which "even the lowly skillet will be a work of art."[644] But at decade's end, *Time* reported that "there still remained vast, unexplored regions of ugliness and inefficiency" for designers to tackle.[645]

Returning GIs tended to favor homes that resembled the ones they'd grown up in. Victorian "gingerbread"—rosebuds, griffin heads, and curlicues—was back, only now manufacturers were calling it "eye appeal." The cognoscenti (still critical of streamlining, which eschewed such excesses) were quick to give the travesty a different name: borax.

Borax, wrote Richard Gump, head of the eponymous, upper-crust San Francisco emporium, meant anything made to "seem what it isn't."[646] The term may have derived from Twenty Mule Team Borax, a popular household cleaner given away as a premium with the purchase of cheap Depression-era furniture. Or perhaps because borax—an extraordinarily versatile compound good for everything from killing roaches, stopping engine leaks, and making indelible ink to curing snakebite and coloring fires with a green tint—was simply too democratic.

Weary of the Museum of Modern Art's declarations about Bauhausian "integrity," Dreyfuss told a conference of Canadian designers, "There are only three museums I'm interested in—Macy's, Marshall Fields and the May Co."[647]

"To the recently arrived Polish miner or the Croat foundry worker still obsessed by the misery of his Balkan background," proclaimed Loewy from his perch atop the pinnacle of Good Taste, borax "represents materialistic splendor. They buy the stuff ravenously, take snapshots of the family sunken in upholstered American luxury, and mail them to Gdynia."[648]

The citizens of lower Slobovia "may not give a hoot" about principles, like free speech, that make America great, Loewy informed Harvard Business School students, but they go into raptures over "gleaming Frigidaire[s], streamlined bus[es and] coffee percolator[s]," all items the Frenchman had, coincidentally, designed.[649] The same unwashed class that gobbled up borax goodies was, apparently, also ravenous for its opposite.

"Couldn't you just *not* produce borax design?" the head of General Electric's Appearance Design Division was asked.

"You have to sell what the consumer will buy," was the response. Companies can try to "educate" Mr. and Mrs. John Q. Public, but if GE ("Progress Is Our Most Important Product") didn't "follow the trend," a competitor surely would.[650]

Of all the things that exemplified the depths to which Bad Modern could sink—free-form tables, anything Cubist, "elongated" black panthers (Loewy preferred large, cerulean Chinese lions, a pair of which perched outside his penthouse window)—Nubian slaves were a favorite Loewy target. In his autobiography, he lambasted ceramic statues of swarthy, taut-muscled, bare-chested ancients decked out in gold turbans, "ecstatically proffering another black ashtray."[651]

In a speech to the Society of Automotive Engineers, he joked that the public's image of an industrial designer was one who "works best when draped at the edge of a turquoise swimming pool while Nubian slaves in gold sarongs serve chilled nectars in silver cups."[652]

Coincidentally, Loewy's much-sought-after *Time* profile included a photo of his flower-filled "black, beige and bronze bathroom, with its motif of Nubian slaves."[653]

CHAPTER 22

Prodigal Daughter

1950S

I didn't see much of my father but I absolutely adored him.

—BARBARA BEL GEDDES[654]

D uring her second year at Putney's, fifteen-year-old Barbs was expelled for kissing a classmate, despite protests from both sets of parents that the punishment was excessive.

Summer offered a temporary respite. In June 1940, early in Futurama's second season, Norman helped his daughter gain admission, as a stagehand and apprentice, to a Connecticut summer stock company where she learned to impersonate teapots and poison ivy under the tutelage of Lee Strasberg. By the end of the first week, she had a walk-on as a maid in an Ethel Barrymore production of *School for Scandal*. (Ethel, Norman's old boxing match pal, wrote him that Barbara was stealing the honors, with her blessing.) The second week she had a few lines. By the third week, she was playing Amy in *Little Women*.

Come fall, she was enrolled at Andrebrook, an establishment catering to "high-strung" girls. It was an improvement over Putney's, but school was still school. Sister Joan, a studious Barnard grad, took after her mother; Barbs was her father's daughter. Before the first semester was over, she "went missing"; a bedsheet was found hanging from her second-story room. Embarrassed, and hoping to find her, school officials waited several days before notifying her father. Days later, Norman received a call from Broadway theater producer Arthur Hopkins.

"Norman," Hopkins asked, "is it all right if I put your daughter under contract?"

"You mean she's been *rehearsing* with you?"

"For the past week. I'm happy to report that she's eminently satisfactory."

"Is she in your office right now?"

"Yes."

"Arthur, you probably don't know that she disappeared ten days ago. The police in three states are looking for her! I'll be there in five minutes. *Keep her talking.* Don't let her run off."

A gurgling sound at Hopkin's end suggested someone grabbing at their throat in mock strangulation.

"What's the play?" Norman asked.

"*Out of the Frying Pan.*"

"Very appropriate. And the answer is yes."

The teenage runaway had just landed her first Broadway role, as an amiably nitwitted ingénue. Though *Out of the Frying Pan* flopped, it marked the end of Barbs's formal education. She never went back to school.

A USO tour in *Junior Miss* and several more Broadway flops followed, including one in which she spent much of the production as a corpse in a closet. In between, Barbs, five foot three, married a six-foot-two electrical-engineer-turned-actor's-agent (true to her nature, the courtship had been short) and gave birth to a daughter.

In 1945, she made a splash with her first important role, as a well-born Southern girl determined to marry a Negro war hero. The director, Elia Kazan, would remember her as having "a luminous quality."[655] Critics called her "superb" in what was a difficult, exacting role,[656] one that won her the first Clarence Derwent Award for outstanding young performers and the New York Drama Critics Award. After fourteen months on Broadway, it headed for London's Globe Theatre.

Barbs was twenty-three.

"As long as one Barbara Bel Geddes can happen," wrote Billy Rose, referring to young women with show biz aspirations, "a thousand would-be Barbaras are going to keep punching."[657]

"This is a fan letter addressed to you because I don't know your daughter Barbara well enough to write to her," Richard Rodgers wrote Norman. *Carousel*, a Rodgers-Hammerstein collaboration, was in the midst of a long-term sold-out run at Broadway's Majestic. "It's perfectly obvious that of all of your creative jobs, the one you started about twenty years ago is probably the most successful. To be able to sit in a theatre seat and watch the beginning of a brilliant career is one of the most satisfying experiences possible." Hers was a talent "too fine and rare to be wasted in celluloid." It was Rodgers's "fervent hope" that she would stand firm against the inevitable "flood" of Hollywood offers.[658]

The flood was, indeed, in progress. "Everybody had Bel Geddes fever," noted director George Stevens, who was sent cross-country to take a look.[659] Lillian Hellman and Kazan ("she was not a commodity, a can of peas"[660]) echoed Rodgers's advice. Stay in New York and hone your craft. But Barbara wanted "desperately" to work with directors like Frank Capra, George Cukor, and Alfred Hitchcock.

In 1946, armed with her father's stubborn determination, she negotiated an RKO contract unheard of for an untried Hollywood newcomer: no more than two pictures a year, at $50,000 each ($605,000 today) the first year, equal billing with veteran film stars and the right to take on a stage play on alternate seasons plus two pictures a year for other studios.[661]

Her first onscreen role, opposite Henry Fonda in *The Long Night*, was a disappointment. Her second hit gold. As the daughter who writes about her immigrant mother in George Stevens's *I Remember Mama*, Barbs was nominated for an Oscar. Ironically, it was a role she'd auditioned for, without success, on Broadway. *Life* magazine featured her on its cover, lying in a field of wildflowers; at twenty-seven, she could pass for a fresh-faced teenager. She was, readers were informed, a fan of the Great Garbo, "a heavy eater" whose preferred Tinseltown footwear was saddle shoes.

Along with Dad's looks (a pretty, female version), she exhibited some of his wit. When photographers asked for a cheesecake shot, meaning something with a lot of leg exposed, she produced an actual cake. But despite a landmark contract and a large hilltop house, complete with swimming pool and tennis court, Hollywood wasn't quite what she'd expected. Tired of makeup men who complained about the difficulty of shading her "nipple" nose, she went to a plastic surgeon and had it bobbed. "It hurt like the devil."[662]

By 1948, she'd made four movies for RKO and was slated to star in two more, when Howard Hughes, RKO's new owner, told studio production chief Dore Schary to fire her. She had, Hughes announced, "no sex appeal."[663] Schary refused, tendered his resignation, and took one of Barbara's upcoming projects to MGM. Sixty-five years later, *The New York Times* would call Max Ophuls's *Caught* brilliant, thanks in part to Barbara's "unassuming . . . subtly off-speed" performance as a carhop-turned-mad-millionaire's-wife. The unstable millionaire in question was believed to be based on Hughes.[664]

In 1950, she worked with Kazan a second time, on *Panic in the Streets*, with Jack Palance (his film debut) as the villain and Zero Mostel as the terrified underling. Next was *Fourteen Hours* opposite Richard Basehart, a forgettable thriller in which Grace Kelly made *her* screen debut.

Then, still furious, if not demoralized, by Howard Hughes's pronouncement, she abandoned her hilltop mansion and headed "home" to a Manhattan brownstone in the East Eighties. Her five-year celluloid investment hadn't been wasted, but she'd been, as one journalist put it, "a diamond in the wrong setting."[665]

All three leads in *The Moon Is Blue*—Barry Nelson, Michael Rennie, and Barbara—initially turned the project down because Otto Preminger, a former Max Reinhardt aide, had signed on to direct. Unlike the soft-spoken Max, Otto had a reputation for yelling.

Barbara's seventh Broadway play was a light-hearted depiction of extramarital sex in which she played Patty O'Neill, a professional virgin courted by a middle-aged lecher and an amiable young wolf. In the end, she leads the former into a marriage proposal and the latter into a promise of chastity. One magazine described her Patty as "a new scatterbrain . . . pretty as a peppermint stick and just about as profound."[666]

Otto refrained from shouting and the reviews were glowing. In April 1951, eighteen months after Loewy's long-coveted *Time* magazine cover coup, *The Moon Is Blue* landed Barbara's face there, anchored by a Peter Pan collar. "Broadway's Barbara Bel Geddes," the caption announced. "For the prodigal daughter, a prodigious homecoming."

At twenty-eight, she'd mastered the very different demands of serious drama and airy farce, her onstage poise "unshakeable"; in "real life," *Time* informed its tens of thousands of readers, she could cuss like a longshoreman. In the works was an amicable divorce and a new husband, a young stage director.[667]

Dad was given his due.

Barbara's fear of and attraction to the limelight is a legitimate inheritance. For a generation before she entered the theater, her father Norman had rumbled and roared like an earthquake in the foundations of show business, making plans, productions, money, noise, friends and enemies on a gargantuan scale. The example of his unbridled imagination and breezy pageantry taught Barbara early in life that the theatre could be both sheen and shoddy . . .

The big brownstone house . . . in which Barbara spent her early
childhood saw an endless stream of visitors from many worlds . . . In
the welter of productive activity . . . Barbara was comparatively small
potatoes . . . no match for such stupendous enterprises.[668]

When Norman objected to some daughterly quotes in a handful of interviews, his youngest chided him. How could he be angry, given all *his* experience with magazine distortions and misquotations? Especially about remarks made in jest? Besides, she wrote, "Since I have a father so much more famous than I am, it is virtually impossible to be interviewed without being asked questions about him."[669]

•

JOSEPH MCCARTHY'S HOUSE on Un-American Activities Committee hysteria was spreading like the orchestrated, hothouse virus it was.

World War I had promoted "Pinkos" lurking under every rock. The alleged threat had moved one letter down, from Bolshevism to Communism—Emma Goldman replaced by Ethel Rosenberg, Sacco and Vanzetti by the Hollywood Ten, Germany and Japan by Korea. The Cincinnati Reds baseball team felt obliged to change their name to "Red Legs," lest someone misinterpret their allegiance. Public schools banned *Robin Hood* (a children's classic since 1883) for its "take from the rich, give to the poor" message.

Those who came into HUAC's sights found their creative lives ruined. The Committee's blacklist would grow to include the best and brightest— among them Stella Adler, Leonard Bernstein, Aaron Copland, Dashiell Hammett, Lena Horne, Paul Robeson, Pete Seeger, Irwin and Artie Shaw, and Orson Welles. It's worth noting that the entertainment world's king-pins, many of whom—Sam Goldwyn, Louie B. Mayer, David Sarnoff— had two strikes against them (being Jewish and hailing from Russia), were never singled out for persecution.

Elia Kazan, one of Hollywood's most successful and progressive directors, and a one-time "Party" member, cooperated with HUAC as a "friendly witness," offering up twenty-six of his fellows. His friend Arthur Miller responded with *The Crucible*, a play about the Salem Witch Trials that electrified Broadway audiences. HUAC retaliated by denying Miller a passport renewal to attend the play's London opening, then subpoenaed him to testify.

At some point, Barbara Bel Geddes appeared on "the list." There's no evidence her name came from Kazan. More likely it was lifted from *Red Channels*, a pamphlet that targeted film industry "sympathizers." That the accusations were unofficial, unsubstantiated, and made behind a cloak of anonymity made them no less lethal in the climate of the times. Barbara had worked with Lee Strasberg, whose wife Paula was on Kazan's list. That was "evidence" enough.

She'd left Hollywood just in time.

IN 1955, SHE worked with Kazan a third time, cast as the eponymous feline in Tennessee Williams's latest Broadway offering, *Cat on a Hot Tin Roof.* Like Howard Hughes before him, Williams found Barbara "too homespun" to portray Maggie, a sexually frustrated young wife (her second time mastering a Southern lilt), though she was fresh from playing a girl caught up in a tragic love affair in Graham Greene's *The Living Room.* Williams wasn't crazy about Kazan's choice for Big Daddy, either—Burl Ives, a folksinger best known for his rendition of "Blue Tail Fly."

The playwright kept a notebook during rehearsals.

Tuesday, 22 February 1955:

The leading actress inadequate, the play not coming to life enough. I'm tired and a bit drunk and I have a beastly cold—I am making plans for a far-away flight (perhaps as far as Ceylon) the night the play opens in New York!

Saturday, 26 February 1955:

Last night the first run-through—devastating. Bel Geddes improved but Burl Ives acted like a stuffed turkey . . . I feel hellish.[670]

"Everyone saw 'Maggie' as beautiful and slinky and seductive," Barbara Bel Geddes would recall half a century later, "and I'm a bit of a dumpling, well-meaning, the girl you marry but begrudgingly fuck."[671] She had, in fact, shed twenty pounds for the role, in which she mostly appeared in a white slip.[672]

Bel Geddes's insecurity about her looks (pudgy, homespun) was partly why Kazan cast her as a bride caught in the machinations of rapacious in-laws, terrified of returning to the poverty of her past, fighting for her life. He saw Maggie as the type of young woman that a repressed homosexual might marry to assuage his family, the kind of hungry, innocent, romantic young woman who believed that love might be enough, in the end, to win out. "He drove me so deep within myself," the actress said, "that I sometimes left rehearsals—and performances—in tears and exhausted." Kazan, "only Kazan," taught her that theater could be powerful, and acting a noble pursuit.[673]

Brick and Maggie's bedroom was created at a forty-five-degree angle (like Norman's acutely angled set for *Dead End*, it conflated naturalism and expressionism), and a corner jutted out past the proscenium (like Norman's *Hamlet* and *It Happened on Ice* sets), allowing the actors to speak directly to the audience. The lighting suggested realism at some points, interior landscapes at others, and allowed for smooth, rapid transitions between them.

As had happened with Wilder's *Our Town*, *Cat*'s seemingly rocky start bore brilliant fruit. Brooks Atkinson called it "stunning drama . . . limpid and effortless . . . superb." Director Arthur Laurents would write that Barbara "had a good-girl-but-I-want-it kind of sex appeal" that made her Maggie "steam."[674] She also had, beneath the well-scrubbed prettiness, a steely intelligence that she brought to the role, something Putney's hidebound faculty had been incapable of recognizing.

Cat received four Tony Award nominations, including Best Actress (Barbara Bel Geddes) and Best Director (Elia Kazan), plus a 1955 Pulitzer. The production ran for almost two years, the longest of any Williams's play, but when MGM cast the film version, Bel Geddes would be passed over for sultry femme fatale Elizabeth Taylor. MGM's first choice, Grace Kelly, was off to Monaco to become a princess.

•

IN 1957, AFTER a seven-year absence from Tinseltown, with HUAC's efficacy waning, Barbara Bel Geddes signed on with Alfred Hitchcock.[675]

Amongst the Dead would prove to be the most analyzed and intensely debated movie of all time, grist for seemingly endless critiques, analyses, debates, graduate theses, and at least three entire books. Renamed *Vertigo*, it starred Jimmy Stewart as John "Scottie" Ferguson and Barbara as Midge Wood, his best friend from college and former fiancée.

One can't help thinking that Sir Alfred, the obese greengrocer's son from Leytonstone, Essex, would be amused, if not bemused, by all the belated attention to a film that was initially panned. *Time* magazine called it "another Hitchcock-and-bull story in which the mystery is not so much who done it as who cares."[676] Orson Welles deemed it "worse" than *Rear Window*.[677] Hitchcock wouldn't live to see his "least favorite" film restored, reissued, and elevated from disparaged flop to undisputed masterpiece. As a final touch, *Vertigo* would eventually dethrone Welles's *Citizen Kane* as "the best film of all time," a spot the latter had occupied for half a century.[678]

The audience first meets Midge when Scottie pays a visit to her San Francisco apartment. She's a commercial designer of some sort. A drafting table dominates one corner of her living room. There are paints, jars of brushes, and various drawings tacked to the wall.[679]

She sits working on a sketch as they discuss the vertigo (which he mistakenly calls "agoraphobia") that has convinced Scottie to quit his police detective job. As they talk, he spots an odd pink contraption on a table.

SCOTTIE: *"What's this doohickey?"*

MIDGE: *"It's a brassiere. You know about those things. You're a big boy now."*

SCOTTIE: *"Never run across one like that."*

It's brand-new, she explains. "Revolutionary Uplift." No shoulder straps or back, but "it does everything a brassiere is supposed to do. Works on the principle of the cantilever bridge. An aircraft engineer down the Peninsula worked it out in his spare time."

"Kind of a hobby," Scottie suggests. "A do-it-yourself type thing."

The scene, skipped over in the hundreds of pages of critiques written in the decades since *Vertigo*'s release, was a hidden-in-plain-sight homage to the actress's father.

The "Revolutionary Uplift" can be read as a jibe at streamlining, and Scottie's follow-up comment about "do-it-yourself type" hobbies recalls everything from Norman's legendary horse race and war games to his reptile-and-insect forays. "Cantilever" references everything from the eastern span of the Oakland Bay Bridge, on view from Midge's window, to the projecting beams Norman incorporated into his stage sets, retail win-

Barbara Bel Geddes, Jimmy Stewart, and the "Revolutionary Uplift."
(Courtesy: Universal Studios)

dow displays, Toledo factory design, houses, and hotels. The Uplift was also a nod to Howard Hughes (an "aircraft engineer" whose *Spruce Goose* was said to have been inspired, in part, by Norman's Airliner No. 4) and the infamous underwire push-up bra he designed ("in his spare time") for Jane Russell to wear (she never did) in *The Outlaw*. Not coincidentally, the same Howard Hughes who'd sent Barbs running back to New York seven years before.

Interviewed years later, screenwriter Samuel Taylor admitted he created the role of Midge with Barbara specifically in mind.[680] The two were friends, her triumph in *Cat* was still fresh, and he certainly knew who her father was.

As did "Hitch." The costume designer for Hitchcock's previous film, *The Trouble with Harry*, had been Edith Lutyens Bel Geddes, Norman's waiting-in-the-wings fourth wife.

Like Bel Geddes *pere*, Hitchcock was a master of detail; his reach extended from script development, locations, camera angles, and lighting to the color of his actors' clothing. He felt strongly enough about the brassiere segment—which had nothing to do with the story line—to ignore pressure from Paramount's legal department and the U.S. Production Code Administration, both of which pushed to eliminate it "on moral grounds."

And then there were the eyeglasses.

They're part of Midge's costume as the practical, independent butterscotch blonde. Later in the film, in an effort to win Scottie back from her "rival," the enigmatic, eyeglass-less platinum blonde Madeleine (played by Kim Novak), Midge paints a portrait of herself dressed as Carlotta (Madeleine's supposed great-grandmother), but she also paints in the glasses. Except that Midge's slender, everyday pair have been replaced, inexplicably, with a large, square, horn-rimmed pair. Just the kind Norman had taken to wearing in real life. The result is sadly comic.

Like Hitchcock's fleeting onscreen cameos and famous McGuffins, the purpose of the "Revolutionary Uplift" scene is ambiguous and, ultimately, irrelevant. It's a loony touch, something only the cognoscenti would catch and, like the nerdy horn-rimmed glasses, a tongue-in-cheek homage to Norman.

CHAPTER 23

Edith

1950–1958

Who has not sat talking above the noise of sobbing saxophones and the humdrum and clatter of a night club, depressed with sudden loneliness?
—CHARLIE CHAPLIN, *MY AUTOBIOGRAPHY*

S ir Edwin Landseer Lutyens (with a middle name predisposing him to his profession) was considered one of Britain's greatest architects. Edith Addams de Habbelinck would claim his surname as her own for the better part of a century—beyond her short-lived marriage to Archibald, Sir Edwin's nephew, and on through two subsequent husbands, to the end of her life—leading many to believe she was the architect's daughter (he was homosexual and childless). Her stationery read, simply: Edith Lutyens.[681]

Born in Brussels, Edith grew up to become a member of the Belgian women's Olympic fencing team. Or so she always said. Belgium was represented at the 1928, 1932, and 1936 Olympics by Jenny Addams, possibly a cousin. The women's Olympic team event wasn't held until the 1950s.[682]

When World War II erupted, Edith was running a London dress shop. On December 29, 1940, over the course of forty-five minutes, the Luftwaffe dropped 10,000 fire bombs on the city's center, an effort enhanced by a sophisticated new "pathfinder" targeting system. As the populace fled to subway tunnels for shelter, a second phalanx sent 550-pound explosives whistling down.

The Blitz would continue for months, ultimately killing 40,000. But before that happened, "Miss Lutyens," age thirty-four, boarded a blacked-out train for Liverpool harbor clutching a one-way ticket on the White Star Line's *Georgic*. In Manhattan, dressmaking led to costume design, and by September 1942, when she married Moseley Taylor, she was beginning to get commissions for ballet and Broadway. By the end of the decade, she'd designed, or executed the designs of others, for *Around the World in 80 Days, Anne of a Thousand Days, South Pacific,* and *Gentlemen Prefer Blondes.*

•

SEVERAL YEARS INTO their marriage, the Taylors were living, as Edith put it, *"faisaient chambre a part,"* each with their own suite on the third floor of a building backing onto Seventy-Third Street. Edith's Saturday night routine consisted of reading the next day's newspapers with her standard poodles, one black, one white, curled up on the bedclothes. On one such evening, a bitter winter night in March 1950, she woke to the sound of her bedroom door opening and the lights being switched on by two policemen. Edith, who slept *au naturel,* clutched at her sheets as the men stepped to either side of her bed, "like a comic set in a Max Sennett movie," asking in unison, "Do you know this man?"

Moseley stood motionless between them in torn, blood-spattered pajamas, his bare feet blue with cold, his face smeared with grime. One of his hands was limp. Edith donned a dressing gown and wrapped the shivering apparition in a blanket.

Her suite was as far from a mid-century modern industrial-design aesthetic as one could get—black and gold octagonal walls, a ceiling sculpted into the signs of the zodiac, French windows veiled with a shimmering sixteenth-century Italian altar cloth, a heavy eighteenth-century Italian door. Her bathroom, gold and silver chinoiserie papered over dark red, housed a sunken marble tub, a gilded Louis Philippe mirror, and a small chandelier.

"Quite a joint," one officer observed.

"What we want to know," the pair demanded, "is why did you push him out the window?"

By the time Moseley crossed paths with Edith, his fourth wife, little remained of his early promise as Harvard's handsome, dark-haired Golden Boy, star footballer and brilliant oarsman, a WWI pilot awarded the Navy Cross.[683] The scion of a prestigious publishing dynasty—his grandfather founded *The Boston Globe*—Moseley managed the *Globe*'s New York office, a glorified remittance man paid by his father to keep his distance. Though still charming, he'd grown fat, nearly bald, and was a notorious alcoholic.

"What do you mean, 'push him out the window'?" Edith protested.

The coppers seemed averse to her British accent. More to the point, why were they there, and *how did they get in?*

"He's over six foot two and I'm five four. It's two in the morning. You found me fast asleep!"

The police inspected the French windows, the chinoiserie bathroom, the closet (mink jacket, two fox stoles, one red, one black, large leopard-skin bag), the vintage Italian door. An adjacent door, leading to Moseley's rooms, was locked.

"Open it. Let's see if you pushed him out *his* window."

"It's locked from the inside," Edith explained.

"Why would he do that?"

"He drinks and doesn't want me to know."

They threatened to break it down. She insisted she didn't have the key. With a ladder, the police accessed the roof outside Moseley's room; his open window was clearly visible. Climbing inside, they unlocked the bolt (the key was still in it) and looked around.

In the kitchen, over coffee, the officers described having seen Moseley lurching down Second Avenue, even before the radio dispatch came in. Eventually the whole story came together. Drinking in his room, he'd turned right instead of left while heading for the bathroom. A few steps later, he fell out the unlatched window onto the roof and may or may not have grappled with the iron safety ladder. Once on the frozen ground, he'd broken the street-level window of a neighbor's house and climbed inside. Woken by the noise, the neighbor watched Moseley walk out the front door, leaving behind a trail of blood. At some point, his wrist had broken. The neighbor phoned the police, already on the apparition's trail. The janitor was woken up to unlock the Taylor home.

After two hours of explanation, argument, and flattery, Edith convinced the coppers, who by now had ascertained who Moseley's father was, to forgo filing a report. They left with two bottles of scotch. Edith's doctor arrived and set the wrist. Still, Moseley said nothing. Exhausted and demoralized, Edith crawled off to bed, weeping into the dismal gray sunrise.

Weeks later, the Taylors were scheduled to attend a party, something to do with "the beastly *Globe.*" If Moseley went, his domineering father was bound to get word about the wrist; the rest would inevitably follow. If he didn't show, that too would cause trouble.

Dressed in her best black taffeta, a reluctant Edith went alone. Golf-

and poker-playing businessmen, their bridge-playing wives. Still, the caviar and champagne were lovely. Wandering into the library, she arranged her rustling skirt onto a chair. Squatting in the corner of the room was a broad-shouldered man with a large head and enormous tortoiseshell glasses, talking to a young woman. When she left, he walked over and introduced himself.

"His name, I'm ashamed to say, meant nothing to me at the time," she recalled years later. They talked for a long while. When Norman mentioned some plays he was working on, she asked to be considered when it came to costumes.

•

IN 1949, THEIR marriage already rocky, Norman and Anne had collaborated on a "Flow-Motion" line of gold-toned costume jewelry. The year-long contract called for 100 sketches. The idea was that the necklaces, in particular, would conform to a woman's neck and move as she moved.

Aside from the fact that jewelry wasn't his forte, Norman's heart wasn't in it. His first wife had left him, arguably for being too successful; Frances, his great love and creative partner, had died. His Rockefeller office had been lost. Despite having rallied after an enforced eight-month bed rest, he still struggled with health issues. And debts. By the time Flow-Motion debuted—it "does not resemble something from Mars, as might be expected by persons familiar with Mr. Geddes's futuristic thinking," observed the *New York Herald Tribune*[684]—Anne had left him.

What speaks loudest about the marriage is the paucity of material in Norman's otherwise overflowing files. Notes on Anne's extensive renovation of the Park Avenue apartment. An occasional newsclip. A paragraph cut from Fitzgerald's *This Side of Paradise*, with "Anne" written at the top:

*She wants what she wants when she wants it, and she is prone to
make everyone around her pretty miserable when she doesn't get it.*

A letter in her neat, legible hand mentioned "the hopeless horror of my personal situation." She apologized for letting him down by moving out "at a very critical time," but she wanted her health back and was under doctor's orders. "I don't see that a nervous breakdown on my part is going to help."

"If anything, you're overburdened with all the attributes necessary to the building of success," read a second letter, "and when such a person begins directing his energies to pulling against himself, the result is a holocaust, to say the least.

"I think," she'd added, "you'll come out of this with flying colors."

•

IT WAS STILL early. Norman and Edith took a taxi to his apartment for a nightcap. Except for Pushkin, Norman's miniature Schnauzer, a gift from Frances, the place had an abandoned look. They drank more champagne and talked for several hours, about the challenges of New York's theater world, about their dogs, all of whom sported literary names—Nana (after Zola's novel), D'Artagnan, also known as Whitey (one of Dumas's *Three Musketeers*), and Pushkin (notorious for fighting duels and writing verses that got him exiled). They perused large, beautiful books of illustrations that he pulled from the shelves. He explained that his wife had left him, that he had two daughters; she told him she was married. At some point, lulled by drink and the lateness of the hour, her new acquaintance admitted that, desperately unhappy with his life and business troubles, he'd tried to kill himself. Even with that, it was the most pleasant evening she'd spent in a very long time.

Norman was fifty-seven, Edith forty-three. She didn't expect to hear from him again. A poodle pin, inspired by Nana and D'Artagnan, was added to the Flow-Motion line. It was the start of a transforming love affair.

•

THE FUTURE, ONCE exemplified by skyscrapers, was now significantly higher up thanks to the ever-looming thermonuclear threat. "A single flight of planes no bigger than a wedge of geese can quickly end this island fantasy, burn the towers," E. B. White wrote of Manhattan more than half a century before 9/11.[685] An entire generation of children practiced "duck and cover" drills, crouched beneath their desks.

Monopoly, a popular board game about real estate, gave way to Risk, a game of world domination. A new generation of superheroes and villains gained their powers thanks to atomic radiation: Spider-Man, Dr. Octavius, the Hulk, the Fantastic Four, Daredevil, Negative Man—the

Midway's "freaks" reimagined for a post-apocalyptic world. "Watch the skies, keep looking. *Keep watching the skies*," warn the scientists at the end of *The Thing*, the first of the decade's sci-fi movies. *Godzilla* and *Creature from the Black Lagoon* were prehistoric beasts awoken by nuclear blasts. There were nuclear spiders, and ants the size of Cadillacs.

The fear of evil skies was further stoked by Russia's Sputnik launch. Ever prescient, Bel Geddes had, months before, tried to drum up interest in a TV show based on Orson Welles's *War of the Worlds* broadcast. *Find out if a copy of the 1938 radio script can be borrowed*, he wrote in a memo to his secretary. *Don't phone Orson about it. It will give him a good idea.*

Mollifying the fear was the "quick and easy," exemplified by frozen TV meals. (Swanson's aluminum trays would eventually be inducted into the Smithsonian, sharing space with Abe Lincoln's top hat and Dorothy's ruby slippers.) Potato chips, curved like the latest Eames and Saarinen furniture, were "redesigned" with ridges to hold the weight of "dips"—a staple of suburban cocktail parties. There was easy-listening "semi-classical" music, Classics Illustrated Comics (Shakespeare and Dickens as cartoons), Paint-by-Numbers ("A beautiful oil painting the first time you try!"), and Readers' Digest Condensed (truncated, mail-order novels). It was an era of "bread and circuses"—the bread factory-made (nutrition-ally and gastronomically questionable), the circuses designed by Disney—Ringling's intelligent elephants replaced with a cartoon Dumbo, a diverse nation replaced with a well-scrubbed, odor-free, one-size-fits-all Main Street USA. "As for realism," Walt told a reporter on the subject of Disney-land, "you can find dirt anyplace you look. I'm one of those optimists."[686]

In the midst of all this plenty came a growing sense of despair, a mis-ery beneath the skin. "Two cars in the garage, chicken every Sunday," raged Raymond Chandler in *The Long Goodbye* (1953), "and me with a brain like a sack of Portland cement."

Some blamed it on burgeoning corporations. For the first time in the nation's history, one out of every two adult males was a "white-collar" worker, a cog in a "culture" that encouraged an almost desperate lust for conformity. *A large percentage of our lives are lost by playing things too safe*, Bel Geddes wrote in his private notes, *which is absurd, because there's no greater risk than life, from instant to instant.* Case in point, his good friend Margaret Bourke-White, the intrepid, unsinkable Maggie, who'd just been diagnosed with Parkinson's disease.

Even *Popular Mechanics*, not exactly a bellwether of social change, noticed the shift. "It is astonishing how easily the great majority of us fall into step with our neighbors. And, after all, is the standardization of life to be deplored when everyone could have his own helicopter, luxurious household appointments, and food that was out of reach of a Roman emperor?"[687]

Not coincidentally, the first FDA-approved tranquilizer took postwar America by storm. TV superstar Milton Berle, also known as Uncle Miltie, took to calling himself Uncle Miltown; Jimmy Durante and Jerry Lewis praised the efficacy of "the little white happy pill" before millions during Emmy and Academy Awards telecasts. In less than two years, 36 million Miltown prescriptions had been filled. Atarax, Librium, Valium, and other so-called lifestyle drugs soon followed.[688]

John Kenneth Galbraith, one of the world's most widely read economists, wrote at length about Americans' "tense and humorless" pursuit of goods. "To cast doubt on the importance of production is to bring into question the foundation of the entire edifice . . . Nothing could be less welcome."[689]

Industrial designers were, of course, an indispensable element of that pursuit. *Newsweek* estimated that industry was spending "at least $500 million" annually to implement "solutions"—designed by men like "the Big Three"—to help feed the postwar acquisition frenzy.[690] "We believe in obsolescence," wrote George Nelson. Without advertising and a "culture of waste," the economy would grind to a halt. "We support it the way we support the multi-party system, pasteurized milk and a free press."[691]

But was all the leisure designers were helping to shape producing happiness? Henry Dreyfuss wondered. Were they doing their job "only to train a nation of passive participants filling their time with wrestling on TV, comic books and pre-digested digests?" More and more Americans were spending time in mental hospitals, he observed, more and more men were pleading to stay on the job in lieu of a pension.[692]

What, then, shall we do with our hands? Roger Burlingame had asked more than a decade before.

•

TO A FIFTYSOMETHING workaholic with a troublesome heart, America had lost much of its allure. Bel Geddes's Park Avenue apartment had become "a bus stop" for friends, clients, and out-of-town relatives. Endless phone calls, and cocktail and dinner parties left little time left for tackling "really tough" creative jobs.

Norman having spent time in Florida, the Caribbean seemed a natural jump. He began to take "working vacations" that lasted anywhere from days to months. The islands offered a different world—of people, customs, dress, language, philosophy, food, geography, vegetation, architecture, pace. Within a week of arriving, he was relaxed and ready for the "uninterrupted thinking through" of whatever projects were at hand. Dictating machines, typewriters, and drafting tables were rented as needed.

Initially, he'd focused on Bermuda, but by the time he met Edith Lutyens, he'd exchanged his "runaway spot" for less touristy Jamaica, where he discovered "real" Jamaican rum and "genuine" Havana cigars. Once he was ensconced in a hotel within walking distance of a beach, his habit was to rise early, swim, breakfast on a terrace, then set to work in shirt, shorts, and sandals. He'd completed thirty jobs over the course of these southern sabbaticals, "better, and in one-quarter the time, than I could have in New York."

For company, he reacquainted himself with his old friends, the anolis lizards. They were prevalent on the islands, playful, inquisitive, comical, masters of climbing anything vertical, even glass, thanks to little suckers on their feet. He'd mastered a way to catch them without doing harm. At the start of each visit, he arranged half a dozen long branches, stretched diagonally to a ceiling corner in his hotel workroom, and planted some vines in a pot. He kept raw fruit on hand for food and arranged a lamp to beckon night insects. In a crunch, he laid out a teaspoon of guava jelly.

The difference in appearance and habit, attitude and personality in the anolises is astonishing . . . You come to know them, and they come to know you. Twenty-five percent of them are tame within two weeks, jumping on my desk with apparent delight—typical kids, full of the devil—hopping on my hand or on a pencil I'm holding . . .

They're continually jumping to catch flies midair, then landing with complete poise on a branch. I've seen them jump at least ten feet, which is twenty times the length of their body, the equivalent of us jumping out a ten-story window.

Twice a day, he sprayed the leaves and branches with rainwater. *It has to be in drops, so they can get under and touch them with their tongues.*

Toward the end of his stay, he'd open all the screens; they'd run up and down the drainpipes or overhanging palms but always returned. A few hours before his departure, he'd chase them out, always with great reluctance.

•

SOMETIME AFTER THE evening of the black taffeta dress, he spent ten weeks in St. Thomas (a "highly civilized" alternative to "the indignities" of Reno) to obtain a divorce from Anne. In a set of adjacent hotel rooms, he kept a trio of local draftsmen busy "day and night" to meet his deadline on a Vertical Broadcast Studio, a flexible, multiuse facility (minimum 1,500 seats) designed to maximize both "production flow" and profit for NBC. The annual sale of TVs had jumped from 7,000 to 10 million; in three more years, more than half the nation's households would own a second piece of "talking furniture" (the first being radio cabinets). TV stations were busy expanding their reach beyond major cities, but the fledgling medium's studios were based on antiquated methods. Norman's design, he quipped, "would make it easier for television to be a greater menace."

His Vertical Studio would be followed, over the next several years, by a Horizontal Studio (a bent-line stage for NBC's L.A. operation), a ring-stage (for NYC), and the eleven-floor Atlantis, a 60,000-square-foot amphitheater.

But getting paid to develop those ideas was an uphill battle. "I wish you would quit thinking five years ahead. I probably won't even be in my job then," Pat Weaver, NBC's president, had complained after seeing the result of a year's work on one such studio, an expenditure of $75,000. "What I want is something I can use *next year.*" Norman's reply was that Weaver's line of thinking should have started three years before, in which case, NBC would be implementing it *now.*

Despite the fact that he was working as hard as ever, he was, by the spring of 1951, $9,791 in arrears on the East Fifty-Fourth Street penthouse he'd leased as a work space for his draftsmen.[693] It's unclear whether he was still maintaining his for-tax-reasons Connecticut office at $416.67 a month. Rent on his nine-room Park Avenue flat was $438.50 a month (about $4,000 in 2017), and he was still attempting to pay off the outstanding debts he'd taken on following the closing of his Rockefeller Center office, in pursuit of which he'd reluctantly borrowed funds from daughter Barbara and her husband.

"No one need tell you the battle I've been fighting trying to get going without working capital," he wrote to Webb & Knapp, Inc., to whom the East Fifty-Fourth Street rent was owed. "For four years, I've spent two-thirds of my time each week trying to get next Friday's payroll and the emergency bills paid. The balance of time I used as a designer—no time to promote new business. I'm still in that position," but with only one employee.[694] Which didn't stop him from maintaining charge accounts at the Barberry Room, the Colony, Jack & Charlie's 21, Luchow's, the Newport, Le Pavilion, and the Stork Club, something he considered a business necessity. The speedboat and sailboats were long gone.

In the course of this precarious balancing act, Webb & Knapp offered the still-unlicensed Geddes an "alteration" job on one of their buildings, an assignment that earned him an architectural award from the Fifth Avenue Association.

"The Big Three," all of whom had eponymous, brick-and-mortar offices—none of whom indulged an expensive "fickle mistress," made unnecessarily risky professional choices, personally subsidized high-ticket projects they believed in (like the *Life* spreads), or were burdened with health issues—fared better.

Henry Dreyfuss never got his wished-for *Time* cover, but in May 1951, his face graced the front of *Forbes*. At forty-six, he was, along with Teague and Loewy, "one of the giants of his profession—a Depression-spawned, peculiarly American phenomenon that has . . . effected a major transformation in U.S. industry."[695] His newest phone, a desktop, was in the works (to be followed by the colorful Model 500 and the popular-with-teenagers Princess Phone). "At last," he would write, "we are repaying the debt of culture to the Old World with well-designed, mass-produced goods."[696] In an unprecedented reversal, European manufacturers were following America's lead.

Ray Loewy was the focus of a profusely illustrated feature in London's *Architectural Review*, an overview of "the Moloch of mass taste" that America's industrial design profession had wrought. Largely laudatory, it lacked the bravado of Betty Reese–directed pieces. Some of the Frenchman's designs were described as being "unnecessarily crude," of "dubious merit," and, in the case of a particular mass-market radio, "rep-rehensible."[697]

On the flip side, a weighty mid-century design survey, published by England's Cambridge University Press, called Loewy's career "the most remarkable" of all the early industrial designers. The source of his acco-lades, the author acknowledged, was Loewy's autobiography.[698]

•

CRAMPED AND SHABBY after nearly four decades, the Dodgers' stadium at Ebbets Field was in desperate need of renovation, if not rebuilding. The pro-ject had been under discussion for years when, on January 16, 1952, Bel Ged-des met team owner and president of the Brooklyn National League Baseball Club Walter O'Malley for lunch. O'Malley was happy to catch the designer between trips to Jamaica, what he called Norman's "self-imposed tropical Siberia."[699]

Norman's suggestions included a new lighting system, increased capacity by at least two-thirds (to 55,000, with the possibility of Bel Geddes–designed, foam-cushioned aluminum seats), better all-around visibility, a playing field of synthetic grass that could be painted any color, and a heated 7,000-car garage allowing direct patron access to the stadium. Machines would replace manpower as much as possible, thanks to automatic hot dog vending ma-chines (complete with optional mustard) and automatic ticket collectors and gates. Covering it all would be a weatherproof, retractable domed roof (trans-lucent to admit daylight), hung on cables attached to external towers, like the deck of a suspension bridge, a concept borrowed from his pole-less Big Top.

O'Malley was impressed enough with these "radical" concepts to commis-sion a blueprint and display a rendering of the design in his office. Not one to leave the future of what critics were calling "O'Malley's Pleasure Dome" to one man, he also consulted with Buckminster Fuller, whose geodesic domes he admired, and with engineer Emil Praeger.

In March, Bel Geddes traveled with O'Malley to Vero Beach, the team's spring training site, where they spent a week discussing a 5,000-seat prac-tice venue. "The cheapest item is always the brains," Norman wrote in a follow-up letter, "and the better the brains, the smaller the ratio . . . of re-ceived value to money spent."[700] In the end, the Florida project was awarded to Praeger, a decision that O'Malley expected would "touch a vital spot, but I can see," he wrote, "that my comments merely scratched your epidermis."[701] Money and timing may have been deciding factors.[702] Still, the Vero Beach stadium, which would remain largely unchanged for the next thirty years, incorporated many of Norman's concepts. As for Ebbets Field, it seemed that Norman's old pal, Robert Moses, resented Bel Geddes's "being in the picture."[703] It was not to be.

•

EDITH HAD REMAINED married, probably as a financial measure; Moseley likely owned the lease on her costume shop. Following his death, she headed for Belgium to see her mother and sister. Postwar Europe was a mess, "run-down and poverty-stricken," everything enmeshed in a "sort of deadness." (Set in rubble-strewn Vienna, *The Third Man*—with Orson Welles as the villainous Harry Lime—was fiction, but its portrayal of American optimism versus European exhaustion was not.) Trying to explain to her family what it was that Norman "did" proved impossible. "There is nothing equivalent to you, or even remotely similar . . . You are what in Europe we imagine America is, new ideas, vision, strength, work and production . . . the world of liberty and gold . . . I feel fortunate that I know you . . . to be with you is all I want." [704]

The following June, in 1953, the pair were married in a private garden outside Kingston. During a three-month Jamaica honeymoon, they snorkeled, house-hunted, and began compiling a cookbook (Norman was on a restrictive diet) that eschewed butter and sugar but allowed for alcohol.

We get up at the crack of dawn, about half past five, Edith reported back to a friend.

> *Norman goes immediately to work until breakfast . . . then he disappears again until lunchtime. He has a swim, then goes on working until the sun goes down, around six. After dinner, the two draughtsmen come and he goes over what they've done and what they have to do tomorrow . . . I've never seen anybody work like he's working on these jobs. It's absolutely extraordinary . . .*

> *We have three maids, also a gardener and an errand boy, a chauffeur and a cook. All this sounds terribly elegant, but it really isn't. The errand boy goes to the post office, picks up paper on the lawn, buys little things, salt and stamps . . . The cook just cooks, nothing else. The maid cleans . . . The girl who does the washing doesn't iron . . . But all these people earn so very little . . .[705]*

The plan was to throw a wedding announcement cocktail party once they returned, but Walter Winchell and a fellow gossip columnist leaked the news.

"Edith and I have collaborated on the most important blueprint of our careers," Norman was quoted as saying. They planned to live "six months of 1954, and every year thereafter, in the Caribbean, at home to no one but fellow loafers."[706]

In October 1953, the newly minted Edith Addams de Habbelinck Lutyens Taylor Bel Geddes moved into the Park Avenue flat with Nana and D'Artagnan, who took to snapping at the lizards, toads, geckos, and fish that Norman (weary of the empty, echoing nine rooms, and ignoring the restrictions of his lease) had acquired the previous spring—complete with custom-built cabinets, a thermostatically controlled, thirty-gallon "adult" aquarium, and a six-gallon "baby" tank aquarium. There's no record of how Pushkin responded to this complicated domestic expansion. Little did he know.

Four months later, Edith wrote to Christopher Coates of the New York Zoological Society, asking if he could obtain a pair of "Pocket Monkeys," otherwise known as pygmy marmosets. Would the diminutive tropical tree climbers like it in New York City? she wondered. Could they be taught to walk on a thin little leash? Would they reply to a name?[707]

"Koko" and "Kiki," small enough to fit together inside a wineglass, were soon installed in the household, where their antics and posturing, clicks, whistles, and trills, their ability to clearly express likes and dislikes, and their habit of wrapping around a finger (in lieu of a tree branch) were much enjoyed and commented on. Let out of their cage, they raced up, down, and around the draperies, leaping and running along the bookshelves, busy and, to all appearances, quite happy. A primate-friendly veterinarian was located on Sixty-First Street. Sensitive to drafts, they sometimes burrowed under an electric blanket.

Before long, the marmosets were veteran travelers, spending winters in the Caribbean. Ropes for climbing and swinging were rigged in a corner of whichever house was being rented. Once resettled, they were treated to local cuisine—grasshoppers (heads removed), sea snails, and the occasional anolis (Norman's former friends, alive, tails removed), along with favorites like strawberries, bananas, hard-boiled eggs, and bread soaked in milk.

"Well, the thing happened that we were afraid of," Norman wrote from Ocho Rios during a January visit. Edith had returned to New York to work on *Ondine*, featuring Audrey Hepburn as a water nymph, a Broadway production that would earn Edith a shared Tony for costume design.

"Koko escaped out the window!" Like a scene out of "Rollo the Boy Naturalist," Norman resorted to crawling on his knees, butterfly net in hand. His quarry was swinging in the nearby shrubbery. "I told him what a foolish old man he'd been, that he wasn't going to have a girlfriend anymore and there wouldn't be anyone out in the great open spaces he would know. Probably a hawk would grab him, or one of the big two-foot-long tree lizards. He did an awful lot of kicking and squealing . . . it was a close shave."[708]

By 1955, Norman was giving serious thought to reviving *The Miracle* film project, convinced the twelfth-century tale—an exploration of fragility and strength, youthful revolt against discipline—had parallels in the mid-twentieth. A four-hour version had been planned back in 1936, to be directed by Reinhardt's son, but Warner Brothers "shelved" it.

Casting the Madonna role was key to getting a deal up and running. His wish list included Vivien Leigh, Audrey Hepburn, Ava Gardner, Leslie Caron, Claire Bloom, and Gina Lollobrigida. He was keen on getting John Gielgud, who'd appeared in *The Patriot* nearly three decades before, but after pleading guilty to solicitation in a British public lavatory, the actor's passport had been revoked.

In the meantime, there was a 150-room hotel to design atop a 2,000-foot mountain outside San Juan (he and Edith had switched from Jamaica to Puerto Rico). His plan called for a swimming pool surrounded by a tree-shaded beach "the width of Park Avenue;" an ingenious sprinkling and drainage system would rinse the sand clean every night.[709] He was also investigating resort prospects for the very rich in the Bahamas in collaboration, he hoped, with Axel Wenner-Gren. And maybe something in Caracas.

Then there was Belgium, Edith's home turf.

When he read about an American exhibit being planned for the Brussels International Expo, he had two initial thoughts—that the worst thing the United States could do would be to use the opportunity to stress its postwar industrial, scientific, and material superiority, and that his 1939 Futurama might offer a kind of template. The exhibit needed a hopeful tone, something to supplement the Marshall Plan and Voice of America broadcasts and to offset "the flood of vituperative invective" emanating from behind the newly named Iron Curtain.

"It would be a great blow to American prestige if Russia were to come off with the honors," Norman wrote in a ten-page letter to James S. Plaut, an art historian working with the pavilion's architects. And, no, he wasn't hustling for a job. He'd had his "share of the fun and the glory" as part of the Chicago and New York fairs.[710]

The Futurama "brand" had long legs. A decade after the Fair was dismantled, General Motors released the "Futuramic" Oldsmobile convertible, with Hydramatic Drive and Whirlaway (no more gear shifting or clutch pushing), and Kelvinator introduced its "Foodarama" refrigerator. Then in 1953, GM initiated Motorama, an annual traveling auto show. Billed as "more complicated to produce than a circus," it comprised 210 exhibits, 99 trailer trucks, a 72-ton stage, plus an acrobat, singers, and chorus girls who danced on lily pads. From name to concept, it paid tribute to Bel Geddes.[711]

Three years later, *Time* ran a full-page ad featuring a family in a clear-domed car speeding down an electric superhighway, its progress and steering controlled by devices embedded in the road. "No traffic jams. No collisions. No driver fatigue." The same year, *Look* ran a two-page spread featuring supermodel Suzy Parker and "the magnificent luxury of 'Futurama'!" The latest Revlon lipstick was available in "28 fashion genius shades," and like the future, the refillable Van Cleef & Arpels–designed Futurama case would go on, readers were assured, "forever."[712]

A World Peace Futurama could exploit that "brand," displaying future-focused relief maps, "blueprints" for rebuilding decimated cities. Nothing utopian, all very practical and possible. It would show the "little man" of the world, "now circumscribed by national boundaries and hemmed in by exaggerated chauvinism, that *his* national leaders, assisted by American know-how," were making plans for the realization of better lives, "all in a form that a peasant, unversed in the ways of international diplomacy, can easily understand, *because he can see it with his own eyes.*"

Plaut and his minions thought otherwise. The U.S. pavilion would ultimately be home to fashion shows and a Circarama theater that screened Walt Disney's *America the Beautiful.*

Frustrated but undeterred, Norman finished up work on his suspended "Transit Grid" matrix, an idea that traced back to his nights, circa 1916, lying prone atop the Garrick's greasy, vertiginous gridiron. Designed to guide moving theater apparatuses (lighting mechanisms, scenery) precisely overhead, it would eliminate "bottlenecks," just as Futurama's motorways had been designed to do with car traffic.[713]

•

FOLLOWING A VISIT to the Mayo Clinic, the designer was now augment-
ing his salt-free, sugar-free, low-fat diet with a regimen of digitalis and
Diamox (for a congestive heart), Demerol and Optalidon (for pain), nitro-
glycerine (chest distress), Miltown (stress), and Plexonal (sleep), washed
down with Valentine's Meat Juice (a popular quack cure-all). His eyes were
giving him trouble, too.

In February 1956, Norman, Edith, and "the little ones" traveled to Marbella. Norman was thinking that Spain, with its churches, old towns, and
500 or 600 locals for crowd scenes, would make a great *Miracle* film loca-
tion. If funding came through and everything was kept to a schedule, they
might even complete shooting by fall of the following year.

Befriended by John Davis Lodge, U.S. ambassador to Spain (there
was talk of Lodge's daughter auditioning for a *Miracle* part), the Bel Geddes entourage found a house, owned by Prince Alfonso de Hohenlohe-
Langenburg, on a half-mile private beach. It came with a cook, housekeeper, gardener, driver, and secretary.[714] Typewriters, drafting tables, a
desk, and file cabinets were rented and an electric gramophone installed.
"The Spanish are such jovial, pleasant people," Norman wrote his brother,
compared to the "usually morose and sullen" Jamaicans.[715] Spain had the
added virtue of fewer insects, far less book-attacking mildew, and added
distance from America's growing unease.

•

FIFTEEN YEARS HAD passed since Richard Ordynski's unanswered Labor
Day letter. Perhaps humbled by a pharmacy's worth of pills and a chassis
powered by a problematic heart, Norman set out to find his old mentor,
writing to various addresses in New York and Europe.

*Stop whatever you are doing at the moment and merely drop me a
word to tell me that you have received this letter. Then I will write you
more fully about what I have in mind.*[716]

The long-belated olive branch almost certainly had to do with *The Mir-
acle.*
The letters were all returned, addressee unknown.
Over the course of his life, Herr Professor had crossed the Atlantic

twenty-two times, not in search of warmer climes, like Norman, but to escape the bloodbath and aftermath of two world wars, returning, in the end, to Warsaw, where he'd died at age seventy-five.

The New York Times had run a dozen paragraphs, topped with a photo from which a still-handsome, if haggard, face stared out, its expression unreadable.[717] Other obits had appeared in *Billboard, Variety,* and *Opera News.* Honeymooning with Edith in Ocho Rios, Norman had missed them all.

•

BY AUGUST, NORMAN was busy drafting a "Confidential and Urgent" proposal addressed to Ambassador Lodge. Spain, he noticed, had a lot of dry, unproductive land. He had a master plan for controlling and distributing water and mountain snowmelt to create an irrigation system and a series of dams. Beyond creating fertile fields, it would generate low-cost electricity for the entire country, "easily" paying for itself in twenty years. As with the Brussels Expo, he offered his services gratis.

In September, a visiting New York journalist described Koko and Kiki to his readers as about the size of large mice. "They fly through the air at enormous speed, catching insects on the wing."[718] Once back in New York, Norman planned to acquire two more females.

In October, Danton Walker's "Broadway" column reported that Bel Geddes was building an entire twelfth-century village film set and had spent more than $200,000 (something under $2 million today) acquiring the dramatic rights, employing writers, researchers, and architects, and traveling twice to Hollywood.[719] As for health issues, "I just have to keep off roller skates," Norman wrote his sister-in-law.[720] Edith, meanwhile, was making inquiries about renting oxygen tanks and respiratory masks in Spain.

By November, *The Miracle* was on the move. Norman had partnered with Blevins Davis, who'd recently taken a revival of *Porgy & Bess* to Madrid. Christopher Fry was writing a new script, and Columbia Pictures had agreed to finance a Technicolor production to the tune of $5 million and allowing Norman complete artistic control. The finished film would be projected onto a special, seventy-foot-wide "Technirama" screen designed for installation in standard movie houses.

The two remaining hurtles were casting and the Catholic League of Decency.

Despite the fact that *The Miracle* legend had existed in ecclesiastical literature for 900 years and Reinhardt had mounted the spectacle dozens of times across Europe, the League (founded in 1933, thus sparing Max) had "concerns." Norman and his team had already expended considerable money (part of the aforementioned $200,000), energy, and time complying with their ever-shifting stipulations.

More encouraging was that he'd managed to get in touch with the elusive Greta Garbo. At fifty-one, she'd been absent from the screen for fifteen years, despite pleas from numerous quarters for a "comeback." The Swedish Sphinx as the Madonna would be a master stroke. Norman spent ten months getting the script "written to her satisfaction" and another three weeks prepping for a sit-down reading.

The reading took an hour, but to Norman's delight, the meeting lasted nearly six, with the actress asking questions. Pending her approval, filming could start June first.

Meanwhile the League—more powerful than the notorious Hayes Commission, which answered to it—continued to wage what Norman's lawyer called "a war of attrition." It didn't help that a 1950 Rossellini film, also called *The Miracle*, had been condemned by Cardinal Spellman as "subversive." Finally, Columbia withdrew.[721]

Norman turned his attention to a musical version of *Lysistrata*, his undisputed 1930 hit. Estimating a $300,000 layout (in excess of $2.5 million), he contacted George Auric (for music), Ogden Nash (lyrics), Cyril Richard and Moss Hart (directing), Emanuel Balaban (conducting), and veteran producers Harold Prince and Roger Stevens. Edith, who'd given up her increasingly cutthroat costume business, set to work contacting Norman's top-ten wish list (Greer Garson, Maureen O'Hara, Vivien Leigh, Marlene Dietrich), followed by a second tier (Katharine Hepburn, Betty Grable, Ava Gardner, Ginger Rogers, Rita Hayworth).

At the same time, Bel Geddes began collaborating with Thornton Wilder on a script about "the American experience" seen through the lives of several generations, and there were plans to produce a two-act play by Andre Obey in New York in the fall, sets by Norman, costumes by Edith. A pair of restaurant projects were also in the works—an interior for the Playbill Restaurant in Hotel Manhattan and a collaboration with Joe Baum,[722] the flamboyant force behind Rockefeller Center's Forum of the Twelve Caesars, and, later, a $25 million overhaul of the Rainbow Room. Mimi Sheraton called Baum "The Cecil B. DeMille of restauranteurs," but as Baum saw it, he used opulence the same way the Catholic Church did—to inspire belief in nonbelievers.

•

WORKING AT HOME was "rather like living in Grand Central Station," Edith wrote to Barbara on May 6, 1958—by day, the household included a typist, a secretary and a librarian, not counting the dogs, monkeys, lizards, snakes, fish, friends, and clients—"and when we've been away, it has been the same . . . no privacy or quiet."[723] Grand Central, it turned out, was scheduled for demolition. The traditional brick edifices on their Park Avenue block were being razed to make room for state-of-the-art office high-rises, glass and steel towers in the "International Style."

Given ninety days to vacate, Norman had immediately mounted a campaign to rally his neighbors and a handful of deep-pocketed friends. Collectively, they could buy a fourteen-story building he'd found on Fifty-Seventh at Sutton Place. He would custom-renovate the apartments in return for a penthouse. "Thanks for thinking of me," came a typical response from the editor of *Town & Country*, but "for that kind of money, I'd buy a treasure island off, say, Puerto Rico, and never be seen again."[724]

There were no alternates in the offing should the scheme flounder, and little time to make them. Norman and Edith were scheduled to leave on May 20, stopping in Paris to finalize *Lysistrata* details with George Auric, then back to Marbella.

•

ON MAY 8, shortly after his sixty-fifth birthday, just shy of Futurama's theoretical world of 1960, Norman Melancton Bel Geddes was on his way to lunch at Manhattan's University Club with his old friend Paul Garrett. "Paul," he cried out, and then he fell, his heart locked shut.

The following day, three thousand miles way, Alfred Hitchcock's *Vertigo* premiered in San Francisco. It was an apt, if unintentional, farewell homage for a man whose reputation, like *Vertigo*'s in reverse, had gone from "undisputed masterpiece"—Inventor of the Jet Age, Grand Master of Modernism—to someone who'd inhabited the shadows for too long.

Unbeknownst to anyone but Edith, Norman had, several years before, pledged his corneas, a gesture generally fraught with superstition and loathing in the 1950s. A young surgeon from the city's Eye, Ear, and Throat Hospital was immediately dispatched. It was imperative that removal take place within a few hours of death and transplanted within forty-eight. Dropped gently into a pair of cotton-filled glass containers, the corneas were quickly en route to La Guardia via a Red Cross motor transport.[725] Eddie Rickenbacker, now chairman of Eastern Airlines, had arranged that all such donations be flown free of charge to their destinations.

Edith confided to the press that her husband had been obsessed with the thought that his eyes, which had begun to fail, would be useless once he was gone. "I only wish," she said, "he were alive to know."[726]

Though donor and recipient names were kept strictly confidential, *The New York Daily News* revealed that Norman's gift went to a man and woman, one in Texas, the other in Michigan, both of whom "are seeing the world for the first time since early childhood, through the eye[s] of a near-genius who designed the famed Futurama . . . numerous stream-lined trains, ships and airplanes, resort hotels, a ball park, and more than 200 often brilliantly imaginative Broadway stage sets."[727]

Bel Geddes's death was described in a special tribute issue of the newly minted *Industrial Design* magazine as "the simple inevitable finale to an improbable life which was seldom simple and never inevitable."[728] Henry Dreyfuss called him the modern counterpart of a fifteenth-century master craftsman. *Newsweek* called him "idea-bristling," *The New York Herald*, "a jack of many trades [who] excelled in all of them."[729] *The New York Times*, which had published dozens of pieces on him over the years, ran a photo and twenty-paragraph obit, citing "imagination and boldness" as his trademarks.[730]

Raymond Loewy sent Edith roses.

He left no estate to speak of. According to the *New York Journal–American*, Norman's safe contained three $10,000 "rubber" checks "written by a widower everyone imagined wound up loaded after his wife's death."[731] The Park Avenue rent was three months in arrears. Other creditors lost no time, as Edith put it, "swooping down." The loan from Barbara and her husband, which Norman had been unable to repay, had plagued him to the last.

·

HE WAS RESTLESS and shape-shifting, a pacifist fascinated by war, a naturalist who loved technology, a serious prankster, a pragmatic futurist, a private man who was rarely alone. It was only his "deeper interest" in theater and design that had kept him, he once wrote, from becoming a biologist, a path that might have treated him more kindly.

Every innovator's story is riddled with defeats. It's the price for taking risks, going against the tide. Would his name be more recognizable today had he stayed with his first career, as an art director, something he was extraordinarily good at? Had he reined in his polymorphic curiosity and remained an innovative set and lighting designer, career number two? Had he taken the time to "legitimize" himself with an architectural degree? Had he been more conservative in his outlook as an industrial designer, like Dreyfuss and Teague, more covert like Loewy? Been more open to compromise? Less devoted to his fickle, and very expensive, "mistress"? Had his Rockefeller Center office remained open, had he lived longer, had there been a protégé to carry on the legacy?

Having occupied the limelight for decades, the once-wunderkind was seen by many, at the end, as a washout, a has-been, "pretty much on the skids."[732] At best a kind of magnificent failure, as the Airflow had been. It was a fate he shared with F. Scott Fitzgerald, Erik Satie, Edgar Allan Poe, Nikola Tesla, D. W. Griffith, Paul Poiret, and Orson Welles . . . to name a few.

"A dreamer is one who can only find his way by moonlight," wrote Oscar Wilde, another lauded genius who fell, dramatically, out of favor, "and his punishment is that he sees the dawn before the rest of the world."

The world Bel Geddes departed was almost unrecognizable from the one he'd been born into. He could rightfully take credit for having put a hand to that vast difference.

Epilogue

1959. The Kitchen Debates between Vice President Richard Nixon and Soviet Premier Nikita Khrushchev emphasize what U.S. technology had wrought: A nation of increasingly rapacious consumers and a new word—affluenza—to describe a malaise stemming from the dogged pursuit of "more." In the 1960s, Pop art will address an America where "you are what you buy." "Consumption and acquisition are a natural part of the human psyche," *Time* will write in 2015, "and, incontrovertibly, a part of the American condition."[733]

> Frank Lloyd Wright dies at ninety-two, his life having spanned from Appomattox to Sputnik. Six months later, his long-awaited Guggenheim Museum project opens its doors.

1960. Walter Dorwin Teague dies just short of his seventy-seventh birthday. The company he founded survives today as Teague, a Seattle-based design consultancy.

1962. Edith Lutyens attempts to sue the creators of Seattle's 605-foot "Space Needle" restaurant (built for the Seattle World's Fair) for its "startling resemblance . . . in both design and mechanical operation" to Norman's 1932 Aerial Restaurant.

1964. General Motors reprises Futurama as Futurama II at the 1964 New York World's Fair—its backdrop painted by MGM artists, its music composed by a professional jingle writer, its molded fiberglass chairs grouped in threes (to discourage "romance"), and its "message" focused on exploiting the planet's more recalcitrant environments—leveling jungles and strip-mining oceans.

Five months before the 1964 Fair opens, John F. Kennedy is assassinated, giving the reissued I HAVE SEEN THE FUTURE souvenir buttons (cheap tin versions of the original), yet again, an unintentional edge.

The Fairgrounds include a massive, 9,355-square-foot model of New York City, known as the Panorama, with 895,000 individual structures, commissioned by Robert Moses. Bel Geddes's swooping airplane ride has been replaced with a simulated helicopter ride. It's a major hit.

1967. A book on the history of the Ringling family includes a full-page photo of Bel Geddes standing between two members of the circus's press department. The caption identifies him as "Clyde Beatty, noted wild animal trainer."[734]

1972. Healthy and vigorous at sixty-eight, Henry Dreyfuss commits suicide with his wife of forty-three years, who is dying from inoperable liver cancer. Borrowing a page from Rosamond Pinchot, they opt for carbon monoxide poisoning in their garage, a carefully "designed" exit that shocks his contemporaries.

1979. Long retired, Raymond Loewy signs extent design renderings by former employees with his own name, mounts them in an exhibition, and offers them for sale.[735]

1982. Disney's fifty-acre EPCOT Center (Experimental Prototype Community of Tomorrow), also known as Project Future, opens in Orlando, Florida. Like the 1964 New York Fair's Magic Skyride, EPCOT is "specifically inspired" by Walt Disney's personal memory of visiting the 1939 Futurama.[736]

1985. John Ringling North dies at eighty-two. In the 1960s, having sold the family enterprise, he and his younger brother had purchased their father's ancestral home in County Galway and become Irish citizens. In 2006, John Ringling North II, Johnny's son, will abandon his life as a rancher in Ireland to return to his childhood roots, buying the one-ring Kelly Miller Circus.

1986. Raymond Loewy, ninety-two, dies in Monte Carlo. As a result of that longevity, wrote Christopher Innis, "Loewy has been credited with much that Bel Geddes introduced,"[737] an observation reiterated by design historian Jeffrey Miekle and others.[738]

1991. The Raymond Loewy Foundation is established "with the specific objective of preserving Raymond Loewy's image and heritage." (The Foundation's website claims his death caused "a worldwide media frenzy.") As of 2014, plans were being discussed for a Raymond Loewy Museum of Industrial Design, or "The Loewy."

1999. *Futurama*, Matt Groening's multi-award-winning animated sci-fi parody, debuts. Set in the thirty-first century, it's a witty, ironic, postmodern refutation of the 1939 Fair's utopian vision.

2000. *Time* magazine's March 20 cover story, "The Rebirth of Design," begins and ends with references to Loewy, who is cited as the "father of industrial design." There's no mention of Dreyfuss, Teague, or Bel Geddes.

2002. Having outlived her husband by forty-four years, Edith Lutyens dies at ninety-five, but not before establishing the Edith Lutyens and Norman Bel Geddes Foundation, which since 2004 has provided grants to innovative New York City–based not-for-profit theaters, allowing designers to realize their visions.

2005. A heavy smoker, Barbara Bel Geddes dies of lung cancer at eighty-two, having achieved her most popular success as Miss Ellie in TV's *Dallas*, the first soap-opera star to ever receive an Emmy. In 1993, she was inducted into the American Theatre Hall of Fame in recognition of twenty-five years' distinguished service to her profession, an honor she shared with her father.

•

OTTO KAHN'S FIFTH Avenue mansion survives as the Convent of the Sacred Heart, a private Catholic girl's school. Okeha Castle, his Long Island "weekend" place, is a hotel and popular wedding venue listed on the U.S. National Registry of Historic Places. Taliesen West, also on the National

Registry, houses the Frank Lloyd Wright School of Architecture and the Frank Lloyd Wright Foundation. Hollyhock House, a National Historic Landmark, is the centerpiece of Los Angeles's Barnsdall Art Park. John Ringling's pink marble manse is home to the John and Mabel Ringling Museum of Art. Owned by an American nonprofit, Schloss Leopoldskron (where the *The Sound of Music* was filmed), is used for seminars and conferences.

First class stamp series, United States Post Office, "Pioneers of American Industrial Design," 2011.
From left to right: Bel Geddes's Patriot Radio (Emerson), Dreyfuss's telephone (Bell Labs),Teague's camera case (Kodak), and Loewy's pencil sharpener (never produced).
(Courtesy: U.S. Postal Service series, "Pioneers of American Industrial Design," 2011)

Notes

1. The Fred F. French building at Fifth Ave and East Forty-Fifth Street.
2. Two years later, a New Jersey man who happened to share the pilot's surname was charged with drunk and disorderly behavior. Asked by the judge if he was related to Colonel Lindbergh, he replied, "No sir, I don't go swimming in the air." "Well," the judge countered, "I couldn't fine a man with such a distinguished name. Sentence suspended." *The New York Times*, March 6, 1929, p. 31.
3. Perhaps not coincidentally, Lindbergh, Fitzgerald, and Bel Geddes all hailed from the Midwest.
4. According to Gilbert Miller in *The New York World*.
5. *New York Telegram* review by Leonard Haw, January 1927.
6. *Scientific American*, March 1928, pp. 249–50.
7. Bel Geddes to Mrs. Carrie K. Anderson of Saginaw, Michigan, October 19, 1927. Harry Ransom Center, University of Texas at Austin.
8. Letter to the Editor, *The New York Times*, VIX 4:5, January 27, 1927
9. Norman Bel Geddes, *Horizons (in Industrial Design)* (Boston: Little, Brown and Company, 1932), 8.
10. "Up From the Egg," *Time*, October 31, 1949, p. 71.
11. Stuart Chase, *Men and Machines* (New York: Macmillan Company, 1929), 250-251.
12. Bel Geddes, *Horizons*.
13. Quoted in *Steichen the Photographer* by Carl Sandburg (New York: Harcourt Brace,1929), 55.
14. Presentation by Edward Steichen in "Minutes of the Representatives Meetings," June 31, 1928, p. 10. J. Walter Thompson Archives.
15. Bel Geddes, *Horizons*, p. 281.
16. *The Genius: A Memoir of Max Reinhardt by His Son Gottfried Reinhardt* (New York: Knopf, 1979), 270.
17. Yingling was an Americanization of the German surname Jungling.
18. Founded in the aftermath of the Civil and Indian wars, Carlisle's mandate was to forcibly assimilate Native American children from some 140 tribes. Many had literally been stolen from their parents, others were war orphans. Pupils were beaten for speaking their native languages, their hair was cut off (a sign of mourning, it convinced many that their parents had died), their clothes and possessions burnt. They were taught to sleep in iron beds and pray to a Christian god.
19. Unbeknownst to Flora, there was also local talent worth knowing. Twenty-eight-year-old Willa Cather was living in Squirrel Hill teaching high school English and writing theatre reviews for *The Pittsburgh Leader*.
20. Letter from Will de Haw to Bel Geddes, circa 1955. Harry Ransom Center, University of Texas at Austin.
21. Flora may have been less surprised. In a private journal she kept about her children, she noted that even when Norman was four or five, he'd approached whatever interested him with an unnerving seriousness.
22. *The New York Times* interview with Edison, October 2, 1910.
23. Bel Geddes, *Miracle in the Evening* (Garden City: Doubleday & Company, Inc., 1960). Also, Bel Geddes to Hellman, 21.
24. Thunderbird's portrait, by Eulabee Dix, hangs in the Smithsonian's National Portrait Gallery.
25. Seven months before his untimely death at forty-eight, Remington attended a dinner—given for De-Plante's former employer, "Buffalo Bill" Cody—where it was suggested that he create a colossal statue of a North American Indian at the mouth of the Hudson, something to rival the Statue of Liberty Had it come to pass, Thundercloud might well have been the model for it. —*Frederick Remington: A Biography* by Peggy and Harold Samuels, (Garden City: Doubleday & Company, 1982).
26. A dozen years before, photographer Edward Curtis had made a similar journey to document Montana's Blackfeet, though it's unlikely Geddes had had an opportunity to see the resulting, prohibitively expensive, monograph.

27. Bel Geddes, *Miracle.*

28. A condor-like bird, whose body was large enough to accommodate an enormous wingspan and strong enough to lift a young buffalo calf up into the air, would need a large thermal updraft to hold it aloft. Scientists today know what the Blackfeet could not, that there's always a large thermal updraft in front of moving storms, giving the impression that the Thunderbirds brought the storms with them.

29. Quoted in *Edison and His Invention Factory: A Photo Essay.* (Washington D.C., Eastern National Park & Monument Association, 1989), 2.

30. Bel Geddes, *Miracle.*

31. The diagonal axis approach places the stage from corner to corner of the auditorium instead of in the middle of a wall, permitting more (and longer) rows of seats, consequently greater seating capacity, as well as improved sight lines.

32. *Forbes,* July 1930, p. 57.

33. Lise-Lone Marker, *David Belasco: Naturalism in the American Theatre* (Princeton: Princeton University Press,1975), 27.

34. Quoted in *The Life of David Belasco* by William Winter (New York: Moffat, Yard, Vol. 2, 1918), 248.

35. Kenneth Macgowan, *The Theatre of Tomorrow* (New York: Boni and Liveright, 1921).

36. During its eighteen-month run, *InWhich* would contain the germs of many future Geddes ideas.

37. By 1930, 30,000 miniature golf courses were in operation, an overall investment of $125 million, many earning 300 percent a month.

38. Barnsdall to Bel Geddes, November 2, 1915. Harry Ransom Center.

39. Barnsdall to Bel Geddes, November 30, 1915. Harry Ransom Center.

40. Barnsdall to Bel Geddes, December 19, 1915 and January 8, 1916. Harry Ransom Center.

41. Bel Geddes to Bel Schneider, 28 February 1916. Harry Ransom Center.

42. Its inspiration, Dominique DePlante, had died of pneumonia during a winter "posing trip."

43. *Los Angeles Examiner,* July 1919.

44. In August 1914, the original Taliesen was set on fire by a disgruntled employee who proceeded to murder Wright's mistress, her children, and several staff members.

45. *The New York Times,* December 17, 1911.

46. Quoted in *The Oilman's Daughter: A Biography of Aline Barnsdall* by Norman and Dorothy K. Karasick (Encino: Carleston Publishing Co., 1993). No source cited.

47. In 1977, it would merge with Lehman Brothers.

48. Cleveland Amory, *Who Killed Society?* (New York: Harper & Brothers, 1960), 443.

49. ""Otto H. Kahn, "Art and the People," in *Reflections of a Financier: A Study of Economic and Other Problems* (London, Hodder & Stoughton, Ltd. 1921).

50. Letter from Otto Kahn to playwright Em Jo Basshe, quoted in *The Many Lives of Otto Kahn* by Mary Jane Matz (New York: Macmillan Co., 1963), 272.

51. *Newsweek,* July 1, 1933, p.17.

52. Bel Geddes, *Miracle.*

53. The story wasn't entirely true. Geddes had initially written to Kahn while he was still working with Barnsdall's Little Theatre, in search of a sponsor who could help his work "reach the market." Kahn offered, instead, an advance to cover a roundtrip to New York, but having secured a contract with Universal in the interim, Norman turned it down. That October, when he came upon the magazine profile, he explained to Kahn that Hollywood's moguls weren't ready for what he had to offer.

54. Bel Geddes, *Miracle.*

55. Charlotte Himber, *Famous in their Twenties,* "All the World His Stage," (Freeport: Books for Libraries Press, 1942, reprinted 1970).

56. About $5,000 today.

57. Bel Geddes to Hellman, p. 29.

58. "Strife in Opera at La Nave: Scene and Song Clash." November 1919. Newspaper clipping, source unknown. Harry Ransom Center.

59. Bergman to Hellman, p. 46.

60. *George Gershwin,* edited by Merle Armitage (London: Longmans, Green & Co., Ltd., 1938). The poem, written for Armitage's book, was registered for copyright in 1970 as an unpublished work.

61. To the chagrin of the shocked elite, both matches proved long-term and happy.

62. Bel Geddes, *Miracle.*

63. "sherry with the soup" —Mrs William Morris to Hellman, p. 25.

64. *The New York Times,* January 25, 1920.

65. *A Project for the Theatrical Presentation of The Divine Comedy of Dante Alighieri by Bel Geddes* (New York: Theatre Arts Inc., 1924). Foreword by Max Reinhardt; photographs by Francis Bruguiere. A slightly different version appeared in Bel Geddes's Miracle.

66. Thornton Wilder to Bel Geddes, October 13, 1932. Harry Ransom Center.

67. George E. Bogusch, *Norman Bel Geddes and Art of Modern Theatre Lighting* (Baltimore: John Hopkins University Press), 1972.

68. Norman Bel Geddes, "The Theatre of the Future," in *American Stage Designs: An Illustrated Catalogue of the Models, Drawings and Photographs* (exhibited at the Bourgeois Galleries in New York), April, 1919.

69. "Stage Designs on View at Cambridge." *Boston Transcript*, November 21, 1932.

70. According to costume designer Lucinda Ballard, his students paid to work for him in return for which he gave a weekly lecture through a hangover. Then again, she was angry for having been fired from his production of *Lysistrata*, presumably for having asked for overtime pay.—Ballard to Hellman.

71. Carl Van Vechten, *The Splendid Drunken Twenties: Selections from the Daybooks 1922-1930*. Wednesday, July 26, 1922 entry. (Champaign: University of Illinois Press, 2003), 9.

72. *New York Times* clip, n.d.

73. Stokowski to Bel Geddes, March 28, 1931. Harry Ransom Center.

74. Preliminary sketches show that Norman's figures were caricatures of prominent society couples gleaned from *Vanity Fair*. See VF, Aug 1919, p. 34; Dec 1920, p. 42.

75. "The Egotist," reprinted in *An American in Paris: Profiles of an Interlude Between the Wars* by Janet Flanner (New York: Simon and Schuster, 1940), 217–225.

76. "New York Has No Laughter and No Young Girls." *New York Times*, October 19, 1913.

77. Emily Post, *Etiquette* (New York: Funk & Wagnalls Co., 1923).

78. *Flapper Magazine*, November 1922, p. 2.

79. "Misunderstood, Says Poiret, So He Will Leave," *Evening Globe*, September 1922.

80. Woollcott review.

81. Edmund Wilson, *The Twenties* (New York: Farrar, Straus and Girous, 1975), 189.

82. *Evening Globe*, Dec 21, 1922, review by J.S. Kaufman.

83. "Curves, Angles, Blocks, Layers," by Gerstle Mack. *New York Times*, April 28, 1923.

84. Billy Rose, *Wine, Women & Words* (New York: Simon & Schuster, 1948).

85. *Time*, January 28, 1924, p. 16.

86. Barnsdall to Bel Geddes, n.d. Harry Ransom Center.

87. Peter F. Ostwald, *Vaslav Nijinsky: A Leap into Madness* (New York: Carol Publishing Group, 1991).

88. For this and all subsequent shows, Bel Geddes factored cigars and whiskey into his budget for the stagehands. For a designer, stagehands were essential friends.

89. A private school for the daughters of affluent, socially prominent families. Future alumni would include Jacqueline Bouvier Kennedy, Estee Lauder, Queen Noor of Jordan, Vera Wang and Ivanka Trump.

90. Diana Cooper, *The Light of Common Day* (New York: Houghton Mifflin & Co., 1959) 133.

91. Stieglitz and O'Keefe to Bel Geddes, May 18, 1924. Harry Ransom Center.

92. Will de Haw to Bel Geddes, Toronto, n.d. Harry Ransom Center.

93. *The Blue Book of the Screen*, edited by Ruth Wing. (Hollywood: The Blue Book of the Screen, Inc. 1923), 381.

94. Bel Geddes, *Miracle*, 317.

95. *Autobiography of Cecil B. De Mille*, Prentice-Hall, Inc. 1959, p. 262.

96. From Frank Elliot's review in *The Motion Picture News*, September 27, 1924.

97. Margaret Farrand Thorp, *America at the Movies* (New Haven: Yale University Press, 1939), 69.

98. Charles Chaplin, *My Autobiography* (New York: Simon and Schuster, 1964).

99. Mercedes de Acosta, *Here Lies the Heart* (New York: Arno Press, 1975), 157.

100. Bel Geddes, *Miracle*, p. 333 and 339.

101. *Report of the Commission to International Exposition of Modern Decorative and Industrial Arts in Paris, 1925* (U.S Dept. of Commerce).

102. Katharine Murphy was featured in *Ballet Mechanique* as "the girl on the swing." Dudley also shot erotic sequences of her that were subsequently deleted.

103. Judging by a letter from Kommer, it wasn't Norman's first extramarital encounter. Written in the spring of 1924, early in "The Miracle's" run, it spoke of Norman's "amorous adventure . . . a deep-dyed secret that all the nuns are chanting from the tops of your cathedral." —Kommer to Bel Geddes, May 18, 1924. Harry Ransom Center.

104. Quoted in *Eva Le Galliene: A Biography* by Helen Sheehy (New York: Knopf, 1996), 125.

105. *Sir, The death of my poor Erik Satie has hit me very hard. He was buried this morning and since yesterday, my head has been empty. Excuse me. I wanted very much to see you and speak with you.* —Pasted into Bel Geddes's *Jeanne D'Arc* prompt book. Harry Ransom Center.

106. "The Geddes Production" by Stark Young. *New Republic*, November 11, 1925, pp. 305–306.

107. "Poe Taunts Filmmakers Evermore," by Terrence Rafferty. *New York Times*/Arts & Leisure, April 22, 2012, p. 12.

108. *Hollywood stars need someone to look after them*, Hayward had told Norman, an all-around personal assistant to help organize their lives. He'd offered Norman a partnership but babysitting movie stars wasn't of interest. Soon after hopping a train to L.A., Hayward was "handling" Greta Garbo, Boris Karloff, Fred Astaire, Judy Garland. Katharine Hepburn, Henry Fonda, James Stewart, even Ernest Hemingway and Dashiell Hammett.

109. Barry Ulanov, *A History of Jazz in America* (New York: Da Capo Press, 1972, originally published 1952), 114.
110. "Geddes at the Fair." *Time*, March 4, 1929, p. 46–48.
111. Bel Geddes, *Miracle*, 230.
112. "Geddes at the Fair." *Time*, March 4, 1929, p. 46–48
113. "Games Worth the Candle—Sport for Kings and Good Fellows Fought Out in the Playing Fields of Mr. Geddes" by Ruth Pickering. *Arts and Decoration*, February 1933, pp. 18+.
114. Bel Geddes's War Game anticipated the French military board game, *La Conquete du Monde* (1957), which in turn inspired Parker Brothers' highly successful *Risk!* (1959)
115. More than $226,000 today.
116. "Games Worth the Candle," ibid.
117. "In Spite of Kellogg Treaty, War Again Rages on Wide Front," by Edwin C. Hill. *New York Sun*, January 12, 1931.
118. Newsom to Hellman, p. 48.
119. Newsom to Hellman, p. 48.
     All-white tires had appeared on the earliest automobiles—white is rubber's natural color—but carbon black was soon added to increase traction, endurance and ease of cleaning.
120. Bel Geddes, *Horizons*, p. 260.
121. *Forbes*, July 1930.
122. Raymond Loewy, *Never Leave Well Enough Alone* (Baltimore: John Hopkins University Press, 2002; originally published by Simon & Schuster, New York, 1951) 57-58.
123. Quoted in *Women's Wear Daily*, November 1929.
124. Edward Bernays, *Crystallizing Public Opinion* (New York: Boni and Liveright, 1923) and *Propaganda* (New York: Horace Liveright, 1928).
125. "Recent Economic Changes in the United States." Compiled by the President's Conference on Unemployment. Washington D.C., 1929.
126. From "The Problem of the Used Car," reprinted in *20,000 Leagues Under the Sea or David Copperfield*, by Robert Benchley (New York: Henry Holt and Company, 1928), 131.
127. "A New Profession," 1949 talk by George Nelson, reprinted in *Problems of Design* (New York: Whitney Publications, 1957).
128. *Time*, Oct 31, 1949, p. 71.
129. Raymond Loewy, *Industrial Design* (Woodstock: Overlook Press, 1979), 8.
130. Lawrence Loewy quoted in "Threads/Brief History," by Sara James. *Vogue Magazine*, September 2007.
131. Loewy, *Never Leave*, 91.
132. Walter Dorwin Teague Jr., *Industrial Designer: The Artist as Engineer* (Lancaster, PA: Armstrong World Industries, Inc., 1998), 72.
133. Fleischman, "Women: Types & Movements" by Doris E. Fleischman, in *America as Americans See It*, edited by Fred J. Ringel (New York: Literary Guild, 1932), 115.
134. Mrs. Christine Frederick, *Selling Mrs. Consumer* (New York: The Business Bourse, 1929).
135. Stuart Case, *Men and Machines* (New York: Macmillan Co., 1929), 230.
136. Egmont Arens and Ray Sheldon, *Consumer Engineering: A New Technique for Prosperity* (New York: Harper and Brothers, 1933), 232–233.
137. "Conduct Your American Advertising on the American Plan" by David Leslie Brown. *Printer's Ink*, April 18, 1929.
138. *The New Yorker*, October 8, 1928.
139. *Pencil Points Magazine*, 1934. Harry Ransom Center.
140. Vogue, n.d., Harry Ransom Center.
141. Bel Geddes & Co. record of phone conversation from H. D. Bennett, February 20, 1929. Harry Ransom Center.
142. *Fortune*, July 1930, pp. 51–57.
143. "Odd Business, This Industrial Design." by Seymour Freedgood. *Fortune*, February 1959.
144. *Fortune*, July 1930, pp. 51–57.
145. Bel Geddes, *Horizons*, p. 221.
146. Bennett to Bel Geddes, August 1, 1929. Harry Ransom Center.
147. "Bel Geddes," *Forbes*, July 1930, pp. 51–57.
148. Bel Geddes, *Horizons*, pp. 113–14.
149. Bel Geddes, *Horizons*.
150. Paul Frankl, *Machine-Made Leisure* (New York: Harper and Brothers, 1932), 166–67.
151. Colonel W. A. Starrett, *Skyscrapers and the Men Who Build Them* (New York: Charles Scribner's Sons, 1928), 3.
152. Even if they were, like cathedrals, often tortured with "pastry cook" embellishment—cherubs, cornices, festoons and cornucopias, their ceilings coffered, sub-coffered and gilded.
153. A complete Bel Geddes Skyscraper Cocktail Set sold at auction in 2001 for a reputed $13,000.

154. Raymond Hood profile in *The New Yorker*, April 1931.
155. *Time*, March 4, 1929, p. 46.
     Eighty-four years later, Michael Graves would be called "a modern-day Michelangelo" for his gleaming, toast-shaped toaster and bell-ringing tea kettle.
156. "Stage Designs on View in Cambridge." *Boston Transcript*, November 21, 1932.
157. "Theatres for the Machine Age: Norman Bel Geddes Leaves Honored Traditions by the Wayside" by H. I. Brock. *New York Times*, November 16, 1930.
158. O'Neill to Richard Madden, April 3, 1920. *Selected Letters of Eugene O'Neill* (New Haven: Yale University Press, 1988).
159. Brock, "Theatres for the Machine Age."
160. "Are You Afraid of the Unexpected?:" by M. K. Wisehart. *American Magazine*, July 1931, pp. 71+.
161. Rockefeller Center's Music Hall (dubbed Radio City after its first tenant, RCA) would open on Dec 27, 1932, its general hall plan and the entire lighting of its auditorium and stage predicated on principles devised by Bel Geddes. A Geddes-esque mechanism (though he likely had nothing to do with it) produced rain, fog, and clouds via steam drawn from a nearby Con Edison generating plant.
162. *Time*, March 4, 1929, p. 46.
163. Dictated memo, "Geddes Explanation of Architectural Registration Difficulties." February 2, 1940. Harry Ransom Center.
164. Letter from Bel Geddes to architect Walter Harrison, April 27, 1956; *Theatre Arts Monthly*, April 1932; See also *Horizons*, pp. 173–80.
165. Bel Geddes to Hellman.
166. "Sales Promotion at the Brevoort," by Margaret Fishback, *New Yorker*, April 30, 1932.
167. Cunningham to Bel Geddes, June 4, 1932, Harry Ransom Center.
168. R.L. Duffis in *New York Times Book Review*, December 18, 1932.
169. Joseph H. Jackson, in *San Francisco Chronicle*, January 1933.
170. Lillian C. Ford in *Los Angeles Times*, Jan. 1, 1932.
171. *Victoria Daily Times*, n.d., 1933. Harry Ransom Center.
172. John de la Valette, intro to *The Studio Year Book of Decorative Art*, quoted *Industrial Art Explained* by John Gloag (London: George Allen & Unwin, Ltd., 1934), 98.
173. *Saturday Review of Literature*, December 31, 1932. Harry Ransom Center.
174. Frank Lloyd Wright, *An Autobiography* (New York: Duell, Sloan and Pearce, 1932), 525.
175. Bel Geddes to Wright, September 27, 1934. Frank Lloyd Wright Archives, Taliesen West, Scottsdale, Arizona.
176. Earhart to Bel Geddes, Sept. 15, 1933, Harry Ransom Center.
177. Earl Newsom to Frank Gledhill of Pan Am, Dec 5, 1933. Harry Ransom Center.
178. Summary by E. A. Paxton of meeting with Pan Am executives, November 21, 1933. Harry Ransom Center.
179. *Vogue* clipping, no date. Harry Ransom Center.
180. "Glamor Crossing: How Pan Am Dominated International Travel in the 1930s." blog.longreads.com.
181. Norman Bel Geddes & Co memo, March 6, 1946. Harry Ransom Center.
182. Published in *Fortune*, June 1930.
183. Bel Geddes, *Miracle*, p. 327.
     Sergei Eisenstein's *The Film Sense* (1942) included diagrams illustrating his method of preparing a manuscript prior to shooting, exact breakdowns, shot by shot—a "measured matching" of the visual, music phrases, pictorial composition, and actors' movements from one scene to next.
184. Sergei Eisenstein, *Immortal Memories: An Autobiography*, translated by Herbert Marshall (Boston: Houghton Mifflin Co., 1983).
185. Eisenstein, *Immortal*, 144.
186. Arthur and Martha Cheney, *Art & The Machine: An Account of Industrial Design in the 20th Century* (New York: Whittlesey House, 1936), 100.
187. Henry Dreyfuss, *Designing for People* (New York: Viking Press, 1974; originally published by Allworth Press/Design in Management Institute, 1955), 75.
188. The prototype, which never went into production, was prominently displayed in Loewy's office. Young Budd Steinhilber, later a designer in his own right, was charged with making sure his boss' pencils were needle sharp. "Frankly," he'd recall in 2012, "it didn't work worth a damn." The idea that "form trumps function . . . is intellectual bullshit. —"The Streamline Era: A Personal View," by Budd Steinhilber, FIDSA. IndustrialDesignHistory.com, September 1, 2012.
189. "Parts of a New Civilization" by Renford G. Tugwell, *The Saturday Review of Literature*, April 13, 1940, pp. 80–81.
190. Antoine de Saint Exupery, *Wind, Sand and Stars* (New York: Reynal & Hitchcock, 1939), 66.
191. Two years after *Horizons* debuted, an eleven-page article on streamlining and aerodynamics in *The Atlantic Monthly*, under Bel Geddes's byline, expanded on streamlining theories, noting that streamlining was already being haphazardly and arbitrarily applied.
192. "My Stove's in Good Condition" by Lil Johnson, circa 1936.

193. Fleischman, "Women: Types & Movements."
194. David E. Kyvig and Ivan R. Dee, *Daily Life in the United States, 1920-1940: How Americans Lived Through the Roaring Twenties and the Great Depression* (Chicago: Ivan R. Dee, 2000), p. 195. The J. Walter Thompson survey was conducted in 1936.
195. Frederick, *Selling Mrs. Consumer.*
196. Silas Bent, *Slaves by the Billion: The Story of Mechanical Progress in the Home* (New York: Longmans, Green & Co., 1938), 14–34.
197. Earnest Elmo Calkins, *Business The Civilizer* (Boston: Atlantic Monthly Press/Little, Brown & Co., 1928), 14.
198. British Commercial Gas Association advertisement, 1930.
199. "Three Women," General Electric Co., 1935. 35-millimeter promotional film.
200. Frederick, *Selling Mrs. Consumer,* 154.
201. "Norman Bel Geddes" by Arthur Shawn, *Outlook* Magazine, February 12, 1930, p. 273.
202. CNN/Money/Fortune, Jan 24, 2000.
203. "Story of a Stove" *Fortune,* Febriary 1934, p. 42.
204. Bel Geddes, *Horizons,* pp. 228–30.
205. Arens, *Consumer Engineering,* p. 174.
206. Dreyfuss, *Designing for People,* p. 65.
207. "Story of a Stove," *Fortune.*
208. "Story of a Stove," *Fortune.*
209. "Both Fish and Fowl" by George Nelson, *Fortune,* February 1934, pp. 40–98.
210. Dreyfuss to Hellman, pp. 25–27.
211. Cheney, Art & The Machine, p. 66.
212. According to *Forbes.*
213. Loewy, *Never Leave,* pp. 115-18.
214. Philippe Tretiack, *Raymond Loewy* (New York: Assouline, 2005), p. 9.
215. Loewy, *Never Leave,* p. 128. See also Loewy, *Industrial Design,* pp. 98–100.
     In his ghostwritten "Both Fish and Fowl" (*Fortune,* February 1934), George Nelson wrote of an un-named refrigerator designer who "proudly" designed a foot lever as a door opener, something a house-wife "with her hands full of dishes" could easily open. "Result: housewife, balancing on one foot, is caught off balance by the heavy door and sent to the floor with her dishes."
216. Loewy, *Never Leave,* 128.
217. "Durable Goods Go To Town" by Francis Sill Wickware, *Forbes,* Nov 15, 1936, p. 34.
218. Arthur J. Pulos, *American Design Ethic: A History of Industrial Design to 1940* (Cambridge: MIT Press, 1988), 358.
219. "Public Relations or Industrial Design? Loewy and His Legend" by Stephen Bayley. p 232, in *Raymond Loewy: Pioneer of Industrial Design,* edited by Angela Schonberger, (Munich: Prestel, 1990). See also: Bagley's monograph *The Lucky Strike Packet* by Raymond Loewy (Design Classics, Frankfurt-am-Main, Verlag form, 1998), 34–35.
220. Loewy, *Industrial Design,* 13.
221. Email to the author from Lynn Catanese, Curator of Manuscripts and Archives, Hagley Museum and Library, Wilmington, Delaware, December 16, 2009. Additional Loewy papers, housed at the Library of Congress, date only from 1960 through 1976.
     Finished designs were always signed with Loewy's name and everything else, including the staff's best work, was destroyed. —Former Loewy employee Jay Doblin in *I.D. Magazine,* Special Raymond Loewy Issue, Nov/Dec, 1986, pp. 42–43.
222. Cheney, *Art and the Machine.*
223. Loewy, *Industrial Design,* 13.
224. Sam Vining, quoted in *Fortune,* May 1940.
225. Transcript of interview with Selma Robinson for *PM magazine,* January 27, 1942, Harry Ransom Center.
226. *The Batavia News,* January 29, 1933. Harry Ransom Center.
227. Bel Geddes, *Horizons,* p. 43
228. "The World Today" ("Machine-Made Art" supplement), *Encyclopaedia Britannica,* 1935.
229. "On the Sun Deck" by John M'Clain. *New York Sunday,* January 16, 1933. Harry Ransom Center.
230. Newsom to Hellman, p 48.
231. Unidentified magazine article, Harry Ransom Center.
232. "Durable Goods Go to Town" by Francis Sill Wickware. *Forbes,* November 15, 1936, pp.32–36, 70.
233. Dreyfuss, *Designing for People,* p. 62 and 136.
234. Wickware, "Durable Goods."
235. Loewy, *Never Leave,* 138–141.
     Designer-turned-historian Carroll Gantz wrote that Loewy's overall vision for the GG-1 had already been created by former Westinghouse design head Donald Roscoe Dohner. "Loewy's claim triumphed due to his own aggressive self-promotion." —*Industrialization of Design: A History from the Steam Age*

to Today by Carroll Gantz (Jefferson, N.C.: McFarland, 2011), 159.

236. Austrian to Hellman, p 41.
237. "Railroad Futurity." *Business Week*, June 10, 1933.
238. "The Artist Turns Big Shot" by Gretta Palmer, in *Today*, June 1935.
239. George Howe papers, Avery Library, Columbia University archive, New York City.
240. Bel Geddes to Bourke-White, May 27, 1935. Margaret Bourke-White archive, Syracuse University.
241. Bel Geddes & Co memorandum, dictated by N, dated February 4, 1940. "Wright might not admit it but we have the drawings in our possession to prove it." Harry Ransom Center.
242. $100,000 in Depression-era dollars.
243. Bel Geddes offered to buy *Comic Supplement*, a production he'd worked on that Flo Ziegfeld was unhappy with, for $100,000.
244. *Scratch Pad: A Chronicle of Common Sense.* Chesla C. Sherlock, editor. Vol. III, No. 1, July 1934, pp. 7–10.
245. Pemberton to Hellman.
246. Bel Geddes to Hellman.
247. "Private Lives," by Edwin Cox, April 25, 1941. Nationally syndicated series.
248. Cheesecloth bags, detailed instructions, and a map of "best places" were provided.
249. Bel Geddes to Southwest Zoological Supply House (Oct. 10, 1934) and Florida Reptile Institute (Nov. 14, 1934).
250. Dr. G. Kingsley Noble to Bel Geddes, July 18, 1934. Harry Ransom Center.
251. Bel Geddes to Mr. R. T. Berryhill, Sr., December 11, 1934. Harry Ransom Center.
252. Pemberton to Hellman.
253. Ray Schechter (General Biological Supply House, Chicago) to Bel Geddes, April 14, 1934.
254. An inveterate inventor, Birdseye also came up with a modernized whaling harpoon and a process for making paper out of sugarcane stalks.
255. Bel Geddes to Beebe, June 7, 1935. Harry Ransom Center.
256. G. Kingsley Noble to Bel Geddes, June 12, 1935. Harry Ransom Center.
257. *Dead End* by Kingsley.
258. Ben Webster to Hellman, p. 35.
259. $180,000–$216,000 per year today.
    In the eighteenth century, Joseph Hayden reputedly spent an entire year's salary on a particularly rare tropical bird.
260. "I Salute Walter P. Chrysler," *Saturday Evening Post*, December 16, 1933, p. 31.
261. When *The Green Hornet* returned as a TV series in 1966, the Black Beauty returned as a Chrysler Imperial, modified to fire rockets. Ford's 1955 Futurama-esque one-off, the Lincoln Futura—a cross between a Manta Ray and a Mako shark—would end up as the Batmobile in the 1966 *Batman* TV series.
262. "Chrysler: Beauty is No Chance Creation." *Literary Digest*, January 12, 1929, p. 36.
263. *Saturday Evening Post*, September 30, 1930, p. 45.
264. Frederick Lewis Allen, *The Big Change: America Transforms Itself, 1900–1950* (New York: Harper & Row, 1952), 147-148.
265. Vincent Curcio, *Chrysler: The Life and Times of an Automotive Genius* (Oxford University Press, 2000), 541.
266. Curcio, *Chrysler*, 542.
267. "The History of the Airflow Car" by Howard S. Irwin. *Scientific American*, August 1977, pp. 98–106.
268. The result of studies of "the exact 'periodicity of movement' most restful to human nerves."
269. Minutes of meeting, Bel Geddes & Co., October 19, 1933, p. 3 +.
270. Christopher Innes, *Designing Modern America: Broadway to Main Street* (New Haven: Yale University Press, 2005), 160–161.
271. Bel Geddes to Harvey S. Firestone Jr., January 22, 1948. Harry Ransom Center.
272. Articles appeared in *Automobile Topics* (December 16 and 30, 1933; January 6, 1934); *Automobile Trade Journal* (January 1934); *American Automobile* (April and June 1934); and *Fortune* (April, May and June 1934).
273. Alec Woollcott posed for a similar ad in *Collier's*.
274. "Fashioned by Function," a 13-minute Chrysler Sales Corp. promotional film.
275. "The Phantom Fleet of the Highway" by H. W. Magee. *Popular Mechanics*, June 1938.
276. Frederick Lewis Allen, *Since Yesterday: The 1930s in America* (New York: Perennial Library, 1986, copyright 1940), 182.
277. Paul Merchant on Harcourt Brace Jovanovich letterhead to Bel Geddes, September 29, 1934. Harry Ransom Center.
278. Carl Breer and Arthur J. Yanik, *Birth of Chrysler Corporation and Its Engineering Legacy* (Detroit: Society of Automotive Engineers International, 1994), 159.
279. Westinghouse had purchased Tesla's AC current patent.
280. Topsy's demise can be seen on YouTube.

281. Jan Cohn, *Creating America: George Horace Lorimer and the* Saturday Evening Post (Pittsburgh: University of Pittsburgh Press, 1989), 218+.

282. Upon receiving a personalized copy of *Horizons*, General Motor's Charles Kettering wrote Bel Geddes in thanks, adding, "We [at GM] have . . . read the papers of many of our theoretical people who still insist on treating the automobile as though it were an airplane,. Nature did a very good job in streamlining birds and fishes, but it did not go to much trouble when it came to creatures that run on the ground." Harry Ransom Center.

283. *Saturday Evening Post*, March 31, 1934, pp. 32–33; April 28, 1934, pp. 38–39; May 12, 1934, pp. 40–41; May 26, 1934, pp. 40–41; June 1934, pp. 36–37.

284. One casting was "gifted" to the Metropolitan Museum of Art. Norman gave another to Margaret Bourke-White.

285. Watch the cliff drop, Bonneville Flat, and safety glass tests in *Trials of Triumph* on YouTube.

286. *Saturday Evening Post*, December 16, 1933.

287. In 2015, under the catchphrase "Dare Greatly," General Motors would adopt a line from Herman Melville (perhaps inspired by Apple's highly successful "Think Different" campaign) to promote its latest model Cadillac: "It is better to fail in originality than to succeed in imitation." Neither GM's investors nor its Board of Directors believe a word of it.

288. Bel Geddes would continue to work for Chrysler, supply "study" models for General Motor's 1939 Buick Series 40, a small-model Plymouth, and contribute to the "streamlined" 1941 Nash.

289. "The Great Packager," by John Kobler. *Life* magazine, May 2, 1949, pp. 110+.
    Loewy's MAYA principle ("Most Advanced Yet Acceptable"), combined with his allegedly superior European aesthetic, was designed as a bulwark against just such hugely expensive faux pas, though it's unclear when, exactly—before the Airflow's demise or after—he coined the acronym.

290. Loewy, *Never Leave*.

291. Headline in Hupmobile ad, *Harper's Magazine*, 1932.

292. "Big Car Enemy Raymond Loewy Showed Detroit the Way" by Sarah Booth Conroy. *Pittsburgh Post Gazette*, March 24, 1975.

293. "Loewy's automobile designs of the 30's were invariably awkward and timid, no match for the brio of Bel Geddes' streamlining," an *Art in America* critic would observe in the 1970s. Only with the advent of WWII would Loewy "master the automobile by mating it with the war plane—fusing cockpit and bullet nose with four wheels." —"Loewy and the Industrial Skin Game" by Richard Pommer. *Art in America* , March-April 1976, p. 44–48.

294. Innes, *Designing Modern America*, 157.

295. In 2009, a 1934 Airflow sold for $44,850. In 2012, a Tucker Torpedo went for $2,915,000.

296. Pulos, *American Design Ethic*, 384.

297. "Bel Geddes Denies Spending Too Much on Plays" by Robert Coleman.

298. The league was principally financed by Reis's husband, a businessman in the underwear trade who became a Geddes War Game devotee.

299. *New Yorker*, n.d. Harry Ransom Center.

300. Film footage of *Hamlet's* rehearsals, both on and backstage, is housed in Bel Geddes's archive.

301. Dreyfuss to Hellman, p. 26.

302. *New York Telegraph* review by Whitney Bolton, n.d. Harry Ransom Center.

303. Ethan Mordden, *All that Glittered: The Golden Age of Drama on Broadway, 1919–1959* (New York: St. Martin's Press, 2007).

304. Bel Geddes, *Miracle*, 163-164.

305. "Dead End" by Stark Young. *New Republic*, November 13, 1935, p. 21.

306. *New York Times*, October 29, 1935.

307. "Dead End" by Stark Young.

308. *New York World Telegram*, February 17, 1937.

309. Anais Nin, *Fire: The Previously Unpublished, Unexpurgated Diary, 1934–1937* (New York: Harcourt Brace & Jovanovich, 1995), February 1, 1936 entry.

310. Nin, *Fire*, March 9, 1936 entry.

311. Nin, *Fire*, March 9, 1936 entry.

312. Nin, *Fire*, March 9, 1936 entry.

313. Nin, *Fire*, March 9, 1936 entry.

314. Nin, *Fire*, March 9, 1936 entry.

315. *Adrian Daily Telegram*, n.d. Harry Ransom Center.

316. Rebecca West to Gordon N. Ray, n.d., Pierpont Morgan Library, quoted in *Anais Nin: A Biography* by Deirdre Bair (G P Putnam's Sons, NY, 1995. p. 563, note #32.

317. Nin to Bel Geddes, 1936. Harry Ransom Center.

318. "His Models Never Moved" by Lloyd Lewis. *The Chicago Daily News*, Sept 12, 1936.
    His motivation for claiming Belmont as a middle name at this juncture is unclear. The press often asked for an explanation of his "odd" name, and an explanation required acknowledging his first wife and failed marriage.

319. "Iron Men" Open at the Longacre" by Douglas Gilbert. Unidentified newspaper clip.
320. "'Iron Men' Swell Construction Job" by Burns Mantle, *Daily News*, October 20, 1936.
321. Kommer to Bel Geddes, May 18, 1924 Harry Ransom Center.
322. Bel Geddes, confidential report to the Guggenheim Foundation, November 17, 1941. Harry Ransom Center.
323. Reinhardt, *Genius*, 28.
324. Reinhardt, *Genius*, 246.
325. Meyer Weisgal, . . . *So Far: An Autobiography* (New York, Random House, 1971), 117.
326. Reinhardt, *Genius*, 248.
327. Reinhardt, *Genius*, 258.
328. *Genius*, p. 256.
329. Brooks Atkinson, *Broadway* (New York: Macmillan,1970), 343.
330. Weisgal, *So Far*, 126.
331. Reinhardt, *Genius*, 297.
332. Frances, by now a partner in Bel Geddes & Co., helped finance both *Eternal Road* and *Dead End*. And in 1937, she put $3,000 into the business.
333. *New York Times*, January 8, 1937, p. 14.
334. "Tourism Boom Gives Broadway Record Season," by Michael Paulson, *New York Times*, May 27, 2015, p. 1+
335. Weisgal, *So Far*, p. 125.
336. Wilder to Lady Sibyl Colefax. November 2 1932. *The Selected Letters of Thornton Wilder* (New York: Harper Collins, 2008), 257.
337. Wilder to Bel Geddes, August 4, 1936. Harry Ransom Center.
338. Wilder to Bel Geddes, September 10, 1937. Harry Ransom Center. Wilder was being disingenuous about "distractions." He was involved with a sexual trophy hunter named Samuel Stewart, whom he'd met through Stein.
339. Wilder to Stein/Toklas, December 6, 1937. Letters of Gertrude Stein & Thornton Wilder (New Haven: Yale University Press, 1996), 197–198.
340. Burton Davis in *The Morning Telegraph*, 1926. Harry Ransom Center.
341. "Milestones," *Time*, August 22, 1932.
342. *NY World Telegram*, May 12, 1934. Bel Geddes also designed the Hotel Pennsylvania's Street Grill.
343. Theresa Collins, *Art, Money and Modern Time* (Chapel Hill, University of North Carolina Press, 1968).
     Despite having saved the Met from bankruptcy, Kahn was, for years, denied an opera box, his right as Director, because he was Jewish, a religion that meant little to him. His children had been christened at St. Bartholomew's.
344. Bibi Gaston, *The Loveliest Woman in America: A Tragic Actress, Her Lost Diaries, and Her Granddaughter's Search for Home* (New York: William Morris, 2008), 218.
345. Quoted in *Jed Harris: The Curse of Genius* by Martin Gottfried. (Boston: Little, Brown & Co., 1984), 169.
346. Jed Harris, *A Dance on the High Wire: Recollections of a Time and a Temperament* (New York: Crown Publishers Inc. 1988), 66.
347. Maurice Zolotow, *No People Like Show People* (New York: Random House, 1947), 224–225.
348. Gottfried, *Jed Harris*, 168-169.
349. Wilder to Stein, February 1,1938. Quoted in *The Enthusiast: A Life of Thornton Wilder* by Gilbert A. Harrison (New Haven: Ticknor & Fields, 1983) 183.
350. George Woods to Hellman.
351. Among Bel Geddes's voluminous papers is a note, on torn foolscap, in his handwriting, briefly describing the evening.
352. *Montreal Gazette*, Jan 22, 1938.
353. George Kivel and Grace Robinson in *New York Daily News*. Jan 24, 1938.
354. *New York Daily Mirror*, January 25, 1938.
355. *New York Times*, January 25, 1938, pp. 1 + 4.
356. *New York Post* [cover story,] January 24, 1938.
357. Gaston, *Loveliest Woman*, 244.
358. Rosamond's "unexplained" exit would remain in the Pinchot family's midst, enormous, unacknowledged, un-exorcised, even into the next generations, wreaking havoc—botched suicide (father Amos), heroin overdose (cousin Edie Sedgwick), successful suicides (both of Sedgwick's brothers), murder, execution-style (half-sister, Mary Pinchot Meyer, JFK's mistress, who also keep a diary that "went missing")—in the best ancient-Greek-tragedy tradition.
359. Wilder to Dwight Dara, February 1938. In *Selected Letters of Thornton Wilder* by Thornton Wilder and Jackson R. Bryer (New York: Harper Perennial, 2009).
360. Louise Spence to Hellman.
361. "A fantastic farm where ashes grow like wheat into ridges and hills and grotesque gardens . . ." *The*

Great Gatsby by F. Scott Fitzgerald.

362. A San Francisco World's Fair was also in the works, but Old Guard organizers there were opting for a Beaux Arts "ancient walled city." Northern California, the yet-to-come bellwether of personal computers and artisanal foods, would play Queen Victoria to New York's David Bowie.

363. Nylon's discoverer, forty-one-year-old William H. Carothers, swallowed cyanide in a hotel room, convinced that DuPont, his employer, didn't want to promote it. Seventeen years later, Alan Turning—the Father of Computer Science—would bite into a cyanide-laced apple (after having been forced to endure "chemical castration" to "cure" his homosexuality). He, too, was forty-one.

364. The jet engine was invented simultaneously by British cadet Frank Whittle and Hans von Ohain, a German; neither was aware of his counterpart. According to Von Ohain, jet-powered planes would have obliterated the Luftwaffe, and WWII might have been prevented, had the British government backed Whittle's work (for which he was knighted).

365. *Man and the Motor Car*, edited by Albert W. Whitnet. (New York: National Conservation Bureau, 1938).

366. View it at www.archive.org/details/WeDriver1936.

367. Spence to Hellman, p 34.

368. Harting to Hellman.

369. There would, in fact, be 61 million by 1960.

370. Seventy-seven years later, *Time* magazine's cover story with focus on self-driving automobiles. In it, Bel Geddes's radio-controlled Futurama cars are credited with anticipating the DARPA (2005), the Tesla Model 5 (2015), Google's prototype (2015) and the Mercedes-Benz F015, among others. —*Time*, March 7, 2016.

371. Futurama's floating airport harks back to a design in *Horizons*.

372. Forty-five years later, echoes of Bel Geddes's question, one that ultimately changed the game, could be heard in Steve Jobs's bid to lure John Sculley away from Pepsico to join forces with Apple.
    1938: "Do you think it's advisable for General Motors to admit it hasn't had a new idea in five years?"
    1983: "Do you want to spend your life selling sugar water or do you want a chance to change the world?"
    1938. 1983. Grist for numerologists.

373. Raymond Loewy would claim a similar laurel, making a point of wearing a Legion d'Honneur red rosette on his lapel. His name doesn't appear on the Legion's recipient lists.

374. Bel Geddes to Paul Merchant, Sept 29, 1934. Harry Ransom Center.

375. Teague to Fred Black, April 29, 1938. (Acc 544, Box 15, Ford Motor Co. archives, Edison Institute, Dearborn, Michigan).

376. Frederick Lewis Allen, *Since Yesterday: The 1930s in America* (New York: Harper Perennial, 1986). Between August 1937 and March 1938, General Motor's stock dropped from 60 1/8 to 25 1/2.

377. Though the busboy story was true, his disparagement of the Waldorf wasn't entirely honest. Norman was a perennial guest at the Waldorf's lavish, invitation-only, black-tie, strictly stag, annual Xmas party thrown by hotel manager Lucius Boomer—ten-pound tins of caviar and immense spreads of cold shellfish and canapes were followed by a lavish meal of wild game, entertainment by Duke Ellington, ballroom dancers Tony and Sally de Marco, or The Russian Choir. Norman went out of his way to arrange his schedule in order to attend. "I remember seeing television there five years before I ever saw it anyplace else."

378. Whalen's salary was believed to have been $75,000 a year plus $25,000 for expenses. —"Barnum in Modern Dress" by Elmer Davis, *Harper's Magazine*, October 1938, p. 459.

379. According to Earl Newsom, who went into partnership with the designer in 1935, Norman had a mistress while married to Frances—Consuelo Flowerton, a Ziegfeld Follies veteran who played the First Theban Woman in *Lysistrata*. (Newsom to Hellman, p. 48) The liaison, like the dalliance with Anais Nin, may have begun when Frances's illness worsened.

380. Gaylord Farm Sanatorium.

381. *Time*, April 9, 1951 (cover story).

382. Barbara Bel Geddes to Norman Bel Geddes, Harry Ransom Center.

383. Mrs. Hinton, headmistress, to Bel Geddes, 1938. Quoted in Bel Geddes's unpublished notes.

384. Buel was the son of Bel's sister, Florence.

385. Bel Geddes to Frank and Florence Worley, September 26, 1938. Harry Ransom Center.

386. Miles Davis and Quincy Jones, *Miles: The Autobiography* (New York: Simon & Schuster, 1990).

387. Billy Rose, *Wine, Women & Words* (New York: Simon & Schuster), 35.

388. "Case History of GM Intersection," Norman Bel Geddes & Co., 1941. Harry Ransom Center.

389. Bel Geddes to Waite, September, 1938. Harry Ransom Center.

390. Bel Geddes to Waite, September 28, 1938. Harry Ransom Center.

391. The tables were mounted on iron pipe, the legs threaded so that they could be inclined or leveled, and aligned to hairline accuracy by a turn of a joint.

392. Harting to Hellman, p. 40.

393. Norman had the satisfaction of out-mastering the master. F.L. Wright's 12-by-12' *Broad Acre* model had been crafted by Taliesen interns. Norman's 35,000 square foot *World of Tomorrow* required a team

of thousands.

394. Frank Harting to Hellman, p 40.
395. Eeva-Liisa Pelkonen, *Eero Saarinen: Shaping the Future* (New Haven: Yale University Press, 2010), 103. Years later, Saarinen would marry Norman's former apprentice, costume designer Aline Bernstein, formerly novelist Thomas Wolfe's mistress.
396. "Seeking Forms that Function," by Janet Mabie. *Christian Science Monitor*, July 8, 1939, p.6+.
397. *Minutes of Meeting*, April 5, 1938. Norman Bel Geddes & Co. in-house memo. Harry Ransom Center. A futuristic novel by Jules Verne (d. 1905), set in 1960 (the same year as Futurama), was discovered in the 1990s in a safe. *Paris in the 20th Century* predicted elevators, fax machines, and horse-less vehicles. Verne's publisher had refused to print it on the grounds that it was too farfetched and had a subversively pessimistic view of technology.
398. Recounted in a letter to Frances Waite, November 1, 1938, Harry Ransom Center.
399. Bel Geddes to Waite, October 26, 1938. Harry Ransom Center.
400. Bel Geddes to Waite, November 1938. Harry Ransom Center.
401. Frank Harting to Hellman, p. 40.
402. Bel Geddes to Waite, December 13, 1938, Harry Ransom Center.
403. Bel Geddes to Waite, Jan. 16, 1939. Harry Ransom Center.
404. Harting to Hellman, p. 40.
405. Natasha Rambove to Waite circa 1939. Harry Ransom Center.
406. Norman had suggested that visitors be carried up the ramps and into the building in GM automobiles, but "Kettering killed that, just out of orneriness."
407. *Time*, February 1, 2010.
408. The total cost included salaries for seventy-five maintenance men for the four buildings, "unseen" technicians, and a crew of artists who "touched up," cleaned and vacuumed *Futurama*'s roads each evening. Dust blowing into the diorama from outside would require a permanent staff of 30 to replace, wash or vacuum vast areas of foliage and fields. The continuous window panes, through which some 28,000 visitors peered every day, would be cleaned each evening by a team of four. Also on salary were watchmen who patrolled the building every 45 minutes, 24/7.
409. David P. Billin and David P. Billin Jr., *Power, Speech, Form: Engineers and the Making of the Twentieth Century* (Princeton: Princeton University Press, 2006).
410. Bel Geddes to Hellman.
411. "Seeking Forms that Function" by Janet Mabie. *Christian Science Monitor*, July 8, 1939. p.6+.
412. "Magic Carpet in Futurama" by Waldemar Kaempffert, *New York Times*, Sept. 10, 1939, section II, p. 8.
413. 'Motoring at 100 M.P.H.' *Business Week*, Sept. 9, 1939, p. 27.
414. "Case History of the GM Intersection," Norman Bel Geddes & Co.1941. Harry Ransom Center.
415. "The World's Largest Rendering" by Robert Henri Mutrux for *Pencil Points*, draft, May 24, 1939.
416. Harold N. Simpson to Bel Geddes, Oct 18, 1939. Harry Ransom Center.
417. "The Skyline: Genuine Bootleg," by Lewis Mumford, July 29, 1939. Reprinted in *Sidewalk Critic: Lewis Mumford's Writings on New York*, edited by Robert Wojtowicz. (New York: Princeton Architectural Press, 2000), 234.
418. From "The World of Tomorrow" by Morris Bishop. *The New Yorker*, March 18, 1939, p. 20.
419. "In Mr. Whalen's Image," *Time* (cover story), May 1, 1939, p. 72.
420. "World: As You Enter." *Time*, June 26, 1939.
421. Reported in July 1939 by Commander Howard M. Lammers and Mrs. Mary H. Ellis. (NYPL's World's Fair files.) The Fair maintained extensive files on "Morality and Nudity Issues."
422. *Life*, July 29, 1940, pp. 50–55.
423. Rube Goldberg's Concession Application dated June 18, 1937, New York Public Library/World's Fair archive.
424. Bel Geddes to Grover Whalen, n.d. Harry Ransom Center.
425. *Time*, January 28, 1924, p. 16.
426. Metro Goldwyn-Mayer memo dated February 23, 1938 in the Arthur Freed Collection (NYPL).
427. *Morris Gest's Little Miracle Town: With the World's Greatest Midget Artistes* (New York: New York World's Fair Publications, 1939), 2.
428. Norman Bel Geddes & Co. memo, "Geddes' Explanation of Architectural Registration Difficulty," February 4, 1940. Harry Ransom Center.
429. Beryl Austrian to Hellman.
430. Quoted in *The Wonderful Future That Never Was* (New York: Hearst Books/Sterling Publishing Co., Inc., 2010).
431. "Painting the Town." *Esquire*, May 1939, p. 22.
432. Bel Geddes, *Miracle*.
433. "Table Talk," *Cue Magazine*, April 15, 1939, p. 40.
434. Bel Geddes to Theodore Backer, January 30, 1939 and May 14, 1941. Harry Ransom Center.
435. "Norman Bel Geddes Gets Very Boyish Idea as He Sees a Dream Reflected in Mirrors," by Michel

York. *New York Post*, May 11, 1939.

436. "Bel Geddes' Mirror Show," in *World's Fair Official Guidebook*, p. 35.

437. *Bel Geddes, "Very Boyish Idea."*

438. Bel Geddes to Waite, February 21, 1939, Harry Ransom Center.

439. Bel Geddes to Waite, January 10, 1939. Harry Ransom Center.

440. Some of Bragdon's ideas on geometric pattern would be adapted by Buckminster Fuller.

441. *Official Fair Guidebook*, p, 35.

442. "Fashions of the Future," *Vogue*, February 1, 1939.

443. Introductory notes to *Things to Come: A Film by H.G. Wells* (New York: Macmillan Co., 1935), xiii–xiv.

444. "Tomorrow's Daughter." *Vogue*, February 1, 1939, p.61.

445. George Woods, to Hellman, p. 58. *Crystal Lassies* was set up as a separate corporation. Frances, Worthen Paxton, and George Woods also contributed.

446. It was believed that collapsing an infected lung allowed it to "rest" so the lesions could heal, a method known as the "pneumothorax technique."

447. Bel Geddes to Frances Waite, n.d. Harry Ransom Center.

448. The *New York Post's* Westbrook Pegler, quoted in *Only a Paper Moon: The Theater of Billy Rose* by Steve Nelson (Ann Arbor: UMI Research Press, 1987), 108–109.

449. NYPL World's Fair archives.

450. "Two Fair Showgirls Arrested for Nudity." *New York Times*, June 20, 1939, p. 16.

451. Undated, unidentified newspaper clipping. Harry Ransom Center.

452. Interview in unidentified newspaper clipping. Harry Ransom Center.

453. Clipping, n.d. Harry Ransom Center.

454 Bel Geddes to Hellman circa 1940. Harry Ransom Center.

455. *Official Guide Book, The World's Fair of 1940 in New York: For Peace and Freedom* (New York: Rogers, Kellogg-Stillson, Inc.), 138.

456. Tirza's "Grapes of Bath" routine had her showering in red "wine" amid dry ice and flashing lights while a narrator told the myth of Bacchus.

457. Interviews with 1,000 men and women conducted by SM-Ross Federal at the Fair's exits.

458. *New Yorker*, May 11, 1940.

459. "Is There a Ford in Your Past?" by Clive Barnes. *Dance Magazine*, August 1997.

460. *New York Times* editorial, "Super Highways," Jan 28, 1940.
Moses's staff had been privy to a day-long private viewing of the Shell Oil "City of the Future" model in 1937, the same year Moses was drafting plans to reconfigure Brooklyn, a subject on which he and Bel Geddes corresponded.

461. Unfortunately, the Casino was also a favorite of flamboyant N.Y. mayor Jimmy Walker, whose agendas conflicted with Moses's.

462. "Fair's Theme Song Has Its Premiere, Geddes Assails Moses." *New York Times*, February 3, 1940, p.9.
"I have never fully recovered from my astonishment over the enormous success of your show," Moses would write Bel Geddes about Futurama a decade later. "Even Billy Rose with his bathing girls couldn't touch it. Your show did a whole lot for motoring even if it temporarily embarrassed us who were suddenly expected to produce in quantity, and on a usable basis, all the extraordinary gadgets you exhibited . . . some of which I still continue to believe are out of this world." Harry Ransom Center.

463. "Norman Bel Geddes at Work," by Morton Eustis. *Theatre Arts Monthly*, December 1940, p. 880.

464. Sales between 1938 and 1940, based on figures quoted in *My Years with General Motors* by Alfred P. Sloan, Jr. (New York: Doubleday & Co, Inc., 1964), 214.

465. According to a Norman Bel Geddes & Co. in-house memo, the Fair's total attendance was 44,929,000, compared to Yankee Stadium's seasonal attendance of 1,000,000.

466. *New York World's Fair, 1939*, an illustrated, limited edition (New York: Lord & Thomas and the NY World's Fair, Inc., 1936).

467. Gallup poll quoted in *Culture as History* by Warren I. Sussman (New York: Pantheon Books, 1973), 268.

468. *NY Sunday News*, June 18, 1939.

469. Letter from Dr. Leo Weiselberg, Jackson Heights, NYC, October 21, 1939. Harry Ransom Center.

470. Garrett to Bel Geddes, January 2, 1940. Harry Ransom Center.

471. Norman Bel Geddes & Co. memorandum, July 3, 1940. Harry Ransom Center.

472. "Goodbye, Folks," by Eugene Kinkead. *New Yorker*, May 31, 1941, p. 32.

473. "We Hope We Set A Boy To Dreaming." *New York Times*, October 28, 1941.

474. Quoted in several books as a remark Balanchine made to journalists.

475. *The London Observer*, October 8, 1961.

476. "The Changing Circus," *New York Times*, November 25, 1940.

477. *NY Herald Tribune*, November 21, 1940.

478. *The Minneapolis Star Journal*, Dec. 13, 1940.

479. "Streamlining the Circus" by Norman Bel Geddes. *New Republic*, January 20, 1941, p. 80-81.

480. With the arrival of the French bikini in 1946, Americans who accepted similar attire on circus per-

formers would be outraged to see it on the beach.

481. North and Aussey had met in Paris the previous Christmas Eve (1939) during a blackout. That the Nazis were about to invade France may have had something to do with her decision to marry a foreigner she barely knew.

482. "Careful There, Mr. Geddes." *New York Herald Tribune*, November 21, 1940.

483. "Four Steel Towers Support New Poleless Circus Tent." *Popular Science*, March 1941, p. 97.

484. *New York Herald Tribune*, January 25, 1941.
Bel Geddes got a retainer of $1,000 a month (approx. $16,900 in today's currency), plus all expenses.

485. "Theatre: Menagerie in Blue," *Time*, April 21, 1941.

486. *New York Journal–American*, n.d. Harry Ransom Center.

487. When all available seats sell out, straw is spread on the ground to accommodate an additional 2,000 people.

488. "A Circus in Red, White and Blue" by Mark Murphy. *New York Post*, April 8, 1941,

489. From a review in the *New York Sun*, April 8, 1941.

490. For the 1941-42 season, Bel Geddes received $30,000, plus an additional 10 percent should his costs exceed $15,000.

491. Porter to Bel Geddes on Waldorf Astoria letterhead, January 24, 1941, Harry Ransom Center.

492. Berlin to Bel Geddes, February 11, 1941, Harry Ransom Center.

493. Bel Geddes to Roy Chapman Andrews, October. 8, 1941. Harry Ransom Center.

494. Robert Moses to Bel Geddes, October 14, 1941. Harry Ransom Center.

495. Bel Geddes to Moses, October 15, 1941. Harry Ransom Center.

496. Moses paid Bel Geddes $3,500. (approx. $53,500 today) for the preliminary report. Fourteen years later, the American Museum of Natural History's "fossils" (Moses's term) remained adamantly against a multi-million dollar overhaul, but the good news, Moses wrote, was that thanks, in part, to Norman's report, the shows had improved and annual attendance steadily increased. —Moses to Bel Geddes, October 4, 1955. In the summer of 1949, Moses would invite Bel Geddes to lunch to discuss revamping New York's botanical gardens.

497. Robert Edward Sherwood, *Here We Are Again: Recollections of an Old Circus Clown* (New York: Bobbs-Merrill Co., 1926).

498. North managed to get Stravinsky to agree to a fee even lower than the modest one Stravinsky himself suggested. In return, North got sixteen bars, without harmonies or orchestration. These Stravinsky agreed to add for an additional fee; then more measures were required, to avoid too much repetition. There was also a matter of orchestration. In the end, North was foiled at his own game, with Stravinsky collecting more—reputedly $1,500—than he'd originally requested.

499. Connie Clausen, *I Love You, Honey, But the Season's Over* (New York: Holt, Rinehart & Winston, 1961), 30-31.

500. Clausen, *I Love You, Honey*, 25.

501. Vera Zorina, *Zorina: An Autobiography* (New York: Farrar, Straus & Giroux, 1986), 260.

502. Quoted in "Entr'acte: Stravinsky and the Elephants" by George Brinton Beal, *Concert Bulletin of the Boston Symphony Orchestra*, January, 1944.

503. John Murray Anderson, *Out Without My Rubbers* (New York: Library Publishers, 1954), 219.

504. Caption in the official *Route Book for the 1941 Season*.

505. "It's Still One a Minute," by Meyer Berger. *New York Times Magazine*, April 13, 1941, p. 10+.

506. Interview with Frederick Franklin, Sept, 2, 2001. NYPL's Performing Arts archive.

507. From the official program notes.

508. Zorina, *An Autobiography*, 260-261.

509. "Circus Opens Amid New Brilliance," *New York Times*, April 10, 1942, p. 14.

510. *Variety*, April 15 and April 22, 1942.

511. Loewy, *Never Leave*, 243.

512. Russell Lynes, *The Tastemakers: The Shaping of American Popular Taste* (New York: Harper & Brothers, 1954), 319.
The 1954 bestseller would belatedly join Ringling's "streamline" detractors, calling Bel Geddes' imprint a supreme example of lowbrow art, corrupted. By wrapping the circus up "in pink middlebrow sequins," he'd destroyed its "special flavor of authenticity."
But *The New York Times* would have the last word. In 2006, 64 years after half a hundred massive, pink-clad elephants last swayed, stomped and posed on blue sawdust, in an artificial blue twilight, the ballet was described as "one of the strangest and loveliest high-meets-low moments in American culture." —"The Big Dance," *New York Times*, May 14 2006.

513. "Talk of the Town." *The New Yorker*, April 29, 1939.

514. Marguerite Manners, daughter of actress Laurette Taylor.

515. Hellman was paid $1,650 for the profile (approx. $26,500 today).

516. *The New Yorker*, May, 25, 1940, p. 25.

517. "Silly Milly" strip. *New York Post*, February 4, 1941.

518. *St. Louis Times*, October 10, 1940.

519. "Bel Geddes Termed Jack of All Trades" by John Ferris. *The Republic*, Phoenix, January 8, 1940.
520. According to James Thurber.
521. "Bel Geddes Denies Spending Too Much on Plays" by Robert Coleman, *The Daily Mirror*, March 8, 1937.
522. Hellman p. 58.
    "He's the original 'thinking guy' in this industrial design business," Woods added. "He does just as he goddamn pleases all the time."
523. Janet Flanner, James Thurber, and others described Harold Ross as "the most blasphemous good talker on record."
524. Ordynski to Bel Geddes, February 26, 1941. Harry Ransom Center.
525. Ordynski may have been partly Jewish, further incentive to leave Poland.
526. "Captain Fritz," *Life*, June 26, 1939, p. 69. Also "Letter to the Editor, *Life*, July 17, 1939, p. 2., from J. H. D. Jr.
527. Reinhardt, *Genius*, 361.
528. Salzburg Seminar Website, www.sjsu.edu.
    According to one account, "Princess" Stephanie, who knew Max, was able to secure sixteen crates of Reinhardt's books, porcelain, silver, and furniture from the schloss and ship them to California.
529. There are a couple of undated translations by Ordynski in The Billy Rose Collection, NYPL: *The Sailor*, a comedy in three acts by Jerzy Szaniawski, and *Little Kitty and the Big Politicians*.
530. Ordynski to Bel Geddes, September 1941. Harry Ransom Center.
531. Norman and Dororthy Karasick, *The Oilman's Daughter: A Biography of Aline Barnsdall* (Encino: Carleston Publishing Co.,1993), 88. J. Edgar Hoover's minions had continued to keep close tabs on Barnsdall, noting it all down. Ordynski would remain in touch with his daughter and, later, her off-spring, his grandchildren.
532. According to *Caught in the Act*, the unpublished memoir of set designer George James Hopkins, Hopkins and Ordynski were lovers during the filming of Theda Bara's *Cleopatra*, which would have been just after Ordynski left Los Angeles and just before Bara starred in *Rose of Blood*, based on Ordynski's story. See William J. Mann's *Behind the Screen: How Gays and Lesbians Shaped Hollywood 1910-1969* (New York: Viking, 2001), 28–31.
533. Unedited transcript, interview with Selma Robinson for *PM magazine*, January 27, 1942,
534. *Scientific American*, March 1938, p. 134.
535. "Helena Rubinstein's Beauty Salons, Fashion, and Modernist Display," by Marie J. Clifford. *Winterthur Portfolio*, vol. 38, 2003, pp. 83–108.
536. In 1975, Rubinstein's bed sold at auction for $10,000.
537. Kaufmann's father, a wealthy department store owner, commissioned F. L. Wright's "Fallingwater," which Edgar would inherit in 1955. The family referred to it as "Rising Mildew." Edgar would inherit the house in 1955.
538. "Borax, or the Chromium-Plated Calf" by Edgar Kaufmann. *Architectural Review*, 104, No. 62, August 1948, pp. 88–93.
539. Borax/Kaufmann, ibid.
540. By Sgt. Philip Reisman. Reprinted in *Shaping America's Products* by Don Wallance, (New York: Reinhold Publishing Corp., 1956), 25.
541. Dreyfuss, *Designing for People.*
542. Loewy disliked the word "streamline" because "it had entered the language of industrial design before he came along." —"Design and the American Dream" by Elizabeth Reese., in *Raymond Loewy: Pioneer of American Industrial Design* by Angela Schonberger (Munich: Prestel Publishing, 1990), 39.
543. Loewy, *Never Leave.*
544. Roger Burlingame, *Engines Of Democracy: Inventions & Society in Mature America* (New York: Charles Scribner's Sons,1940), 526-527.
545. Harold Van Doren, *Industrial Design: A Practical Guide* (New York: McGraw-Hill Book Co., Inc., 1940), 16.
546. Janet K. Smith, *Design: An Introduction* (Chicago/New York, Ziff-Davis Publishing Co., 1946), ix.
547. Valley Upholstery Corp to Bel Geddes, November 9, 1944. Harry Ransom Center.
548. Ira M. Pink to Bel Geddes, November 30, 1941. Harry Ransom Center.
549. Bacardi had just converted a nineteenth-century penitentiary into a bottling plant, its "execution by hanging" site redesigned as a tourist cafe. They wanted something "more unusual" than a "regular" architect might supply, something "equal to the imagination [of] Futurama." —Norman Bel Geddes & Co., "Minutes of a Meeting," October 1941. The project was ultimately cancelled.
550. Bel Geddes patent #2,317,682 awarded April 27, 1943 The U.S. government considered using it as a hospital tent.
551. In an attempt to recoup some of the substantial personal monies he'd invested in both the models and various military projects, Bel Geddes eventually sold his flotilla (with some ships replicated in sterling silver) to a New York attorney.
552. "Amphibious War: Bel Geddes Model Explains Land and Sea Attack." *Life*, November 16, 1942, pp.

115–23.
553. "Model Ships Show World's Navies." *Popular Science Monthly*, Nov. 19, 1941, p. 116.
554. Correspondence between Commander C. G. Moore, U.S. Navy, and Bel Geddes, Oct 24, Dec 13 and Dec 18, 1941. Harry Ransom Center.
555. Edison, Norman's adolescent hero, spent considerable time working on submarine detection methods during WWI.
556. "Coral Sea: Norman Bel Geddes' Models Reenact Naval Battle." *Life*, May 25, 1942, pp. 21–25.
557. "The Bel Geddes Method" by Andrea Gustavson. In *Norman Bel Geddes Designs America*, edited by Donald Albrecht (New York: Abrams, 2012), 353.
558. *Architectural Forum*, March 1944, pp. 4–5.
559. Loewy, *Never Leave*, 148.
560. Loewy, *Industrial Design*, 52.
561. Loewy, *Never Leave*, 180
562. *Architectural Forum*, March 1944, p. 150.
563. During World War II, helped along by generous tax breaks and other government incentives, GM assembled the tanks, plus vehicle and weapon camouflage, trucks and "firepower" for fighter planes;75 decades later, evidence surfaced that GM's German-based factories were dependent on the forced labor of thousands of POWs, from civilians to concentration camp prisoners. —"Ford and GM Scrutinized for Alleged Nazi Collaboration" by Michael Dobbs. *The Washington Post*, November 30, 1998; Page A01. *The Schneider Report* is available from the National Archive.
564. Alfred P. Sloan, Jr., *Adventures of a White-Collar Man* (New York: Doubleday, 1941), 198.
565. *Saturday Evening Post*, September 27, 1940.
566. George Nelson, *Tomorrow's House*, introduction by Howard Myers, publisher of *Architectural Forum* (New York: Simon & Schuster, 1945).
567. The U.S. military kick started the country's oral hygiene.
568. Norman Bel Geddes & Co in-house memo, October 13, 1943. Harry Ransom Center.
569. The play was *Sons and Soldiers*, with Stella Adler, Karl Malden, and Gregory Peck. Peck had worked as a barker at the 1939 Fair's Speedway, where he earned extra cash allowing customers to climb up the Speedway's scaffolding for a commanding view of Norman's gyrating *Crystal Lassies* next door. —*Gregory Peck: A Charmed Life* by Lynn Haney, Da Capo Press, 2006, pp. 72–73.
570. A sentimental choice, as Helene was the name of his mistress-turned-second-wife.
571. Gottfried Reinhardt to Bel Geddes, November 24, 1943. Harry Ransom Center.
572. Kevin Maney, *The Maverick and his Machine: Thomas Watson Sr. & the Making of IBM* (New Jersey: John Wiley & Sons, Inc., 2003), 336.
573 Maney, *The Maverick*.
574. Computer Oral History Collection, Hopper interview, January 7, 1969. Smithsonian Archives, National Museum of American History, p. 7–8
575. "The Egotist," reprinted in *An American in Paris: Profiles of an Interlude Between the Wars* by Janet Flanner (New York: Simon and Schuster, 1940), 217–225.
576. Lauded design advocate Ralph Caplan compared Loewy to Bob Hope, a comedian known for his one-liners. "No matter how marvelous the execution," the one-liner is, like styling, "always sharply limited in range." —*On Design* by Ralph Caplan, 1982 edition, p. 140.
577. Interview with Betty Reese, *ID: The Magazine of International Design*, Special Raymond Loewy Issue, November/December, 1986, pp. 40–42.
578. "Looking Backward to the Future, by Raymond Loewy as told to B. Smith Reese" *Collier's*, November 13, 1943, pp. 13+.
579. "Sherry with Egg" (Talk of the Town), *The New Yorker*, December 23, 1944, pp. 12–13.
580. Reese in *ID*, Special Raymond Loewy Issue.
581. "Perpetual" as in self-winding, a 1931 breakthrough. $1,075 is approx. $8,450 today.
582. Evert Endt, "A Frenchman in New York: Loewy's Debut in the United States." in *Raymond Loewy: Pioneer of American Industrial Design* (Munich: Prestel,1990), 26. Also, Endt interview in *ID: The Magazine of International Design*, Special Raymond Loewy Issue, November/December, 1986, p. 47.
583. Reese in *ID*, Special Raymond Loewy Issue.
584. Mimeographed Raymond Loewy & Associates Christmas program, Cooper Hewitt archive, New York City.
585. *The Smyth Report* by Henry De Wolf Smyth. "The Official Report on the Development of the Atomic Bomb Under the Auspices of the United States Government." Appendix 6. War Department Release On New Mexico Test, July 16, 1945.
586. "One Hell of a Big Bang" by Studs Terkel, *The Guardian*, August 6, 2002.
587. Rambova to Bel Geddes, October 15, 1941. Harry Ransom Center.
588. "Your New World of Tomorrow, by F. Barrows Colton, *National Geographic*, October 1945, pp. 385–410.
589. "Tomorrow Has Arrived," by Waldemar Kaempffert, *American Magazine*, March 1941, p. 45+.
590. "New Foods to Tempt Your Palate" by S. S. Block. *Science Digest*, October 1944.

591. Norman V. Carlisle and Frank B. Latham, *Miracles Ahead: Better Living in the Postwar World* (New York: Macmillan Co., 1944), 10, 264.

592. Transcript for *PM* magazine interview, Jan 27, 1942. Harry Ransom Center.

593. "Tomorrow's Consumer." Talk presented on December 9, 1943 at 39th Annual Meeting of the American Society of Refrigerating Engineers, Philadelphia, PA.

594. Leave it to French. As late as 1957, Roland Barthes would describe plastic as "magical . . . the stuff of alchemy."
One notable exception was hosiery. Nylon, requisitioned for parachutes and ropes, had been unavailable since 1942. With the war over, nylon stockings re-appeared, causing a frenzy (40,000 women reputedly lined up in Pittsburgh, 30,000 in New York City) similar to their initial release six years before.

595. "The Building of the Great Pyramid: Bel Geddes Models Reconstruct for Encyclopedia Britannica How the Biggest Pyramid was Built," *Life*, December 3, 1945, pp. 75–80.

596. "Future Toledo: The Model Gives Citizens a Prophetic Look at the Wonderful City They Could Have in 50 Years." *Life*, September 17, 1945, pp. 87–94.

597. Minutes of a General Meeting with RCA executives, Norman Bel Geddes & Co. December 30, 1942,

598. The contract also stipulated designs for post-war electric refrigerators, washing machines and a company trademark, to be produced within two years, with a minimum payment (including overhead) of $4,000 a month, to be followed by designs for irons, vacuums, water heaters and kitchen ranges.

599. According to Stephen Bayley, the three IBM employees who invented the transistor insisted that the device had absolutely no commercial application. Then came Sony. IBM's Watson had had similar thoughts about the future of computers.

600. Innes, *Designing Modern America*, 284.

601. "Construction/Comeback," *Time*. December 13, 1948, pp. 96–98.

602. Norman Bel Geddes to Barbara Bel Geddes and her husband Windsor Lewis, August 7, 1952. Harry Ransom Center. (Approx. $1,800,000 today.)

603. Norman Bel Geddes & Co office memo, February 3, 1947. Harry Ransom Center.

604. Joint Bel Geddes–Noyes patent, #160139, filed January 8, 1949, awarded September 19, 1950.

605. "The Forgotten Pioneer of Corporate Design" by Jesse Scanlon, *Bloomberg Busniess*, January 29, 2007.

606. Quoted in *Terrible Honesty: Mongrel Manhattan in the 1920s* by Anne Douglas (New York: Farrar Straus & Giroux, 1995), 471.

607. "Raymond Loewy: Designs for Living" by Woodrow Wirsig. *Look*, February 14, 1949. pp. 78+.
Viola Erickson Uzzell would eventually become a vice president in Loewy's firm.

608. "The Great Packager" by John Kobler. *Life*, May 2, 1949, pp. 110+

609. "The Artist Turns Big Shot" by Gretta Palmer," *Today Magazine*, June 22, 1935 p. 7.

610. Longtime Loewy employee Jay Doblin guessed that "25 percent of everything," e.g. business revenues, wound up in the boss's pocket. Doblin interview in *ID, The Magazine of International Design*, Special Raymond Loewy Issue, November/December, 1986, p. 44.

611. Reese in *ID*, Special Raymond Loewy Issue.

612. Loewy, *Never Leave*, 174-175.

613. Letter to the Editor from Walter Dorwin Teague, *Life*, May 23, 1949, p. 14.

614. Letter to the Editor from RL, *Life*, May 23, 1949, p. 14.

615. Loewy, *Never Leave*, 376.

616. Russell Flinchum, *Henry Dreyfuss: The Man in the Brown Suit* (New York: Cooper-Hewitt/Smithsonian/ Rizzoli, 1997), 26. See also Bel Geddes' obituary by Dreyfuss in *Industrial Design Magazine*, June 1958, pp. 48+.

617. *Time*, October 31, 1949 issue.

618. "Modern Living." *Time*, October 31, 1949, pp. 68+.

619. "Up From the Egg." *Time*, October 31, 1949, p. 68.

620. According to Dreyfuss's friend, Donald Holden. In Flinchum's *The Man in the Brown Suit*, p. 79.

621. In contrast to Bel Geddes, Loewy "couldn't draw a straight line." Epstein, Betty Reese, Jay Doblin and Evertt Endt, all longtime employees, remarked on the fact that Loewy, "incapable of sharing the limelight," "grabbed 100 percent of the credit" for everything, on every project, stamping completed work with his name. His staff, "largely young and hungry," went along with it. —*ID: The Magazine of International Design*, Special Raymond Loewy Issue, November/December 1986.

622. Adrian Forty, *Objects of Desire: Design & Society from Wedgwood to IBM* (New York: Pantheon Books, 1986), 244.

623. Coca Cola's growing ubiquity would inspire two French wine growers to promote a bill banning it as a health hazard. The reason? Coke's popularity in Loewy's homeland was lowering wine sales. —*Time* (cover story), May 15, 1950.

624. An unromantic mix of corn syrup, phosphoric acid, sulfite, ammonia, caffeine, carbon dioxide and coloring, plus vanilla, cinnamon, and sugar.

625. Design critic/historian Stephen Bayley goes so far as to say that Loewy's "overweening" vanity would be recognized today as "narcissistic personality disorder." In *Design A-Z* (New York: Firefly Books, 2010), 204.

626. "Public Relations or Industrial Design? Loewy and His Legend," by Stephen Bayley. In *Raymond Loewy: Pioneer of Industrial Design*, edited by Angela Schonberger (Munich, Prestel, 1990, published in conjunction with exhibit by the same name in Berlin), 238.

627. Philippe Tretiack, *Raymond Loewy* (New York: Assouline, 2005), 15.

628. Raymond Loewy quoted in John Kobler's "The Great Packager," *Life*, May 2, 1949, pp. 110+.

629. Loewy's much-praised 1963 Avanti sports car was "primarily the work of [veteran car designer] Bob Andrews, who received no recognition for it." —*Industrial Designer: The Artist as Engineer* by W. Dorwin Teague Jr., 1998, p. 217.
     According to Stephen Bayley, the Avanti made no money and had no impact on the market. "Loewy constantly made huge claims for it but they were not ones that could be substantiated on any way." "Public Relations or Industrial Design?," 1990.

630. "He Changed the Look of America: Raymond Loewy Started Industrial Design and Transformed Attitudes about Art and Industry." by Victoria Donohoe, art critic. *Philadelphia Inquirer.* August 25, 2002.

631. www.britannica.com/biography/Raymond-Loewy.

632. No mention is made of Loewy's "slimming" work or involvement in the white Applied Color Label. —"Master of Design: Coke and the Legacy of Raymond Loewy" by Ted Ryan Dec 21, 2012. www.coca-colacompany.com/history/master-of-design-coke-and-the-legacy-of-raymond-loewy.
     Loewy would file a patent (#1012DD7E) for a Coca Cola dispenser on June 7, 1940, and one for a "cooling cabinet" (USD147370S) on December 21, 1945.

633. Loewy submitted 25 or 30 different sketches. www.fanta.co.uk/en/history.

634. Fanta proved to have very long legs. It remains available in 188 countries (it's huge in Thailand) in some 70 flavors. Loewy's ringed glass bottle, with its "twin peak" logo on the shoulder, remains in production in some markets.

635. "Construction/Comeback," *Time*, December 13, 1948.

636. "Construction/Comeback."

637. In the range of $11 million today.

638. "Heatter Hails New Copa City As Impossible Dream Come True." *Miami Beach Florida Sun*, December 24, 1948.

639. *Variety*, December 28, 1948.

640. Radio broadcaster "Mr. Mutual" on the Copa City opening. Harry Ransom Center.

641. "Bel Geddes in Boca Raton." www.newyorksocialdiary.com, "Palm Beach Real Estate Roulette," August 18, 2010.
     "Construction/Comeback."

642. In 1956, Schine sold his Boca properties to Arthur Vining Davis, former president of Alcoa, for $22 million. By 2010, all but three Bel Geddes houses had been demolished; the rest were about to be.

643. Bel Geddes to Dreyfuss, May 9, 1951. Harry Ransom Center.

644. Van Doren, *Industrial Design*, 84.

645. *Time*, October 31, 1949, p. 73.
     One new design that caught on like wildfire after debuting at Saks Fifth Avenue was The Boater, the first disposable diaper, invented, not surprisingly, by a housewife.

646. Richard Gump, *Good Taste Costs No More* (New York: Doubleday & Co, Inc., 1951), 105.

647. "The Silent Salesman in Industry," address by Henry Dreyfuss to the 81st meeting of Canadian Design Conference, Toronto, May 28, 1952,. Also quoted in *Forbes*, May 1, 1951, p. 20.

648. Loewy, *Never Leave*, 291.

649. Loewy speech, 1950, quoted in *Never Leave*.

650. Quoted in "Borax, or the Chromium-Plated Calf" by Edgar Kaufmann. *Architectural Review*, 104, No. 62, August 1948, pp. 88–93.

651. Loewy, *Never Leave*, 289–290.

652. Loewy, *Never Leave*, 306.

653. *Time*, October 31, 1949, p. 68.

654. "Actress BBG Has Died." August 10, 2005,"Today: Pop Culture," msnbc.com.

655. *Kazan on Directing*, Preface by Martin Scorsese (New York: 2009), 297.

656. *Herald Tribune, World-Telegram.*

657. Billy Rose, *Wine, Women and Words* (New York: Simon & Schuster, 1946), 177.

658. Rodgers to Bel Geddes, Oct 8, 1945. Harry Ransom Center.

659. "Rising Star." *Time*, April 9, 1951, p. 80.

660. "Rising Star." p. 80.

661. "Rising Star." p. 80.

662. "Rising Star." p. 83.

663. Some believed that Hughes's reason had more to do with the fact that he'd hated *Deep Are the Roots*.

664. "Prisons of Gender and a Generation" by J. Hoberman. *New York Times*, July 9, 2014.

665. "Rising Star," p. 83.

666. "Broadway's Best Girl." *This Week* (nationally syndicated Sunday newspaper magazine), June 17, 1951 cover story.

667. "Rising Star." p. 83

668. "Rising Star." p. 79

669. Barbara Bel Geddes to Norman Bel Geddes, December 14, 1951. Harry Ransom Center.

670. Tennessee Williams, *Notebooks*, Edited by M.B. Thornton (New Haven: Yale University Press, 2007).

671. "Barbara Bel Geddes: Too Far Within," on JamesGrissomblogspot.com. Grissom quotes from interviews conducted for his book, *Follies of God: Tennessee Williams and the Women of the Fog* (New York: Knopf, 2015). This quote was deleted in the published text.

672. A clip of Barbara Bel Geddes in the play (in the slip) is available on YouTube.

673. Grissom, *Barbara Bel Geddes.*

674. Arthur Laurents, *Original Screenplay By: A Memoir of Broadway & Hollywood* (New York: Knopf, 2000), 143.

675. It's worth mentioning that 1956 saw the release of Hitchcock's *The Wrong Man*, a HUAC-like allegory about a bass player at the Stork Club whose life gets turned inside out by unjust accusations.

676. Quoted in *The Films of Alfred Hitchcock* by Robert A. Harris and Michael S. Lasky. (New Jersey: Citadel Press, 1976), 186.

677. Henry Jaglom, *My Lunches with Orson: Conversations Between Henry Jaglom and Orson Welles* (New York: Picador, 2013).

678. In 2012, the British Film Institute announced the results of its once-a-decade poll of 846 critics, programmers, academics and distributors, conducted by *Sight & Sound Magazine.*

679. Hitchcock, like Bel Geddes, Sr., did a stint early in his career as a draughtsman and advertising director.

680. Dan Auiler, *Vertigo: The Making of a Hitchcock Classic*, foreword by Martin Scorsese (New York: St. Martin's Press, 1998), 51.

681. Even Canadian scholar Christopher Innes, who interviewed Edith toward the end of her life, described her in print as "the daughter of the British imperialist architect Sir Edwin Lutyens." Innes, *Designing Modern America*, 22.

682. *David Wallechinsky's Complete Book of the Olympics* (New York: Penguin, 1984).

683. Author's interview with Adolphus Andrews (Moseley's son-in-law), San Francisco, CA, June 30, 2009.

684. "Jeray Jewlery By Bel Geddes Put on Display" *New York Herald Tribune*, June 15, 1950.

685. "This is New York," by E. B. White. *Holiday Magazine*, summer 1948.

686. Walt Disney: Giant At The Fair" by Gereon Zimmermann. *Look*, February 11, 1964, p. 32.

687. *Popular Mechanics*, 1950, quoted in *The Wonderful Future That Never Was* by Gregory Benford & the Editors of *Popular Mechanics* (New York: Hearst Books/Sterling Publishing Co., Inc., 2010).

688. "The little white happy pill" may have inspired Aldous Huxley's "soma" in *Brave New World.*

689. John Kenneth Galbraith, *The Affluent Society* (New York: Houghton Mifflin Harcourt, 1958), 141.

690. "Designers—That Extra Touch." *Newsweek*, November 23, 1957, pp. 105–108.

691. "Obsolescence," in *Problems of Design* by George Nelson, 1957, p. 45.

692. Dreyfuss, *Designing for People.*

693. An account sheet entitled "Capital Requirements as of 4/20/49" listed his weekly office overhead at $1,380, plus $150 weekly for expenses, not counting his own salary.

694. Handwritten draft to Bill Zekendorf of Webb & Knapp, Inc. n.d. Harry Ransom Center.

695. *Forbes*, May 1951.

696. Dreyfuss, *Designing for People*, 88.

697. "Popular Art Organized: The Manner and Method of Raymond Loewy Associates" by Alec Davis. *Architectural Review*, London. November 1951, Vol. 110, pp. 319–26.

698. Michael Farr, *Design in British Industry: A Mid-Century Survey* (London, Cambridge University Press, 1955).

699. O'Malley to Bel Geddes, March 31, 1952. Harry Ransom Center.

700. Bel Geddes to O'Malley, December 15, 1952. Harry Ransom Center.

701. O'Malley to Bel Geddes, December 17, 1952. Harry Ransom Center.

702. The project was completed in 55 days for $3,000.

703. Bel Geddes to O'Malley, February 10, 1954. Harry Ransom Center.

704. Lutyens to Bel Geddes, 1952. Harry Ransom Center.

705. Edith to "Suzanne," July 9, 1953 from Ocho Rios, Jamaica.

706. "Society Today" by Charles Ventura. *New York World Telegram & Sun*, October 1953.

707. Lutyens to Christopher Coates, October 29, 1953. Harry Ransom Center.

708. Bel Geddes to Lutyens, January 25, 1954. Harry Ransom Center.

709. "Society Today" by Charles Ventura. *New York World Telegram & Sun*, n.d. Harry Ransom Center.

710. Bel Geddes to James S. Plaut, March 4, 1956. Harry Ransom Center.

711. At the same time, Bel Geddes was attempting to get GM interested in doing something for Futurama's twentieth anniversary (1959), especially in view of President Eisenhower's proposed ten-year

Highway Construction Program. (GM would decline, but in 1956, Motorama would be built around a "Highway of Tomorrow" theme.)

712. *Time*, December 10, 1956. *Look*, October 16, 1956.

713. Norman's Transit Grid patent was filed on April 2, 1957, complete with a baker's dozen of technical drawings. It was approved (patent # 2,943,579) on July 5, 1960.

714. Edith's unhappiness with the housekeeper and her husband, and her demands that they leave the premises "immediately"—they'd been in the family's employ for years—did not sit well with Max's wife, the Princess.

715. Bel Geddes to Dudley Geddes, Ft. Myers, Florida, July 13, 1956. Harry Ransom Center.

716. Bel Geddes to Ordynski, May 15, 1956. Harry Ransom Center.

717. *New York Times*, August 16, 1953, p. 77.

718. "Noel Whitcomb Says" column, *Daily Mirror*, September 20, 1956, p. 2.

719. "Broadway" by Danton Walker, October 26, 1956.

720. Bel Geddes to Mary Geddes, Ft Meyers, Florida, July 13, 1956. Harry Ransom Center.

721. Three years later, in 1959, Warner Bros. would take up the mantle, choosing "face" actors (Carroll Baker and Roger Moore) and the screenwriter from Bob Hope and Bing Crosby's comedic "road" pictures. A huge flop, the film had the added misfortune of being released the same week as MGM's *Ben Hur*, one of the most successful film epics of all time.

722. The Baum–Bel Geddes collaboration may have been on The Four Seasons restaurant.

723. Lutyens to Barbara Bel Geddes, May 6, 1958. Harry Ransom Center.

724. Henry Blackman Sell to Bel Geddes, April 15, 1958. Harry Ransom Center.

725. "I've Given Away My Eyes: Speed and Skill Turn Dream into Real Vision," by Eckert Goodman. *New York Daily News*, Sept 10, 1958.

726. "News Series on Eye Bank Hits Thousands of Hearts" by Eckert Goodman. *New York Daily News*, September 19, 1958.

727. Goodman, "I've Given Away My Eyes."

728. *Industrial Design*, June 1958, pp. 48+.

729. *Newsweek*, May 19, 1958. *New York Herald*, May 9, 1958

730. *New York Times*, May 9, 1958.

731. "Smart Set" column, *New York Journal–American*, June 20, 1958.

732. Geoffrey Hellman papers, Fales Library, New York University.

733. "The Joy of Less" by Josh Sanburn. *Time*, March 23, 2015, pp. 44–50.

734. Gene Plowden, *Those Amazing Ringlings and Their Circus* (Idaho: Caxton Printers Ltd., 1967), opposite p. 96.

735. John Epstein, longtime Loewy employee and eventual VP of the firm, in *ID Magazine*'s special Raymond Loewy Issue, November/December 1988, pp. 38–39. "He was an utter egomaniac . . . a heated protest over attribution marred the last decade of his life . . . It was quite sad but I know that it was deeply resented by many designers . . . I still adored the man, even though he was terribly self-centered."

736. Innes, *Designing Modern America*, 151–152.

737. Innis, pp. 270–71.

738. "In the final analysis, the power of the Loewy legend owed most to the simple fact of longevity." Jeffrey Meikle, "Loewy," *ID: Magazine of International Design*, Special Raymond Loewy Issue, November/December 1986, pp. 28–47.

Selected Bibliography

Allen, Frederick Lewis. *The Big Change: America Transforms Itself 1900–1950*. New York: Harper and Row, 1952.

Bel Geddes, Norman. *Horizons (in Industrial Design)*. Boston: Little, Brown, 1932.

Burlingame, Roger. *Engines of Democracy: Inventions and Society in Mature America*. New York: Scribner's, 1940.

Chase, Stuart. *Men and Machines*. New York: Macmillan, 1929.

Cheney, Sheldon, and Martha Chandler Cheney. *Art and the Machine: An Account of Industrial Design in the 20th Century*. New York: Whittlesey House, 1936.

Le Corbusier. *Towards a New Architecture*, translated by Frederick Etchells from the 13th French edition, 1931. New York: Dover Publications, 1986.

Cross, Gary S., and Robert N. Proctor. *Packaged Pleasures: How Technology and Marketing Revolutionized Desire*. Chicago: University of Chicago Press, 2014.

Dreyfuss, Henry. *Designing for People*. New York: Allworth Press/Design Management Institute, 1955/2004.

Frankl, Paul T. *Machine-Made Leisure*. New York: Harper and Brothers, 1932.

Frederick, Christine. *Selling Mrs. Consumer*. New York: Business Bourse, 1929.

Giedion, Siegfried. *Mechanization Takes Command: A Contribution to Anonymous History*. New York: W. W. Norton, 1948.

Holme, Geoffrey. *Industrial Design and the Future*. London/New York: The Studio Limited, 1934.

Innes, Christopher. *Designing Modern America: Broadway to Main Street*. New Haven, CT: Yale University Press, 2005.

Kobler, John. *Otto the Magnificent: The Life of Otto Kahn*. New York: Scribner, 1988.

Leach, William. *Land of Desire: Merchants, Power, and the Rise of a New American Culture*. New York: Vintage, 1993.

Lebergott, Stanley. *Pursuing Happiness: American Consumerism in the Twentieth Century*. Princeton, NJ: Princeton University Press, 1993.

Loewy, Raymond. *Never Leave Well Enough Alone*. New York: Simon and Schuster, 1951; reissued by John Hopkins University Press, 2002.

Meikle, Jeffrey. *Twentieth Century Limited: Industrial Design in America, 1925–1939* 2nd ed. Philadelphia: Temple University Press, 2001.

Mumford, Lewis. *Technics and Civilization*. New York: Harcourt Brace Jovanovich, 1934.

Nelson, George. *Problems of Design*. New York: Whitney Publications, 1957.

Petroski, Henry. *Invention by Design: How Engineers Get from Thought to Thing*. Cambridge, MA: Harvard University Press, 1996.

Pulos, Arthur J. *American Design Ethic: A History of Industrial Design to 1940*. Cambridge, MA: MIT Press, 1983.

Read, Herbert. *Art and Industry: The Principles of Industrial Design*. New York: Harcourt, Brace, 1938.

Reinhardt, Gottfried. *The Genius: A Memoir of Max Reinhardt by His Son Gottfried Reinhardt*. New York: Alfred A. Knopf, 1979.

Sayler, Oliver M., editor. *Max Reinhardt and His Theatre*. New York: Brentano's, 1924.

Schutte, Thomas F., editor. *The Uneasy Coalition: Design in Corporate America*. Philadelphia: University of Pennsylvania Press, 1975. (The Tiffany-Wharton Lectures on Corporate Design Management.)

Sheldon, Roy, and Egmont Arens. *Consumer Engineering: A New Technique for Prosperity*. New York: Harper and Brothers, 1932.

Susman, Warren I. *Culture as History: The Transformation of American Society in the Twentieth Century*. New York: Pantheon, 1984.

Teague, Walter Dorwin. *Design This Day: The Technique of Order in the Machine Age*. New York: Harcourt, Brace, 1940.

Van Doren, Harold. *Industrial Design: A Practical Guide*. New York: McGraw-Hill, 1940.

Walker, Stanley. *The Night Club Era*. New York: Frederick A. Stokes, 1933.

Acknowledgments

B el Geddes's autobiography, *Miracle in the Evening*, was based on a hodgepodge of overlapping drafts written years after the facts and published posthumously. (Focused exclusively on theater, it ends when he's still in his thirties.) His collaborations with Max Reinhardt were conducted through a translator. Given that, I've taken a certain amount of creative license when reconstructing conversations.

Unattributed Bel Geddes quotes are from his unpublished writing. Cecil B. DeMille's comments in chapter 7 are based on his "What Psychology Has Done to Pictures," in *The Blue Book of the Screen*, edited by Ruth Wing (1923). The story of how Edith Lutyens and Bel Geddes met is based on Lutyens's unpublished notes.

SINCERE THANKS TO:

The Edith Lutyens and Norman Bel Geddes Foundation;

Helen Baer and Rick Watson at Harry Ransom Center for the Humanities, University of Texas at Austin;

The Max Reinhardt Archive at Binghamton University; the Margaret Bourke-White and Walter Dorwin Teague archives at Syracuse University; the Otto H. Kahn archive at Princeton University; Fales Library, NYU (Geoffrey Hellman papers); the Henry Dreyfuss archive at Cooper-Hewitt/Smithsonian Library; the Raymond Loewy archive at Hagley Museum and Library, Wilmington, Delaware; the Performing Arts archive of the New York Public Library; the Butler and Avery libraries at Columbia University; and the Pike County Historical Society, Milford, Pennsylvania (Chief Thundercloud);

The Link+ library system, through SFPL, that sent dozens of out-of-print materials practically to my doorstep;

Hatchfund for facilitating a research grant;

To Sarah Miller and Dana Randall of the Convent of the Sacred Heart school, New York City, for allowing me to tour the former Otto Kahn mansion;

To friends, family, and occasional others who offered support: Elsa Cameron, James McGrath Morris, Jacki Lyden, Jessica Hagedorn, Louis Spagna, Martin Muller, Mark Belton, Richard Defendorf, John Geresi, Teme Levbarg, Arthur Hanna, Dagny Lux-Hanson, Susie Hoimes, John Solt, Tatiana Kliorin, Mark Miller, Hannah Kahn, Ziril Szerlip, David Szerlip, Anna Swan (of the London's Biographer's Club), Carlos Pulido, and Stephen Bayley;

And to those who offered refuge along the way: Ricardo Noreiga (Malinalco), Victor di Suvero (Santa Fe), Mary Pieratt (Chico), Bibi Gaston (Connecticut), Mercedes Palau-Ribes (Majorca), the Corporation of Yaddo (Saratoga Springs), and Villa Lena (Tuscany).

This book would not have seen the light without my agent, Jill Marsal, who believed (despite considerable obstacles) from the start, and editors Mark Krotov and Ryan Harrington, who made it a personal mission to help "resurrect" Bel Geddes.

Sections of this book appeared, in somewhat different form, in *The Believer, Le Believer* ("Best of The Believer," Paris), *The Paris Review Daily,* and *Berfrois* (London).

Index

About the Author

B. ALEXANDRA SZERLIP has been a two-time National Endowment for the Arts Writing Fellow, a Yaddo fellow, and runner-up for Britain's Lothian Prize for a first biography in progress (the first American ever to make it to the finals). She has contributed to *The Paris Review Daily* and *The Believer*, among other publications. Born and raised on the East Coast, she currently lives in San Francisco.